Handbook of Green Engineering Technologies for Sustainable Smart Cities

Green Engineering and Technology: Concepts and Applications

*Series Editors: Brujo Kishore Mishra, GIET University, India
and Raghvendra Kumar, LNCT College, India*

Environment is an important issue these days for the whole world. Different strategies and technologies are used to save the environment. Technology is the application of knowledge to practical requirements. Green technologies encompass various aspects of technology which help us reduce the human impact on the environment and create ways of sustainable development. Social equability, this book series will enlighten the green technology in different ways, aspects, and methods. This technology helps people to understand the use of different resources to fulfill needs and demands. Some points will be discussed as the combination of involuntary approaches, government incentives, and a comprehensive regulatory framework will encourage the diffusion of green technology, least developed countries and developing states of small island requires unique support and measure to promote the green technologies.

Handbook of Green Engineering Technologies for Sustainable Smart Cities
Edited by K. Saravanan and G. Sakthinathan

Green Engineering and Technology
Innovations, Design, and Architectural Implementation
Edited by Om Prakash Jena, Alok Ranjan Tripathy, and Zdzislaw Polkowski

Machine Learning and Analytics in Healthcare Systems
Principles and Applications
Edited by Himani Bansal, Balamurugan Balusamy, T. Poongodi, and Firoz Khan KP

Convergence of Blockchain Technology and E-Business
Concepts, Applications, and Case Studies
Edited by D. Sumathi, T. Poongodi, Bansal Himani, Balamurugan Balusamy, and Firoz Khan K P

For more information about this series, please visit: www.routledge.com/Green-Engineering-and-Technology-Concepts-and-Applications/book-series/CRCGETCA

Handbook of Green Engineering Technologies for Sustainable Smart Cities

Edited by
K. Saravanan and G. Sakthinathan

CRC Press
Taylor & Francis Group
Boca Raton London

CRC Press is an imprint of the
Taylor & Francis Group, an **informa** business

First edition published 2022
by CRC Press
6000 Broken Sound Parkway NW, Suite 300, Boca Raton, FL 33487–2742

and by CRC Press
2 Park Square, Milton Park, Abingdon, Oxon, OX14 4RN

Library of Congress Cataloging-in-Publication Data
Names: Saravanan, Krishnan, 1982– editor. | Sakthinathan, G., editor.
Title: Handbook of green engineering technologies for sustainable smart
    cities / edited by K. Saravanan, G. Sakthinathan.
Description: Boca Raton : CRC Press, 2021. | Series: Green engineering and
    technology : concepts and applications | Includes bibliographical
    references and index.
Identifiers: LCCN 2021000825 (print) | LCCN 2021000826 (ebook) | ISBN
    9780367554989 (hbk) | ISBN 9781003093787 (ebk)
Subjects: LCSH: Smart cities. | Sustainable engineering.
Classification: LCC TD159.4 .H3635 2021 (print) | LCC TD159.4 (ebook) |
    DDC 307.76028/6—dc23
LC record available at https://lccn.loc.gov/2021000825
LC ebook record available at https://lccn.loc.gov/2021000826

ISBN: 978-0-367-55498-9 (hbk)
ISBN: 978-0-367-55499-6 (pbk)
ISBN: 978-1-003-09378-7 (ebk)

Typeset in Times
by Apex CoVantage, LLC

In Memoriam

Dr. G. Sakthinathan ME, PhD was born on June 20, 1976. He was an Associate Professor in the Department of Manufacturing Engineering, College of Engineering, Guindy, Anna University, Chennai. He had been Dean of the Regional Campus of Anna University, Tirunelveli and Thoothukudi. Dr. Sakthinathan inspired many students and staff by his kind approach and practical teaching methods. He was inspired by nature and devoted himself to several environmental development activities. He along with the district administration was involved in the cleaning of River Thamirabarani (South Tamilnadu, India) and had rejuvenated several irrigation tanks and ponds in the districts of Tirunelveli, Thoothukudi and Thenkasi. For all his environmental contributions, Dr. Sakthinathan received the Abdul Kalam Seva Award and was named the "Waterman of Tirunelveli". He also received an award from the government of India for improving water management activities. Dr. Sakthinathan passed away on May 10, 2021. He is survived by his wife, son, and daughter. He was a beautiful soul never to be forgotten. He will be remembered for his kindness, mentoring, and caring for nature. River Thamirabarani salutes you!!

Contents

Preface

Industrial Revolution 4.0 empowers smart cities with Internet technologies and ICT infrastructures. Smart things, or the Internet of Things (IoT), play a major role in the development of smart sustainable cities. Several IoT applications are implemented for urban growth in the fields of education, waste management, transport, surveillance, energy management, e-governance, water distribution and environmental technologies. The drastic changes from these technologies' adoption and implementation should not harm the urban ecosystem. However, instead of IoT improving the quality of life for urban citizens, possible environmental hazards may create issues in urban areas. The application of green engineering principles is thus necessary to prevent waste, to improve the natural ecosystem and to maximize the efficient utilization of resources in smart cities. Smart cities' applications of green engineering techniques are combined to form the green smart city for sustainable urban growth and a better urban lifestyle.

Waste and materials management, energy management and renewable energy are detailed in separate chapters. Green IoT for the sustainable growth of smart cities is explored. ICT technologies such as artificial intelligence, machine learning, big data and IoT are also presented in this book. A case study for "A Green and Sustainable University Campus Dwelling" at NIT Rourkela is explored with different green life cycle indicators. The impact of smart city projects in sustainable green cities is discussed. "Privacy and Security Issues in Green Smart Cities" with the use of ICT components is presented in the book.

Editors

K. Saravanan is a Senior Assistant Professor, Department of Computer Science and Engineering at Anna University, Regional Campus, Tirunelveli, Tamilnadu. He has published papers in 14 international conferences and 27 international journals. He has also written six book chapters and four books published by international publishers. He is an active researcher and academician. He is engaged in consultancy services for government organizations and smart city projects. Also, he is a reviewer for many reputed journals with Elsevier, IEEE, etc. He has trained and interviewed engineering college students for placement training and counselling. He has conducted many ISTE workshops in association with IIT Bombay and IIT Kharagpur. He is also coordinator for IIRS (Indian Institute of Remote Sensing) e-outreach programme. He has delivered guest lectures in many seminars/conferences at reputed engineering colleges.

G. Sakthinathan is an Associate Professor, Department of Manufacturing Engineering, College of Engineering Guindy Campus, Chennai, India. He has carried out a multitude of real-time projects and brought significant changes through different technological interventions. He has many publications to his credit. He is engaged in various research projects in different government organizations and research institutions. Further, he is also engaged in consultancy services for government organizations and smart city projects. He is a reviewer for numerous reputed journals. He has conducted trainings in many upcoming technologies for the benefit of academicians, researchers and students. He has also delivered several motivational lectures for graduates and students. He has been recognized with awards and appreciation by the government and private organizations for his technological contributions to various societal problems. He is also involved in developing solutions for the welfare of society and students' communities. He belongs to various advisory boards for colleges and organizations.

Contributors

Dr. M. Balamurugan
Jain University Global Campus, Bengaluru
Karnataka, India

Dr. J. V. Bibal Benifa
Department of Computer Science and
  Engineering
Indian Institute of Information Technology
Kottayam, India

Dr. Karan Chandrakar
Indian Institute of Technology
Delhi, India

Dr. Vidya S. Dandagi
Department of Master of Computer
  Application
KLE DR. M.S. Sheshagiri College of
  Engineering & Technology
Belagavi, India

Dr. Kesavaraja Duraipandy
Department of Computer Science and
  Engineering
Dr. Sivanthi Aditanar College of
  Engineering
Tiruchendur, India

Dr. Aboobucker Ilmudeen
Department of Management and
  Information Technology
South Eastern University of Sri Lanka
Oluvil, Sri Lanka

Dr. J. Colins Johnny
Department of Civil Engineering
University VOC College of Engineering
Thoothukudi, India

Dr. P. Karuppasamy
Department of Chemistry
Anna University Regional Campus
Tirunelveli, India

Dr. T. Ananth Kumar
IFET College of Engineering
Tamil Nadu, India

Dr. T. Deva Kumar
National Engineering College
Kovilpatti, India

Dr. S. Dilip Kumar
Sri Krishna College of Technology
Coimbatore, India

Dr. V. Kishore Kumar
IFET College of Engineering
Tamil Nadu, India

Dr. Manish Kumar
Electrical Engineering Department,
  School of Engineering and
  Technology
Central University of Haryana
Haryana, India

Dr. K. Suresh Kumar
IFET College of Engineering
Tamil Nadu, India

Dr. Vikas Kumar
Chaudhary BansiLal University,
  Bhiwani, Haryana, India

Prof. Madhusree Kundu
Department of Chemical Engineering
National Institute of Technology,
   Rourkela
Odisha, India

Prof. Manju Lata
Chaudhary BansiLal University
Bhiwani, Haryana, India

Dr. Amiya Kumar Mallick
IIT Kharagpur

Mr. Navdeep Mor
National Institute of Technical Teachers
   Training and Research
Chandigarh, India

Dr. S. Narendiran
Sri Krishna College of Technology
Coimbatore, India

Dr. K. Padmavathi
Department of Computer Science
PSG College of Arts and Science
Coimbatore, India

A. Mohideen Pathumuthusabana
Anna University Regional Campus
Tirunelveli, India

Dr. S. Pitchaimuthu
Post Graduate and Research Department
   of Chemistry
Thiagarajar College
Madurai, India

Prof. S. Porkodi
Department of Computer Science and
   Engineering
Dr. Sivanthi Aditanar College of
   Engineering
Tiruchendur, India

Dr. S. Suja Priyadharsini
Department of Electronics and
   Communication Engineering
Anna University Regional Campus
Tirunelveli, India

Aditya Punia
National Institute of Technical Teachers
   Training and Research
Chandigarh, India

Dr. V. Rajapandian
Post Graduate and Research Department
   of Chemistry
Sri Ramakrishna Mission Vidyalaya
   College of Arts and Science
Coimbatore, India

Dr. R. Rajesh
Sri Krishna College of Technology
Coimbatore, India

Dr. J. Bruce Ralphin Rose
Department of Mechanical Engineering
Anna University Regional Campus
Tirunelveli, India

Mr. Praveen Kumar Sahani
Indian Institute of Technology
Delhi, India

Dr. G. Sakthinathan
Department of Manufacturing
   Engineering, CEG Campus
Anna University
Chennai, India

Dr. T.S. Arun Samuel
National Engineering College
Tamil Nadu, India

Dr. K. Saravanan
Department of Computer Science
Anna University Regional Campus
Tirunelveli, India

Mr. Yash Shah
Department of Chemical Engineering
National Institute of Technology
Rourkela, Odisha, India

Dr. Nandini Sidnal
Department of Computer Science and
 Engineering
KLE DR. M.S. Sheshagiri College of
 Engineering & Technology
Belagavi, India

Dr. S. Sundaresan
SRM TRP Engineering College
Tamil Nadu, India

1 Green Smart Cities
An Introduction

P. Karuppasamy, S. Pitchaimuthu,
V. Rajapandian and K. Saravanan

CONTENTS

1.1 INTRODUCTION

A city which is designed based on its social, economic and environmental impact is called a green or eco smart city [1, 2]. The universally accepted definition for eco-city is "cities that enhance the well-being of citizens and society through integrated urban planning and management that harness the benefits of ecological systems and protect and nurture these assets for future generations" [2]. But a green smart city has not been generally defined till now. The sustainability of smart cities is a major challenge for all countries due to providing the solutions for sustaining the natural environment, powering itself with renewable sources of energy, conceiving its ecological footprint, minimizing pollution by way of composting and recycling or converting waste to energy [3–8]. For example, the Adelaide City Council (ACC) [3] in Australia said a smart city must be fair to all; they are interconnected, democratic and offer a desirable life to the people.

Many countries have adopted smart green resilient (SGR) missions for making green smart cities. SGR is concerned mainly with the ecological problems such as structuring long-term sustainability plans in broad perspectives and guiding smart city transformation. SGR detects a variety of important urban problems and provides solutions for them. An SGR planning approach is categorized by three main elements such as people oriented, contemporary relevance and future proofing [9–12].

In recent years, most citizens around the world have moved to cities due to urbanization, the need to find better jobs and a higher standard of living. More than 3 million people every week since 2009 have moved to cities around the world due to low economic status, increasing population, unavailable resources, and lack of infrastructure development, healthcare and education facilities, etc. Among the world population, nearly 55% are in city areas, and this will reach 70% in 2050 as estimated by the United Nations (UN) [13].

Many countries face challenges related to solid waste management (SWM) such as environmental hazards and the risk to public health because of population growth, the financial burden and the unavailability of efficient and suitable SWM methods [14–17]. The US Environmental Protection Agency (USEPA) has reported that the United States of America produces nearly 250 million tonnes (MT) of solid wastes every year; it is 1.3 billion tonnes (BT) at global level. This will increase and reach 2.6 BT in 2025 [14, 17]. Asian and African countries will produce 5.3% and 97% respectively [15, 17] in 2025.

Waste materials produced from healthcare facilities cause more high-risk related infections and injuries than other types of waste. Due to lack of knowledge for handling and disposing of biomedical waste, this material leads to serious adverse effects to the environment. Annually 0.35 million tonnes (MT) of biomedical waste is produced in India [18–20]. Biomedical waste has not only affected the environment but also has created health problems for health workers and the public. Hence, effective and safe biomedical waste management techniques are needed now.

Electronic waste (e-waste) is one of the solid wastes that create pollution in the environment as well as health problems for human beings. Many e-waste management tools are available such as life cycle assessment (LCA), material flow analysis (MFA), multi-criteria analysis (MCA) and extended producer responsibility (EPR) to manage toxic materials in e-wastes [21]. The circular economy (CE) and integrated solid waste management system (ISWM) may be suitable to manage these solid waste materials globally [22]. It not only provides waste management-related solutions but also it provides safe employment, recycling of waste and enhances the economic status of every country.

1.2   KEY CHALLENGES PERTAINING TO SMART CITIES AT GLOBAL LEVEL

In cities at the global level, establishing sustainability of a green environment includes many challenges, including possible energy crises and water scarcity. Green industrial development, smart building construction, discrete management systems, renewable energy resources, pollution-free air, smart transportation, smart climate adaptive systems, and creation of forest-like cities, etc. each offer their own challenges. Among these, a few of them are explained very briefly [23–26]. The worldwide urbanization rate (in percentage) with smart city challenges by year is depicted in Figure 1.1.

1.2.1   Waste Management

Disposal of solid waste into landfills has a huge impact on the environment and it threatens the safety of the ecological system. If waste in smart cities passes through the water supply, it would make the water unfit for drinking. Sewage waste mainly consists of chemical constituents and microbial organisms and its effect on humans is characterized by physical ill health. Second is sludge, which is a semi-solid precipitate created from waste treatment plants in cities. This type of waste contains higher concentrations of heavy metals, nonbiodegradable organic compounds, nitrogen, phosphorous and pathogenic microorganisms. The sludge, with rich organic matter and nutrients, may be utilized as fertilizer on land for enhancing the soil [25, 26]. The schematic representation of wastewater containing sludge and sewage waste is shown in Figure 1.2.

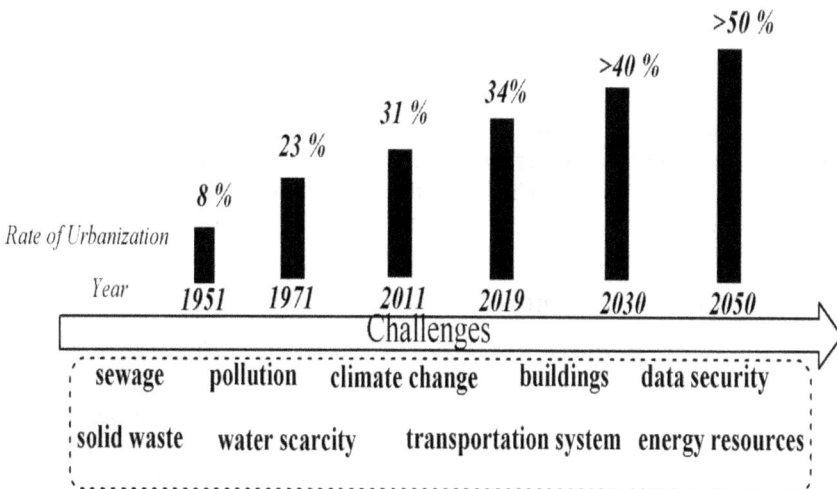

FIGURE 1.1   Rate of urbanization (in %) between 1951 and 2050 at the global level.

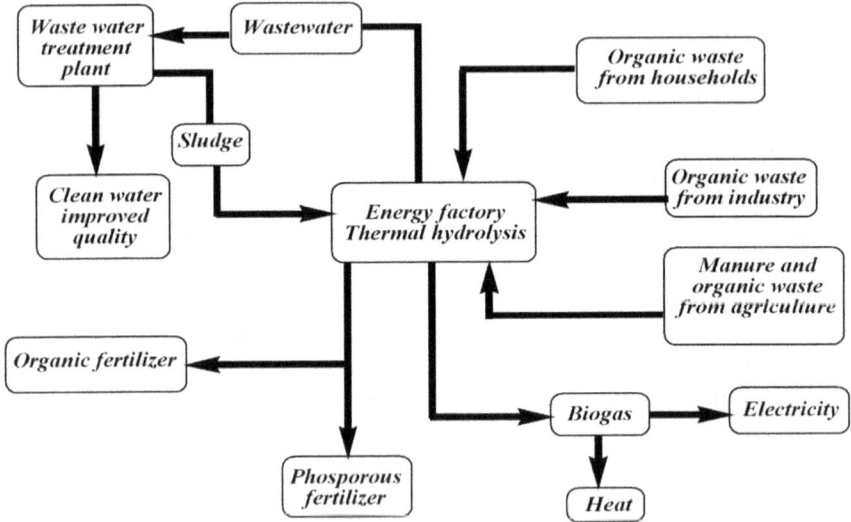

FIGURE 1.2 All-in-one chart for wastewater treatment, energy production and resource recovery.

Managing industrial waste in smart cities is complicated and costly (turnover > $433 billion per annum and engages around 40 million workers). In recent years, recycling waste materials in smart cities has become an essential activity and 12 million employees are currently working in this industry in Brazil, China and United States of America [9, 10]. A recent study in India has reported that of more than 62 million tonnes of waste materials generated per annum, nearly 80% is disposed of in an unhygienic way, which leads to adverse effects to ecological systems.

The municipal solid waste (MSW) in urban cities has been categorized as follows: biodegradable (51%), recyclable (17.5%) and inert (31%) respectively. The recent data clearly pointed out that the continuation of the current scenario in India (nearly 62 million annual generation of MSW) requires 1,240 hectares of land space every year for dumping without further regeneration and recycling processes. It will reach more than 165 million tonnes in 2031 and the requirement of land will become more than 66,000 hectares [12].

1.2.2 CREATING SMART, GREEN AND LIVEABLE CITIES

Due to increasing congestion, six out of ten people will shift their habitat to urban cities by 2030. All the previously mentioned problems create adverse effects on human beings and the environment. According to United Nations Sustainable Development Goals (UNSDG), creating smart cities with sustainability depends on social, cultural, economic and climatic factors, etc. Basic needs such as energy, water and mobility for an increasing population will be met slowly in the fast-growing cities. Smart

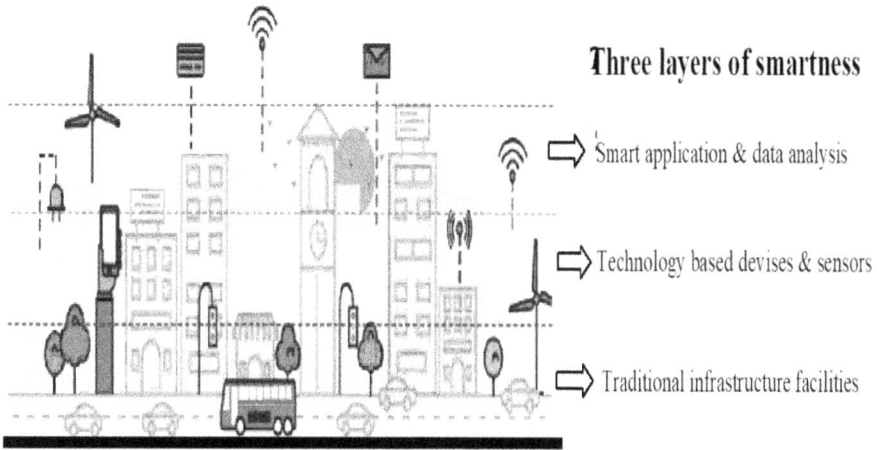

FIGURE 1.3   Pictorial representation of three-layer smartness based on digital intelligence to solve public problems and improve livability in green smart cities.

cities are slowly growing as smart technologies develop to optimize economic activity, energy consumption and environmental impact. Many cities have adopted a holistic approach to enrich 'the good life' for urban citizens in the developing urban areas and to facilitate the transition to inclusive, safe, resilient and sustainable cities and communities [26]. The pictorial representation of three layers of smartness is presented in Figure 1.3 based on digital intelligence to solve public problems and improve liveability in green smart cities.

1.2.3   Moving towards a Circular Economy

The UN reported that the global population increases by more than 83 million people every year. Indirectly this indicates that more than 83 million people per year are ready to share our resources, hence the world's resources are decreasing due to population increase every year. A new and smart economic model must be followed by all people globally. The efficient utilization of resources through a circular economy is the only solution for these challenges [25, 27].

1.2.4   Clean Energy Sources

A conceptual change is taking place every day in the energy-related systems at a global level. Nowadays, fossil fuels are used for energy-related purposes, but one fine day this will be over, leading to an energy crisis among the nations despite technological developments and growing political will. No single clean energy resource is able to power an energy system in isolation. Recently Denmark has generated 99.7% of its power from bio-waste materials using clean energy resource methods [26].

1.2.5   CLEAN AIR

Cities throughout the world are facing many challenges every day. Among them air pollution is one of the most serious challenges, and causes millions of deaths every year at the world level. Air pollution caused by transportation and industries are the main sources in urban cities. To reduce air pollution from transportation, we need to change the mobility pattern of people accessing public transportation and motivate them to utilize electric vehicles and we need to make rules reducing industrial air pollution. Government must encourage new ideas related to reducing air pollution, for example architects designing buildings with good indoor air quality. Since 2018, Denmark has been adopting all these measures and it has a vision of clean air by 2030 [28].

1.2.6   WATER SCARCITY

Drinking water is in inadequate reserve due to industrialization, water pollution and the ever-growing population, etc. A recent report says [28] the consumption of water will be increased 30% by 2030. Many countries have adopted integrated waste water management systems for water purification, recycling and to reduce water consumption. More than 1 billion people have faced water scarcity problems daily due to unavailability of clean drinking water. In addition to this, the very poorest regions of the world lack access to adequate sanitation facilities due to water scarcity. Currently, climate change (droughts and floods) is one of the reasons for the lack of available fresh water. The increase of inhabitants is also a vital part of the water challenges recently. People are suffering not only from the lack of drinking water but also lack of fulfilment of their daily needs, such as energy and food.

1.2.7   CLIMATE CHANGE ADAPTATION

Cities across the world are facing adverse effects from climate change day by day. Climate change due to increasing urbanization is an alarm to society to make our cities greener. The reduction of costs and decrease of ecological impact may be achieved through climate change adaptation plans [29]. Usage of rainwater harvesting techniques is an example of cost-effective resource management in creating green smart cities.

1.2.8   SOCIAL FACTORS OF SUSTAINABLE CITIES

In the past two decades more than 60% of global population occupies cities. There are many social factors affecting the sustainability of smart cities on the global level including population increase, migration of people from one place to another, job scarcity, urbanization, poverty, unavailable resources, lack of infrastructure, challenges for providing services, local political influences, improper utilization of public funds, environmental pollution and climate change [30].

1.2.9 SUSTAINING SMART CITIES BY GREEN IT

Sustainability in smart cities can be achieved using smart technologies. In recent days, superfast computer technology is available, and computer technology is growing so fast that the ordinary person is seemingly always ready to replace their old device. Manufacturers of computer devices consume more than 70% of the world's natural resources. Even though this industry has many advantages, power consumption of computers and their global servers emits CO_2. Data centers' high-energy consumption and resulting surplus emissions of carbon dioxide directly contribute to global warming. But compelling power management systems based on green IT mechanisms in smart cities is a better solution for organizations. These can be created as large data analyzing resources [31].

1.3 GLOBAL-LEVEL INITIATIVES FOR SMART CITIES

Smart cities are expanding their green infrastructure which needs common initiatives and schemes suggested by experts all around the world. This is an emerging area but the experts believe that more research is needed to prove the sustainability of smart cities. Many countries have adopted initiatives or schemes based on their own resources and infrastructure needs for sustainable development of green smart cities, such as energy-based environmental design (EED), sustainable sites initiatives (SSI), green IT-based initiatives, green context-based initiatives and geographical-based initiatives. The pictorial representation of present and future smart city scenarios and its smart changes at the world level are presented in Figure 1.4.

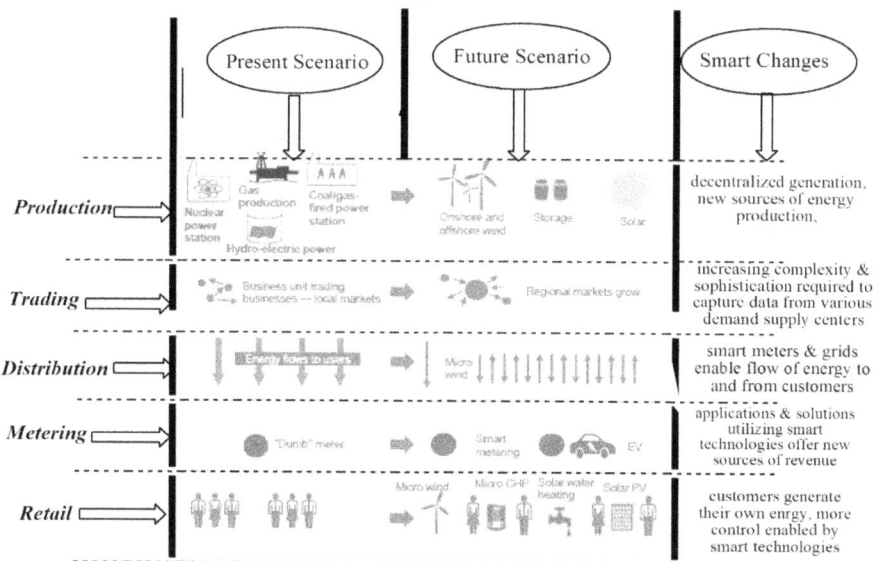

FIGURE 1.4 Graphic representation of present and future smart city scenarios and their smart changes at the world level.

1.3.1 ENERGY-BASED ENVIRONMENTAL DESIGN (EED)

Energy-based environmental design (EED) strategy develops green practices via creation and implementation of globally accepted tools and practices. It is also a globally approved green building system and it has special elements including optimum water, energy, air availability, renewable energy resources, quality of education, innovative ideas and awareness, etc. EED mainly focuses on its design, construction and use for certification because it has used all certified materials for sustainable design. EED promotes sustainable ideas for green smart city development based on energy and efficiency [32].

1.3.2 SUSTAINABLE SITES INITIATIVE (SSI)

SSI has their own guidelines for creating design, construction and maintenance of landscapes. The SSI principles are designed based on the nature, culture, conservation, regeneration, living ecological system to enrich the environment, green ethical and research approach, collaboration and leadership, etc. An idea of SSI helps to promote all types of ecological solutions. SSI's main focus is on protecting and restoring existing hydrologic functions, soil and vegetation for minimizing structure heat by using plants [33]. Transportation in smart cities emits CO_2, CO, NO_2, SO_2, etc. that contribute to environmental pollution. The only way to reduce pollution, is to use eco-friendly vehicles such as bicycles, e-bikes, battery vehicles and fuel cell automobiles. Until now, CO_2 emission is increasing and it is calculated to be one-fourth of the overall energy pollution in the world. Many countries have adopted four types of ideas to design sustainable green smart cities, such as car-free cities, emphasis on proximity, diversity in modes of transportation and access to transportation.

1.4 GREEN IT AND GREEN SMART CITIES

Smart cities can provide a wealth of opportunities for people to enhance their lifestyle and quality of life. In order to meet the increasing population's requirements, governments can introduce a new, smart plan using information technology. IT-based smart cities can use their collective data and analysis to provide remedial solutions for society. With the increase of IT-based products and their applications, our environment suffers a lot from the waste of these products. Thus, a newer Green IT-based technology creates IT commodities for creating smart, greener city applications. The green IT concept was introduced by the US Environmental Protection Agency in 1992. IT aims to reduce the ecological effects from IT-manufactured goods. It helps to ensure sustainability after implementation of green IT in smart cities [34–36]. The trends of green IT shares in smart cities at the global level with challenges are presented in Figure 1.5.

1.4.1 ADVANTAGES OF GREEN IT IN SMART CITIES

Green IT is playing a vital role for the sustainability of green smart cities in many countries due to its wide spectrum of advantages. For example, LCD monitors in

Africa — 23%, 18%
APA
Europe — 34%
America

(bar chart values, top to bottom per bar)

smart retail: 4%, 9%, 53%, 53%
smart agriculture: 4%, 31%, 39%, 30%
smart supply chain: 5%, 12%, 49%, 49%
connected health: 6%, 15%, 55%, 55%
others: 8%, 11%, 50%, 50%
smart energy: 10%, 19%, 54%, 54%
smart automobiles: 11%, 12%, 54%, 54%
smart buildings: 12%, 13%, 53%, 53%
smart industries: 17%, 20%, 45%, 45%
smart cities: 23%, 18%, 34%, 34%

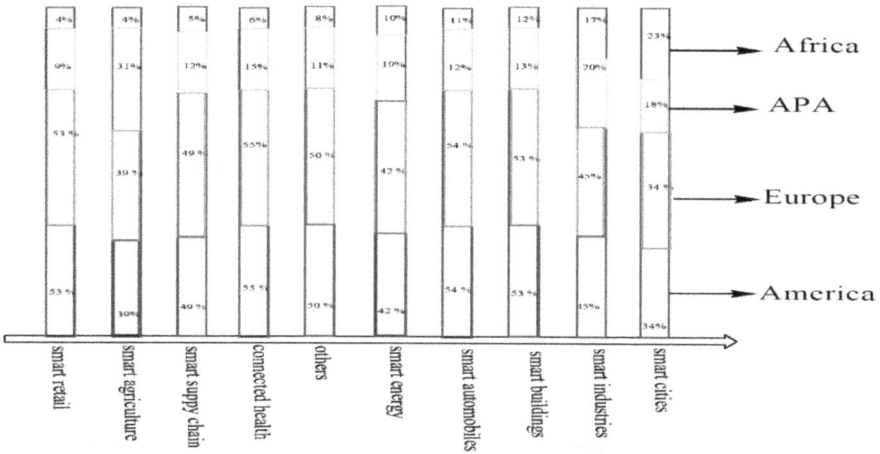

FIGURE 1.5 Trend of green IT shares (in %) in smart cities at the global level; blue color denotes America, rose colour indicates Europe, violet colour indicates Africa, green colour indicate Asian pacific nations.

computers contain minimum amounts of lead (Pb) content and also they require less power than cathode ray tube monitors. New LCD monitors in computers use light-emitting diodes (LEDs) as an alternative of fluorescent bulbs, but because they have a small amount of mercury, they can create health problems for humans due to the toxic nature of mercury. Recent years green IT put into practice technologies such as teleconferencing and telepresence that are useful in reducing the emission of greenhouse gases. Green IT distributes the energy supply as per government requirement. For instance, manufacturers can easily reduce power consumption of the computer [35, 36]. Use of sustainable technology is the need of the hour for creating green smart cities. The green IT is used to create hardware products and software applications that consume fewer resources. The growing demand for computer devices can use the product longevity concept of green IT. The recycling of computer parts will reduce the time and resource wastages for manufacturing new computers and reduce the production of destructive materials. CO_2 emissions could be minimized by using a decentralized system via interconnection of the systems based on green IT.

1.5 GEOGRAPHICAL INFORMATION SYSTEM (GIS)-BASED GREEN SMART CITIES

GIS technology applications are needed to reach climate-smart and green cities practically. It has potential applications for developing green smart cities amidst arising climate change issues. Information and communication technological (ICT) tools give ample support for GIS. GIS offers competent and cost-effective maintenance, scheduling, and growth of green regions on the global level [37]. The advantages of

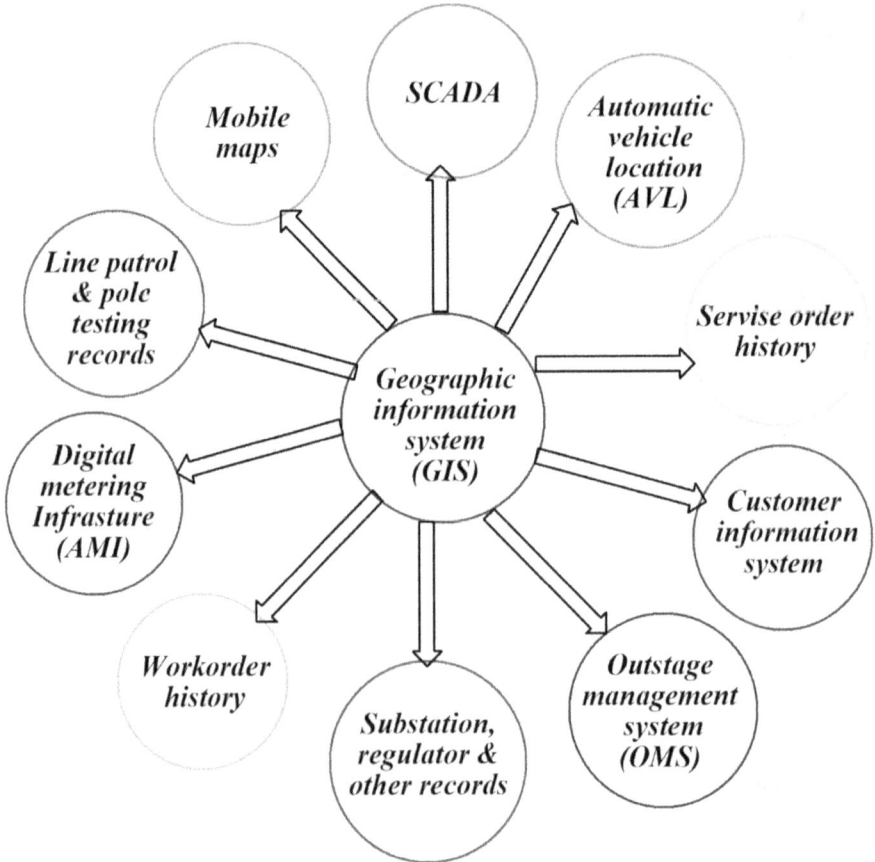

FIGURE 1.6 Schematic representation of various smart city applications using GIS techniques in real life.

GIS are enormous such as maintenance, communications, administrations, decision making, etc. Various applications in daily life using GIS techniques are shown in Figure 1.6.

1.6 GREEN WASTE MANAGEMENT PROCESSES AT GLOBAL LEVEL

1.6.1 BIOCHEMICAL DEGRADATION PROCESSES

Generally, biodegradation processes include refuse reduction using microorganisms, oxidation in the presence and absence of air, and the use of biochemical techniques as shown in Figure 1.7. During oxidation in the absence of air, the organic matter, garbage, leaves and manure, is converted into humus and organic acid and CO_2, but in the presence of air it releases methane as the end product.

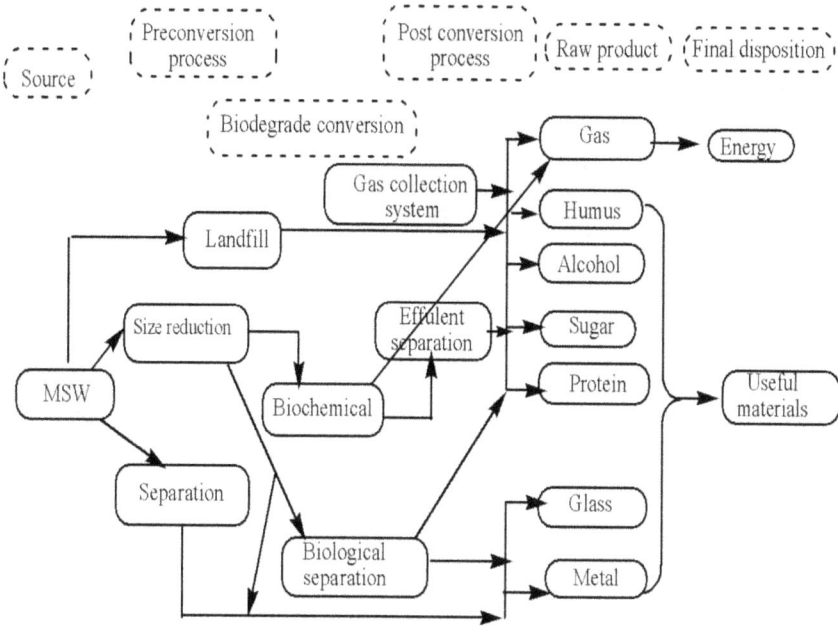

FIGURE 1.7 Biodegradation of refuse.

1.6.2 COMPOSTING

In the presence of microorganisms (bacteria, fungi and actinomycetes) as well as heat and moisture all organic waste matter collapses into humus (i.e., compost). Composting is a simple and benign environmental process, which can be done on a small scale in households and a large scale in industries. Microorganisms alter the wastes into sugars, starches and organic acids under optimized conditions [38]. In recent days, composting technologies have been improved for converting waste to compost in smart urban cities. Nowadays, many types of composting methods are available. Composting has many advantages, such as converting waste into useful fertilizer, increasing the fertility of agriculture land, not requiring high-level equipment, being cost effective, etc.

The window-based composting technique consists of mounding waste matter in order to allow airflow and maintenance of temperature. Mounds are sporadically turned via specific equipment attached with paddles for increasing the porosity and providing for the easy diffusion of air. Generally, wastes are piled beneath a roof to shelter from rain. In aerated static pile composting, a set of connections of pipelines are attached with a blower under the mechanical aeration of waste piles. The blower provides hot air to the compost and then it can create a humus composite material. It takes just six to twelve weeks to complete the process.

Vessel-type composting requires air, humidity and heat for conversion of waste into manure retained inside a vessel. It requires one to four weeks of time to complete the process. The main advantage of this process is it is easy to control the conditions of an environment, and it produces very minimum leachate throughout the process.

1.6.3 VERMITECHNOLOGY

Vermitechnology is an advanced method of converting waste into fertilizer using microorganisms. This type of conversion modifies the surface structure of the fertile land, reduces the air pollution, and engages diminution in populations of pathogenic microbes. It presents nutrients for the fast growth of plants, and produces high harvest yields. Much research was studied using vermicompost as an adsorbent for the removal of heavy metal ions in contaminated water [39–42]. Vermicomposting technology is employed in small- and large-scale industries for converting waste into manure or fertilizer due to its cost effectiveness under optimized conditions. This method depends on waste material quality, pH, heat variation, humidity, exposure to air and microbes, etc. [43]. Vermicompost is a humus material derived from animal dung, and it adsorbs heavy metal ions which are present in an industrial effluent contaminated water. Due to the activity of humus material, sustainable macro and micro nutrients in soil should be maintained. In the oxidized form of humus, that humic acid binds with the positively charged heavy metal ions from the industrial effluent affected water system. The plausible adsorption pathway is depicted in Figure 1.8 [43].

The natural waste water contaminants are used microbes or bacteria and infringement themselves and converted into CO_2 and H_2O using modern biotechnology methods, i.e. bioremediation.

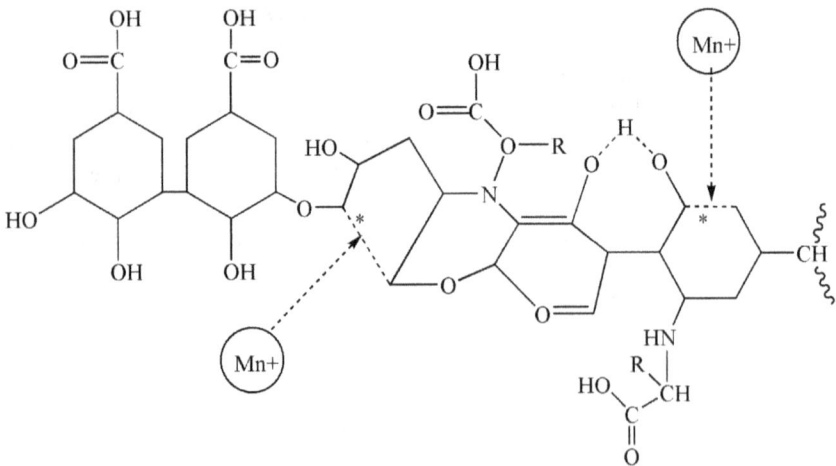

FIGURE 1.8   Vermicomposting mechanism for adsorption of heavy metal ions.

1.7 KEY INFRASTRUCTURE FACILITIES IN GREEN SMART CITY

India is a pioneer in developing countries in sustainable development of smart cities. India has planned to facilitate green smart city task to other nations for enhancing the quality of cities in greener way. The following infrastructure facilities must be included for creating green cities such as sufficient water, electricity, sanitation, waste management, well-organized transport, inexpensive housing for poor people, green IT based digitalization, high-quality e-Governance, ecological sustainability, population well-being, high quality health and education, etc. [44].

1.8 AREA-BASED STRATEGIES OF DEVELOPMENT IN GREEN SMART CITIES

Area-based development in the green smart city promotes city improvement (retrofitting), regeneration (redevelopment) and expansion (green field development). Pan city scheme also provides better solutions for development of green smart cities. Herein three types of area-based development strategies and pan city development in the green smart city are briefly discussed [44].

1.8.1 Retrofitting

This area-based development strategy is applicable for existing settled space containing more than 500 acres. It is based on the well-known areas of the cities and for this technique to become successful, it helps to convert an existing area into an efficient one. It has more intensive infrastructure service levels and an oversized variety of good applications. It may be completed within a short span of time.

1.8.2 Redevelopment

Redevelopment strategies involve the replacement of the present settled areas and to supply redevelopment plans for enhancing infrastructure. It requires more than 50 acres space. Example. Bhendi bazaar project in Mumbai and East Kidwai Nagar in New Delhi.

1.8.3 Green Field Development

Green field development strategy is providing smart solutions in an exceptionally vacant space consisting of more than 250 acres. This innovative plan helps to create affordable housing for poor citizens. It requires developing population in and around the city. Example. Gujarat international finance tec-city (GIFT) in Gujarat.

1.8.4 PAN CITY DEVELOPMENT

Pan city development envisages application of preferred excellent solutions to the current urban-spacious infrastructure. To apply the solutions to the traffic executive system and it reduces average travel time and cost. Example: waste water reprocess and smart metering service.

1.9 CONCLUSION

The global and national level smart city challenges and features briefly discuss the sustainability of various smart solutions. Various existing initiative schemes are suggested to improve smart city related issues based on ecofriendly techniques. Green IT based solutions pave a way for the development of green smart cities. In addition, geographical information systems (GIS) have various applications in smart cities. Increased public awareness of smart city oriented problems and pressure on government sector officials is useful but also every individual has to think and put a step forward to minimize waste to improve their own city.

REFERENCES

[1] Slaper, T.F., Hall, T.J. (2011). *The Triple Bottom Line: What Is It and How Does It Work?* Indiana Business Review. www.ibrc.indiana.edu. Retrieved 2019–10–02.

[2] Richard, R. (1987). *Ecocity Berkeley: Building Cities for a Healthy Future.* Berkeley, CA: North Atlantic Books. ISBN: 9781556430091.

[3] Low, N., Gleeson, B., Green, R., Radovic, D. (2016). *The Green City: Sustainable Homes, Sustainable Suburbs.* Abingdon, UK: Routledge. ISBN: 1136752994, 9781136752995.

[4] Wilfred, L., Christoforos, R., Bruce, C. (2016). *Smart Green Resilient.* Hong Kong: Hong Kong University Press, Ove Arup & Partners. ISBN: 9789881492302.

[5] Hatuka, T., Rosen-Zvi, I., Birnhack, M., Toch, E., Zur, H. (2018). The Political Premises of Contemporary Urban Concepts: The Global City, the Sustainable City, the Resilient City, the Creative City, and the Smart City. *Planning Theory & Practice*, 19(2), 160–179.

[6] Vanolo, A. (2013). Smart mentality: The Smart City as Disciplinary Strategy. *Urban Studies*, 51(5), 883–898. doi:10.1177/0042098013494427.

[7] Viitanen, J., Kingston, R. (2014). Smart Cities and Green Growth: Outsourcing Democratic and Environmental Resilience to the Global Technology Sector. *Environment and Planning A*, 46(4), 803–819. doi:10.1068/a46242.

[8] Wheeler, S.M. (2013). *Planning for Sustainability: Creating Livable, Equitable and Ecological Communities.* London: Routledge.

[9] Taylor, P.J., Derudder, B. (2015). *World City Network: A Global Urban Analysis* (2nd ed.). New York, NY: Taylor and Francis.

[10] Adger, W.N. (2000). Social and Ecological Resilience: Are They Related? *Progress in Human Geography*, 24(3), 347–364. doi:10.1191/030913200701540465.

[11] Coaffee, J. (2013). Towards Next-Generation Urban Resilience in Planning Practice: From Securitization to Integrated Place Making. *Planning Practice & Research*, 28(3), 323–339. doi:10.1080/02697459.2013.78769.

[12] Antrobus, D. (2011). Smart Green Cities: From Modernization to Resilience? *Journal Urban Research & Practice*, 4(2), 207–214.

[13] Ahvenniemi, H., Huovila, A., Pinto-Seppä, I., Airaksinen, M. (2017). What Are the Differences between Sustainable and Smart Cities? *Cities*, 60 (Part A), 234–245. doi:10.1016/j.cities.2016.09.009.

[14] Halbach, T. (2013). *International Trend in Solid Waste Handling.* Minnesota: University Press.

[15] Ahmed, S.A., Alli, B. (2004). Partnership for Solid Waste in Developing Countries: Linking Theories to Realities. *Habitat International*, 28(3), 467–478.

[16] Ayuba, K.A., Abd Manaf, L., Sabrina, A.H., Azmin, S.W.N. (2013). Current Status of Municipal Solid Waste Management Practise in FCT Abuja. *Research Journal of Environmental and Earth Sciences*, 5(6), 295–304.

[17] Salau, O., Osho, S., Sen, L. (2017). Urban Sustainability and the Economic Impact of Implementing a Structured Waste Management System: A Comparative Analysis of Municipal Waste Management Practices Developing Countries. *International Journal of Regional Development*, 4, 1. ISSN 2373–9851.

[18] Patil, A.D., Shekdar, A.V. (2001). Health-Care Waste Management in India. *Journal of Environmental Management*, 63, 211–220. doi:10.1006/jema.2001.0453.

[19] Pandit, N.B., Mehta, H.K., Kartha, G.P. Choudhary, S.K. (2005). Management of Bio-Medical Waste: Awareness and Practices in a District of Gujarat. *Indian Journal of Public Health*, 49(4), 245–247.

[20] Mathur, V., Dwivedi, S., Hassan, M.A., Misra, R.P. (2011). Knowledge, Attitude, and Practices about Biomedical Waste Management among Healthcare Personnel: A Cross-Sectional Study. *Journal of Community Medicine*, 36(2), 143–145. doi:10.4103/0970-0218.84135.

[21] Kiddee, P., Naidu, R., Wong, M.H. (2013). Electronic Waste Management Approaches: An Overview. *Waste Management*, 33(5), 1237–1250.

[22] Ferronato, N., Cristina Rada, E., Gorritty Portillo, M.A., Ionel Cioca, L., Ragazzi, M., Torretta, V. (2019). Introduction of the Circular Economy within Developing Regions: A Comparative Analysis of Advantages and Opportunities for Waste Valorization. *Journal of Environmental Management*, 230, 366–378.

[23] O'Dwyer, E., Pan, I., Acha, S., Shah, N. (2019). Smart Energy Systems for Sustainable Smart Cities: Current Developments, Trends and Future Directions. *Applied Energy*, 237(1), 581–597.

[24] Trindad, E.P., Hinning, M.P.F., da Costa, E.M., Marques, J.S., Bastos, R.C., Yigitcanlar, T. (2017). Sustainable Development of Smart Cities: A Systematic Review of the Literature. *Journal of Open Innovation: Technology, Market, and Complexity,* 3(3), 11. doi:10.1186/s40852-017-0063-2

[25] Elagroudy, S., Warith, M.A. El Zayat, M. (2016).*Municipal Solid Waste Management and Green Economy* (pp. 1–69). Berlin, Germany: Global Young Academy Publisher.

[26] Bishoge, O.K., Huang, X., Zhang, L., Ma, H., Danyo, C. (2019). The Adaptation of Waste-to-Energy Technologies: Towards the Conversion of Municipal Solid Waste Into a Renewable Energy Resource. *Environmental Reviews*, 27(4), 435–446.

[27] Maina, S., Kachrimanidou, V., Koutinas, A. (2017). A Roadmap towards a Circular and Sustainable Bioeconomyrefj through Waste Valorization. *Current Opinion in Green and Sustainable Chemistry*, 8, 18–23.

[28] Achillas, C.H., Vlachokostas, C.H., Moussiopoulos, N., Banias, G., Kafetzopoulos, G., Karagiannidis, A. (2011). Social Acceptance for the Development of a Waste-to-Energy Plant in an Urban Area. Resources, Conservation and Recycling, 55(9–10), 857–863.

[29] Badaloni, F.C., de Hoogh, K., Von Kraus, M.K., Martuzzi, M., Mitis, F., Palkovicova, L., Porta, D., Preiss, P., Ranzi, A., Perucci, C.A., Briggs, D. (2011). Health

ImpactAssessment of Waste Management Facilities in Three European Countries. *Environmental Health*, 10, 53. doi:10.1186/1476-069X-10-53.

[30]   Batty, M. (2013). Big Data, Smart Cities and City Planning. *Dialogues in Human Geography*, 3(3), 274–279. doi:10.1177/2043820613513390.

[31]   Low, L.N. (2005). *The Green City: Sustainable Homes, Sustainable Suburbs* (1st ed.). Sydney: University of New South Wales Press.

[32]   Coaffee, J. (2013). Towards Next-Generation Urban Resilience in Planning Practice: From Securitization to Integrated Place Making. *Planning Practice & Research*, 28(3), 323–339. doi:10.1080/02697459.2013.78769.

[33]   Crnčević, T., Tubić, L., Bakić, O. (2017). Green Infrastructure Planning for Climate Smart and Green Cities. *Spatium*, 38, 35–41. doi.10.2298/spat1738035c.

[34]   Simmhan, Y., Ravindra, P., Chaturvedi, S., Hegde, M., Ballamajalu, R. (2018). Towards a Data-Driven IoT Software Architecture for Smart City Utilities. *Journal of Software: Experience and Practice*, 48(7), 1390–1416.

[35]   Prasad, D., Alizadeh, T. (2020). What Makes Indian Cities Smart? A Policy Analysis of Smart Cities Mission. *Telematics and Informatics*, 101466. doi:10.1016/j.tele.2020.101466.

[36]   Tiwari, A., Jain, K. (2014). GIS Steering Smart Future for Smart Indian Cities. *International Journal of Scientific and Research Publications*, 4(8), 1–579.

[37]   Kandpal, V. (2018). A Case Study on Smart City Projects in India: An Analysis of Nagpur, Allahabad and Dehradun. *Companion Proceedings of the Web Conference 2018, International World Wide Web Conferences Steering Committee*, 935–941.

[38]   Elagroudy, S., Warith, W.A., El Zaya, M. (2016). *Sustainable Solid Waste Management and the Green Economy* (pp. 1–69). Berlin, Germany: Global Young Academy. ISBN: 978-3-939818-65-6.

[39]   Urdaneta, L.M., Marcó Parra, S., Matute, M., Angel Garaboto, H., Barro, C., Vázquez, S. (2008). Spectrochimica Acta Part B: Atomic Spectroscopy, Evaluation of Vermicompost Asbioadsorbent Substrate of Pb, Ni, V and Cr for Waste Waters Remediation Using Total Reflection. *X-Ray Fluorescence*, 63(12), 1455–1460.

[40]   Durán, A.C., Flores, I. (2009). Evaluation of Lead (II) Immobilization by a Vermicompost Using Adsorption Isotherms and IR Spectroscopy. *Bioresource Technology*, 100(4), 1691–1694. doi:10.1016/j.biortech.2008.09.013.

[41]   Jadia, C.D., Fulekar, M.H. (2009). Phytoremediation of Heavy Metals: Recent Techniques. *African Journal of. Biotechnology*, 8(6), 921–928.

[42]   Boni, M.R., Sbaffoni, S. (2009). The Potential of Compost-Based Biobarriers for Cr (VI) Removal from Contaminated Groundwater: Column Test. *Journal of Hazardous Materials*, 166(2–3), 1087–1095.

[43]   Lakshmi, C., Nagarajan, N., Karuppasamy, P. (2017). Bioremediation of Soil from an Industrial Effluent Affected System Using Vermicompost. *International Journal of Current Science Research and Review*, 3(11), 1426–1451.

[44]   Adapa, S. (2018). Indian Smart Cities and Cleaner Production Initiatives: Integrated Framework and Recommendations. *Journal of Cleaner Production*, 172, 3351–3366.

2 Transforming Green Cities with IoT
A Design Perspective

T. Deva Kumar, T.S. Arun Samuel
and T. Ananth Kumar

CONTENTS

2.1 INTRODUCTION

The word 'harmony' means the quality of establishing a pleasing and consistent whole. In a modern city, the principle is "make a habitable place in harmony with good local conditions." With this principle in mind, many agencies in the world presented standards, such as the EPA (Environmental Protection Agency) [1], Building IQ (Building IQ is an agency that helps owners and operators worldwide lower energy use and increase building operations efficiency and tenant comfort), IGBC (Indian Green Building Council) and USGBC (U.S. Green Building Council). The latest approaches to urban planning promote innovative concepts based on the principle

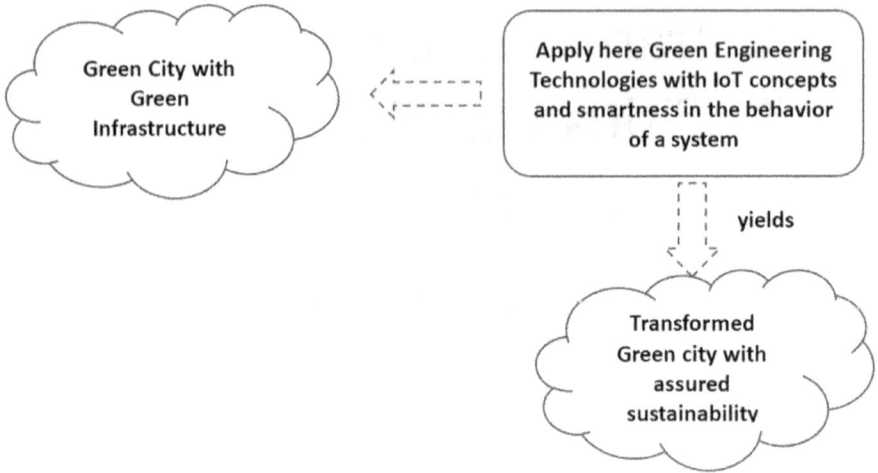

FIGURE 2.1 Overview of green city transformation by IoT.

that "The city will act as a living organism with complex metabolism" [2]. The new concepts meant for this purpose are green urbanism, bio-urbanism, organic urbanism, biophilic city (biophilic means the human tendency of having life associated with nature), eco-city, or sustainable city, and green city.

This chapter describes the introduction of smartness to maintain the sustainability of green cities and describes the procedure for green city development. Here, the smartness is introduced with the adaptation of IoT technologies [3]. Also, the word sustainability relates to the maintenance of the green infrastructure of green cities. Without the smartness introduced by IoT, the sustainability of green cities is impossible. These ideas produce the concept "Green city transformation by IoT." Figure 2.1 shows the overview of green city transformation by IoT. Here, the sustainable green design should consider the impact on the environment and a sense of minimizing waste production and pollution. Also, the design should consider the various inputs for the green city such as energy, water, and food.

2.1.1 GREEN CITY AND ITS BENEFITS

In general, the word 'green' suggests natural resources such as forests, wildlife, and river banks. Green city is a group of individuals, neighbors, staff, and visitors trying to combine economic, environmental, and societal needs to ensure that every member of society and every generation will have a stable, secure, and clean environment [4][5].

The significant benefits of green cities are

- Oxygenation and purification of air.
- Increasing biodiversity by creating semi-natural habitats.
- Mitigation of the heat-island phenomenon.

- Psychological and sanogenesis impact on people.
- A dense cover of plants will reduce soil erosion.
- By using trees to modify the temperature, the number of fossil fuels for cooling and heating is reduced.

By managing Green infrastructure properly, that will become local tourist assets that will enhance the communities' economic benefits.

2.1.2 Components of Green Cities and IoT's Role in Their Transformation

The components of green cities influence their functionality and morphology, which reflects the difference between then and unplanned cities [6]. The main components of any green city are,

- Green and blue oxygen-production areas: These areas contain planted flowers, shrubs, and trees.
- Blue-green corridors: This component integrates the water surface and green areas.
- Green belts: These areas are around the green city, with a view toward protecting natural elements.
- Green walls and vertical gardens.
- Greenhouses: this is a new housing tradition blending energy efficiency with cultural aesthetic values.

The elements mentioned here are the parts of the green infrastructure of a green city. Green infrastructure refers to interconnected elements of green areas and hydrographic elements that contribute to the preservation and enhancement of biodiversity. Furthermore, they maintain the abiotic process within the urban environments close to their natural conditions [7]. Thus, they promote the quality of life and sustainability. Apart from that, green infrastructure contributes to the natural process of keeping water and the air clean and recycling waste. For maintaining some degree of balance in the natural system, it is needed to have engineering systems to create and transport energy, remove and process waste, and to control the storm runoff. Figure 2.2 illustrates the green infrastructure components of any green city. Also, the diagram depicts the various components associated with that green infrastructure. A "green building" is an essential element within the green infrastructure [8]. It creates structures and uses environmentally friendly and resource-efficient processes throughout the lifetime of the building, from design to construction, operation and maintenance. With the green building concepts, if the information technologies are included, the green building is converted into a "living building" [9]. Here, the information technologies enable

- Effective utilization of data and resources.
- Reduced building life cycle cost.
- Creation or establishment of an adaptable, responsive learning space.
- A self-sustaining entity.

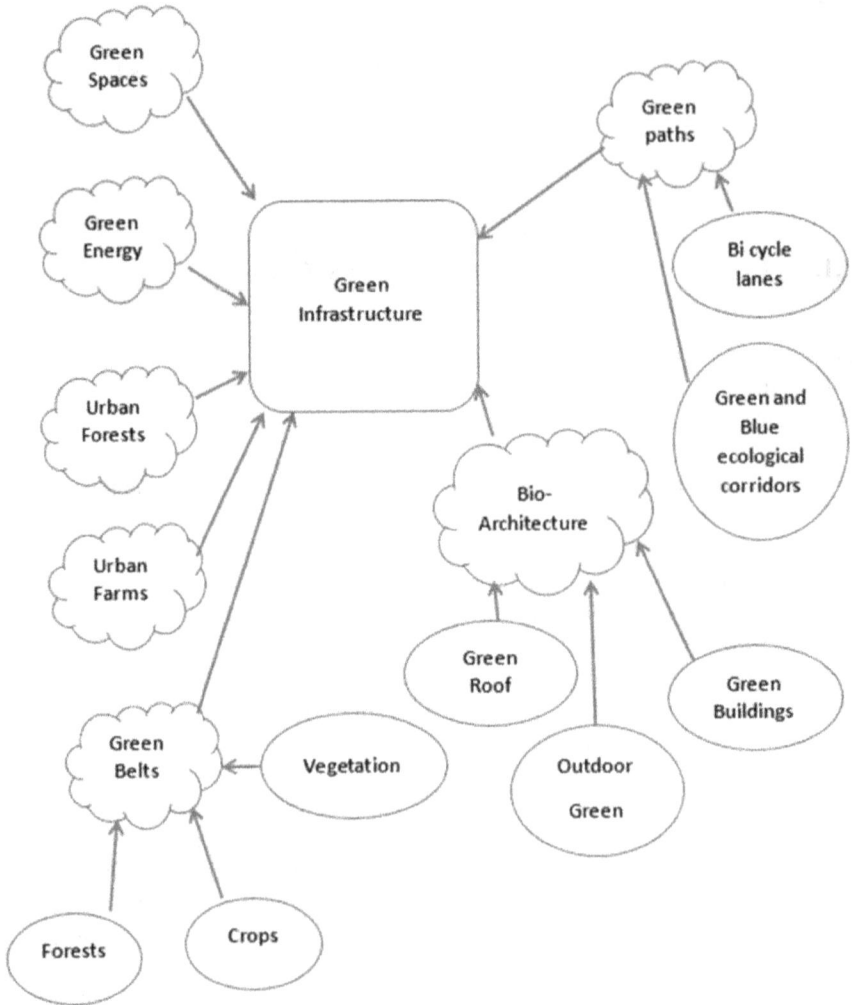

FIGURE 2.2 Green infrastructure components of any green city.

The green building initiative successfully creates awareness of, adoption of, and influences the market for energy-efficient materials and design practices [10]. The focus now is to establish specific standards and compliance for the building design, material preparation, operations tracking, and energy use of the building through the life cycle. The establishment of a green building faces challenges while integrating different systems to achieve desired green factors in order to

- Identify the degradation of equipment for building.
- Be able to operate the building for energy optimization and better comfort.
- Achieve a useful measurement and verification process.

By adopting IT infrastructure with IoT technology, this challenge is easily overcome. IoT enables green buildings to become connected buildings [11]. The information from these connected buildings enables the administrative system to render appropriate services to maintain the buildings' sustainability. Indeed, the general idea is to equip green buildings with sensor-fitted embedded systems and advanced communication infrastructure, which includes IoT technologies-based services transforming green buildings into connected buildings. Connected buildings will generate massive data from multiple systems such as energy, IT access equipment, and security-related elements and operation upon green infrastructure components. Thus, the administrating systems are getting a massive volume of data of different variety and different velocity. The collection of these data is called "big data." Figure 2.3 shows the generalized diagram of green buildings with the sensor-fitted embedded systems connected with the big data system [12].

Using techniques related to artificial intelligence and machine learning, big data are manipulated to render specialized services to green infrastructure to maintain sustainability [13]. These specialized services include maintaining operational efficiencies, energy consumption, and reduction. Also, IoT combined with ubiquitous computing generates enormous data. In particular, they enhance building services. Figure 2.4 depicts the overlapping purpose of building services.

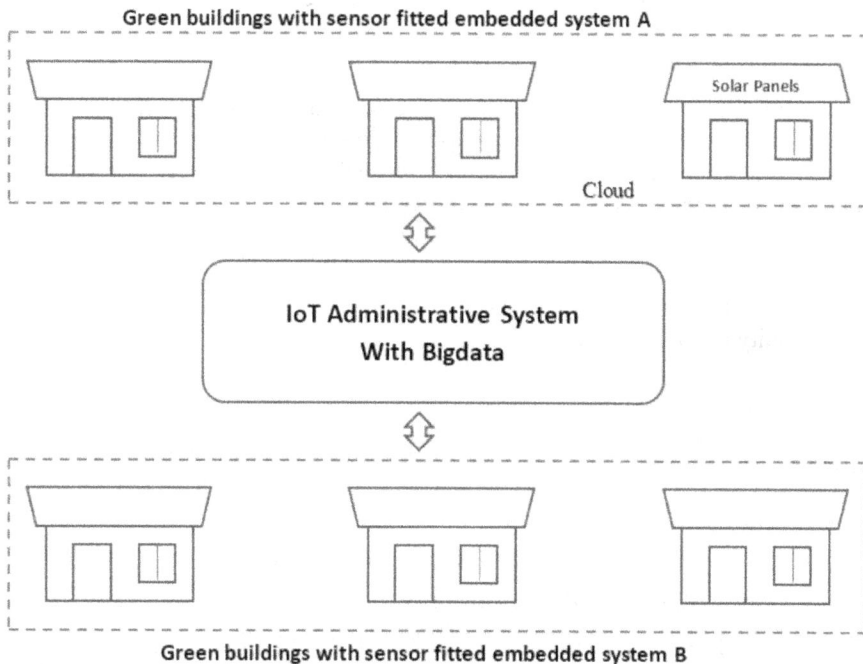

FIGURE 2.3 General block diagram of green buildings into connected buildings.

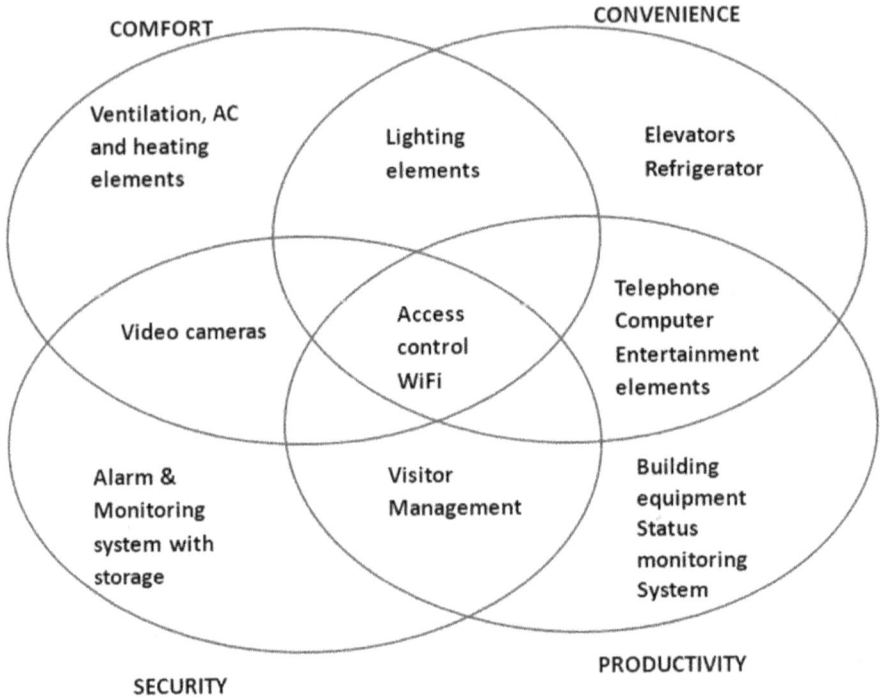

FIGURE 2.4 Overview of building services.

The building services are classified into four categories: comfort-related services, convenience-related services, productivity-related services, and security-related services. While rendering these services, if IoT technologies are included, the following characteristics can be added with green buildings:

- Ability to forecast operation.
- Ability to predict the operation.
- Ability to optimize the operation.

These abilities enable green buildings to become self-estimated, self-regulated, and self-optimized buildings. This transforms a "Sustainable green building" into an "IoT enabled living building" [14].

2.1.3 Objectives of Modern Green City Development and Some of Its Design Considerations

Zero carbon, zero waste, and healthy ecosystem are the three main objectives of modern green city development design [15]. To keep this in mind, some of the planning considerations are noted here.

- Introduce zones that require mixed use of community space.
- Ensure clean water and air.
- Generate electricity only from renewable resources.
- Ensure strict building codes favor green technology.
- Encourage and increase the volume of public transport use.
- Implement efforts and policies to achieve zero waste and to reduce overall water consumption.
- Ensure early access to affordable and healthy food.
- Allow and encourage pedestrian and bicycle usage to achieve zero CO_2 emissions.

IoT technologies play a significant role in achieving nearly zero emissions and zero wastage. In other words, IoT technologies are used to monitor and control the carbon footprints of buildings. Also, they are used to monitor and control waste management schemes within green cities [16].

2.2 TRANSFORMATION OF GREEN CITIES WITH IOT TECHNOLOGY

In green cities, human beings are living with the benefits of green infrastructure. For maintaining the green infrastructure for future utilization, some engineering implementations should be integrated with green infrastructure [17]. One such technology is the Internet of things (IoT). Figure 2.5 shows the block diagram of the transformation of green cities with IoT technology. The subsequent paragraph explores the transformation of conventional "green cities" into "sustainable green cities" using IoT technologies. Smartness using IoT technologies becomes the inbuilt characteristics of the green city with its green infrastructure [18]. Thus, by using IoT technologies, smart environment, smart economy, smart governance, smart mobility, and smart living are developed and made into the inbuilt components of the green city, along with its green infrastructure.

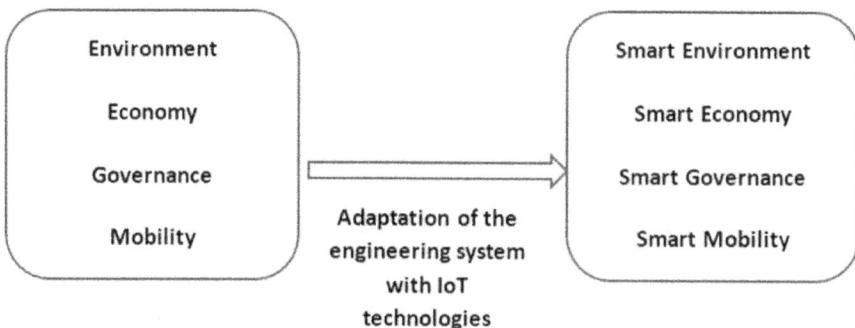

Environment	Adaptation of the engineering system with IoT technologies	Smart Environment
Economy		Smart Economy
Governance		Smart Governance
Mobility		Smart Mobility

FIGURE 2.5 Transformation of green cities with IoT technology.

Thus, the critical implementation of IoT technologies is the establishment of a smart green city. That city includes the following facilities or characteristics using IoT technologies [19].

- The system has less greenhouse gas emissions.
- It maintains greenhouse gas emissions at acceptable levels. Particular examples are green buildings and transportation.
- The system monitors and controls for better resource utilization. Specific examples are water management, building space management, green space management vegetation or farm management, data transport management, and waste management. This last item gives us a specific directive: waste also has to be treated as one of the resources. By adopting this concept, various systems can be developed with the support of IoT technologies, enhancing the design of a sustainable green city.
- Smarter urban transport networks, an upgraded water supply, and more efficient ways to light and heat buildings are needed.

The intelligent system encompasses a more interactive and responding city administration with safer public space that enables the needs of an aging population to be met. In general, greenhouse gas (GHG) control, living buildings, and waste management are the main issues related to green cities [20].

Smart green cities are the future technology that emphasizes the impending digital arrangement to enhance the green environment and that simplifies the authorization of green space planning. Energy management is the most sensational issue that drives the implementation of smart green cities. As the devices used in smart cities are wholly equipped with the IoT, there is a need to ensure the sustainability of such devices. For real-time IoT applications, there exists a demand for sensitivity and latency. Edge computing can provide support to deal with these kinds of issues by providing the most well-organized structure in reducing delays and improving the system's performance. The applications of edge computing and its technologies deal with the development of energy-efficient smart green cities in IoT environments. Along these features an enhanced security mechanism is included in this work [21]. A novel model for the management of soil, water, and the air is proposed by using edge computing technologies. Here traditional approaches are not considered in the handling of water, soil, and air because they mainly focus on dispersed features. The water, soil, and air quality is predicted and analyzed using a smart scheme that incorporates all the quality standard indicators. The sources needed for the analysis are collected directly from the smart city areas.

The quality of the water, soil, and the air is predicted and determined using an improved neural network-based approach. The water, soil, and air quality are calculated by using advanced modeling and analyzing techniques. The obtained value is compared with the historical data and can predict the evolution of water, soil, and air in the smart city areas. The prediction accuracy is improved, and the efficiency is enhanced using optimization techniques—the input for the analysis is obtained from the sensors placed in various locations of smart green cities. The proposed method also concentrates on the privacy and security of the resources in the smart green cities. The data collected for the analysis might be very sensitive for security and the privacy threats because the data are collected mostly from private agencies. For enhancing the security of the system, a differential storage method is used. That includes an integrated blockchain-based approach with Software Defined Networks (SDN), which is done by splitting the network into different channels. The blockchain method will deliver a safety base for data regulation and the transmission of information between the unconventionally working devices in the edge-based smart green city environments. Blockchain helps ensure a safe transmission scheme and thus enhances authentication and protection. The safety measures are inputted directly to the system by including them individually in all sensors placed. The information from the sensors is collected using a four-way handshake method. A single channel that may contain a lot of sensor information is obtained from various sensors placed. The data sharing is made efficient and safe by using the SDN approach. This provides a brilliant edge-based environment that incorporates many functionalities for obtaining reliable resources. This will create a sustainable environment with more security features. The proper handoff mechanism with proper authentication routing strategy is proposed here. This will provide security against all kinds of suspicious attacks and thus improve the network's performance and reduce the response time. Some more parameters considered in this system are the management of latency, sustainability, and interoperability.

2.3 CONTROLLING CO_2 EMISSIONS WITH IOT TECHNOLOGY

Pollution has a significant part to play in destroying cities. Ambient carbon dioxide is driving global warming, which triggers the consequences of climate change, including the melting of ice caps, rising flood levels, the destruction of the environment for wildlife, severe weather, and even more negative side effects harmful to the earth, human lives, and the future, and the implications of this impact are also significant [22]. More than 50% of the world's population is metropolitan, and towns play a significant role in the production of global greenhouse gases (GHG). The Internet of Things (IoT) attracts many innovative real-time services. Big data analytics enable intelligent city initiatives worldwide, which result in marching toward smart city development. Infrastructure development, transport networks, reduction of traffic congestion, waste management, and enhanced quality of life are some of the operational parameters in smart city development. This chapter focuses on the aforementioned parameters for developing a smart city with integrated ICT's in different networks. A description of the existing activities of standard organizations was also added to increase interest. Thus, the briefing about the current smart city infrastructure and its behavior are followed by the case studies. The current synergies and global efforts to support IoT in smart cities are summarized. Finally, to provide potential directions for research, a range of IoT challenges are highlighted.

Smart cities hold that name as long as the components in smart cities act smart with the real-time environment and serve them appropriately in an intelligent manner. Components include items used daily such as mobile phones, gadgets, microwaves, toasters, fridges, air conditioners, vehicles, watches, buses, and trains. The IoT is a smart world that connects everything in the smart world. Motivated by achieving a sustainable smart planet, a variety of green IoT innovations and concerns that further reduce IoT energy consumption are addressed. Notably, an analysis of IoT and green IoT is first explored, and then new developments and a future vision of the cloud paired with the active sensors are examined, which is a new paradigm in green IoT. The IoT gives cities new ways to use the information for traffic control, emissions reduction, better usage of resources, and protection for the public. It is insightful and the latest guide to smart green city solutions. Emerging smart cities will use ICT to tackle urban challenges and improve public well-being. The IoT connects all electronics and with them also gets connected with the things, which are sensor-enabled, with actuators and networking elements with significant ICT development. It also introduces many new techniques and software to become compatible with smart grids and their corresponding infrastructure, intelligent smart homes, schools, and many others. This IoT moves towards a greener society through this integration by reducing the energy and power generated. IoT incorporates radio or wireless communication to get connected. Many smart devices are battery-driven, and battery replacement can be difficult, if not impossible, in many situations. Regardless of the power source, global IoT power consumption with trillions of devices deployed is expected to become significant. Thus, with the continuous development of IoT for use in smart cities, energy efficiency is definitely an essential part of the 'green IoT for smart city' research agenda.

Artificial intelligence (AI) is one of the most challenging advancements within recent memory. Enthusiasm for the utilization of AI for urban development continues developing every day [23]. Green city areas are expected to oversee developing urbanization proficiently and with vitality, maintain a green standard to improve the cost and expectations for everyday comforts of their residents, and raise individuals' capacity to proficiently utilize and embrace cutting-edge ICT [24]. In the smart city community concept, ICT is an essential tool in arrangement plan, choice, execution, and for establishing an extremely profitable administration. Here, the indispensable target is to investigate the job of AI and machine learning (ML) in the development of a smart green city. Demonstrating urban frameworks has intrigued organizers and modelers for a considerable length of time. Various models have been created that depend on science, cell robotization, intricacy, and scaling. The vast majority of these models will generally improve the real world. Today, the use of computer views shows promising potential for understanding the sensitive elements of urban areas, in terms of the changes in the perspective of human brainpower in different fields of science. Computer vision demonstrates improvement in managing several dynamic physical and nonphysical visual transmissions. Computer vision contributes most to developing digital scenarios for the advent of the smart green world. To provide a safe atmosphere in complex areas like galleries, houses, and parking zones, people need to use internet protocol-based addressable cameras. The usage of computer vision mainly helps to identify any movements spotted in those areas with the help of the IoT.

The IoT may permit the interconnection of the physical and the computer-generated things. The application of computer vision in addition to the sensors and IoT paves the way for surveillance, autonomous vehicles, smart agriculture, Industry 4.0 and robotics [25]. Here, the model of algorithms is discussed with the image recognition application, which even identifies the characteristics of an image taken. The proposed method uses sensors to capture the image and processes the obtained image by extracting the desired feature using various algorithms like histogram of oriented gradients (HOG) and scale-invariant feature transform (SIFT), also with face detection. The additional features attached to the smart city via computer vision are the monitoring of the real-time temperature, air pressure, CO_2 level and also the humidity. This can be done by attaching non-wired sensors in the streetlights. The camera is placed at a height similar to the streetlights to detect all activities. The Zigbee module can correspond to the communication of the sensors. Using these decentralized street light monitoring devices will also support the real-time monitoring of traffic conditions and accessing the same through computer vision. This will also support the prioritization of the vehicles intended for some emergency applications (ambulance). Violation of the traffic rules can easily be identified, and the statistics related to this are collected by using an improved algorithm that is flexible for controlling the traffic with the no-collision method. The approach used here is fuzzy-based, which effectively handles the movement of emergency vehicles across the cross-congestion area. Along with this, a new feature is added to find spots for parking using the same computer vision associated with the streetlight-fitted cameras. This uses convolutional neural networks (CNN) to analyze the images taken, and artificial intelligence lends a hand in deciding factors depending upon the gathered data that is stored in the cloud.

2.3.1 Carbon Dioxide (CO_2) and Its Impacts

Carbon dioxide is an important environmental variable; it warms the surface of the earth. However, tracking and managing all these environmental changes is also a tough hurdle. An even more appropriate and efficient approach of prevention and control techniques is long-term monitoring of CO_2 emissions. Some of the CO_2 emissions occur via public transport, industry, and forest fires. Air pollution is an unhealthy, dangerous byproduct, contributing to global imbalances and endangering the natural ecological cycles. Exposure to air pollution triggers many breathing problems such as asthma and bronchitis, raising the risk of life-threatening disorders with considerable health costs.

Here, early detection of forest fire is discussed using an IoT based smart system. It is an evolving technique that connects embedded computing systems to an Internet service platform that enables device communication directly to other devices. IoT involves collaborating with these independent devices to interact rapidly and generate aggregated data without human intervention. Sensor nodes, software modules, and Raspberry Pi computers are needed to make an IoT based smart system.

2.3.2 Examples of CO_2 Emissions

Greenhouse gases that lead to global warming form the principal exhaust from motor vehicles. The major greenhouse gas in automobiles is carbon dioxide (CO_2). The rate of CO_2 emissions from automobiles is related to fuel consumption and quality. The field of road transport relies on oil-based fuel sources. The Green Vehicle Guide

(GVG) ensures that the automobile provides lower CO_2 emissions from the tailpipe. The blanketing impact of CO_2 in the air allows the planet to gradually warm up.

A wide range of industrial processes that do not pertain to energy is responsible for the emission of CO_2. Industrial development processes are the prime determinants of pollutants that modify materials physically and chemically. Several greenhouse gases including carbon dioxide, methane, nitrous oxide and chlorofluorocarbons could be released in those operations. A striking feature of a manufacturing environment that generates a substantial amount of carbon dioxide is cement production. Fire, as an influential producer of global CO_2, was ignored internationally. A few research findings have shown that CO_2 emissions from flames could be up to half the global fossil fuel consumption emissions. In 1997, a wildfire in Indonesia released carbon dioxide emissions into the atmosphere of 13–40% of the global emission of human fossil fuel.

Complexity, connectivity, environmental impact, and power are confronted by the designs discussed in this chapter. There is a wide range of wireless and wired communication standards for various products and communication needs in this framework. IoT technology can influence the carbon emissions consequences of the development and the disposal at the end of all the equipment involved. In standard garbage dumps, many electronic devices may be burnt or deposited, which pollutes soil, air and surface water. All program or hardware malfunction or faults will have massive issues. A loss of power could create many disturbances.

Four main modules form the method of tracking CO_2 emissions: sensing control, data-semantic-storage, decision making and user interface module. The sensor detects carbon dioxide levels released directly into the physical world. Signals are transmitted and communicate when the CO_2 content exceeds anticipated amounts. These are transmitted to the top layer. The sensor control module data is placed in the directory and evaluated productively. It was necessary to translate sensed data into digital formats via ADC and in the format of analog signals. It makes choices using information from the lower semanticized layer. The decision-making system sends the central inspection board notifications that are based on the summarized data. This unit provides essential social network interfaces. The control mechanism is determined by the input and assessment tests. The whole method is divided into two parts: one tracks and controls the concentration of CO_2 generated by the automobiles' emissions and the second monitors the forest ecosystem to forecast the reduction of CO_2 emissions from wildfires.

The sensing modules that can track various phenomena, such as temperature, relative humidity, and smoke, are beneficial to the identification of forest fires. During the fire season, these nodes can be constantly monitored. Also, the system protocol enables sensor nodes to organize themselves into a self-configuration network, thus eliminating manual overhead. Sensors at different forest sites currently use the temperature sensor, the humidity sensor, and the CO_2 sensor. The sensed data is stored in a database and evaluated accordingly with the assistance of Raspberry Pi—the cloud service provides abstract information to make decisions and transmit warnings to the central board. The IoT relates to physical computer access to the Internet. IoT enables these independent devices to communicate efficiently and deliver integrated data without human interference. There are sensor nodes, Raspberry Pi, and an interface module in the network. It is primarily used to forecast forest fires in order to ensure lower emission levels.

Like forest fires, it can also be suited for other applications and could also be monitored and controlled by this methodology. Portable cellular networks such as PDAs, smartphones, and tablet PCs could check for air effectiveness in achieving real-time monitoring of the carbon content at a specific location. It significantly monitors CO_2 emissions.

2.4   GREEN PARKING SYSTEM BASED ON IOT

Intelligent parking systems are divided into different groups of vehicle detection technology and various purposes. Intelligent parking systems assist drivers and operators [26]. The drivers will use the system to find the closest car parks, and parking managers can use the system and the information collected to determine the best parking patterns. Since the demand for parking cannot be anticipated, a flexible pricing approach helps boost income by taking account of customer time and resources. A smart parking system offers other services, for example, smart payment and advance booking, that greatly enhance the driver and operator experience. The intelligent parking system, by increasing traffic controls on vehicles, also helps prevent the illegal use of cars. SPS thus plays a significant role in creating a clean and green atmosphere by reducing pollution from vehicles by spending less time for finding a parking spot. SMS usually consists of several layers of the structure.

2.4.1   CAR PARK OCCUPANCY INFORMATION SYSTEM

The Car Park Occupancy Information System (COINS) uses single-source video sensor technology to monitor vehicle presence and status. It can be recorded in strategically positioned information panels around the parking lot [27]. In particular, COINS is based on a few specific technologies such as counter-based technology, cable-based sensor technology, and wired sensor-based technology. Be aware that the most advanced technology is possible in terms of precise parking conditions without the use of other sensors in any location. The COINS system was generated and simulated in the year 2008. In a multilevel parking lot, the scalability and coverage difficulties cannot make COINS implementation as efficient as other parking systems.

2.4.2   AGENT-BASED GUIDING SYSTEM

An agent-based guiding system (ABGS) simulates each driver's behavior in a dynamic and complicated environment [28]. The agent should describe perceived driving facts and related variables such as independence, behavior, reactivity, adaptability, and social power. For example, in order to find parking, using an agent approach in urban areas was created. In 2008, the authors tried to split the problem of car parking into useful AI techniques. Based on ArcGIS features like SUSTAPARK, known as PARKAGENT, the agent-driven solution was introduced. However, the consequences of the heterogeneity of system entities and driver distribution should be taken into consideration.

2.4.3 AUTOMATED CAR PARKING SYSTEM

Automated parking systems operated by a computer allow carriers to shift their vehicles to a designated bay, lock their vehicles, and carry out remaining automated parking. Stacking cars when there is too little room allows this system to function effectively, using the full available spaces. Vehicle recovery is as simple as entering a given password or code [26]. The method is fully automated, and all processes, including drivers and equipment, have an additional protection and security layer. While initial automatic parking costs are very high, the provided services are reasonable. However, savings of up to 50% compared to conventional high- and time-limited parking modes are required. By using one or more services and sensors, a fast, reliable, and safe parking mode can be implemented with little to no driver–system interactions. A general issue with such a program is that there is no standard building code. For pleasing all the customers, compatibility problems with multiple vehicle models should be discussed and resolved.

2.4.4 GREEN TRANSPORTATION WITH BICYCLE-SHARING SYSTEMS

IoT incorporates Internet, electronic networking, sensing, computation, and built-in devices. This is everywhere for everyone. The sharing of bicycles is a typical and significant IoT system. Chinese bicycle sharing is a significant contributor to green transport, conservation of energy, and environmental health [29]. Currently, IoT is impractical for daily lives. For sustainable and intelligent cities, this can be important. A cycling sharing network is ideal for short-haul journeys, for example, subway and bus stops, shops, post offices, and movies. The network of bicycle sharing would make transport greener and more convenient in a city. In a sustainable and intelligent city, IoT can therefore become an important role. With the introduction of the bicycle-sharing programs, some positive results are apparent. More people opt for public transit and bikes. The sharing of bicycles makes a significant difference in green transport. Bicycle sharing networks have thus contributed significantly to energy conservation and the safety of the environment and need to be widely deployed.

2.5 HOUSEHOLD WASTE MANAGEMENT SYSTEM USING IOT

The concept of smart cities is now very critical to enhancing living standards. The Indian government took the initiative to build 100-plus smart cities. To provide the necessary services such as electricity, water supply, sanitary, recycling, and transportation, the smart town has state of the art technology with sensor networks, cameras, wireless devices, and fast networks such as 5G, IT infrastructure, and data centers [30]. Waste management is a crucial part of urban management, especially where cities need sustainability regeneration. One cannot picture a smart world without a smart waste management system. A town contains markets, banks, agencies, and both small and large houses and communities. Homes collect significant residue. There is organic or inorganic industrial/household waste. A dustbin is the best way to manage household waste awaiting municipal companies. Owing to the routine accumulation of waste, garbage or dust cells are placed in public spaces or in front of homes in towns. Failure to treat waste poses significant health risks, contributing to the spread of infectious diseases and environmental contamination. Some biodegradable waste combinations produce toxic gasses, such as methane; if the dustbin stays for days, that requires immediate action. The segregation and management in the sustainable ecosystem of biodegradable or nonbiodegradable waste is the solution to the fast-growing urban population. There is a mechanism to warn the persons that the bins are being filled up. This will help clean the bin and protect the environment in due course. A city would also be smart if a home community were smart in minimizing waste, saving money, saving water, saving the world.

2.6   REVIEW OF IOT-ENABLED WASTE MANAGEMENT PRACTICES

Regarding waste management, three main objectives are

1. Classification of waste.
2. Quantification of waste.
3. Processing of waste to yield a recycled product for further use or for waste transformation in to another form.

Some waste management practices are aimed primarily at examining waste streams in diverse geographical areas so that waste streams can be separated into particular categories, such as food, paper, metallurgical, and plastic. Also, quantification studies have been studied in different industries such as construction, food, e-waste, forest waste, medical waste, and scraping ships and others [31]. However, waste and waste-quantification studies have covered a range of industries. In the recycled, incinerated, degraded, and composted waste literature, the value was calculated in addition to waste production. Three significant strategies are currently in place, namely minimization (for example, product design), end-of-pipe approaches (for example recycling, separation of waste, incineration, waste efficiency), and environmental recovery. The evaluation approaches such as reducing waste, growing awareness of residential problems, and regulations on waste are considered. Also, practices have been designed to reclaim the interest already embedded in the waste stream through recycling, waste management, incineration, and other actions, such as effective and timely recycling, waste recovery, and waste separations. The final goal is to rebuild the destroyed ecosystem after contamination by air waste streams through environmentally friendly restoration techniques. Despite numerous waste management research projects, the definition of IoT based waste management is very new, and a rising number of research has been published.

In developing countries like India, industrialization and growth have increased rapidly over the last two decades, leading to the question of waste management. Municipal agencies spend 20–50% of the total available budget on solid waste disposal.

Nowadays, various IoT based waste management system have been developed and implemented. In one such example, the smart dust bin, the combination of GSM and electronic control system send SMS to the supervisor to remind the manager of the full dustbin to enable the manager to transfer the garbage. In that system, an ultrasound sensor is used to test the status of the dustbin and the GSM system offers information about whether the dustbin was filled or empty. Thus, the ultrasonic sensor monitors waste levels and ensures immediate dustbin purification when loaded at reference levels. Unattended biodegradable items like food waste often contribute to the production of toxic gasses such as methane [33]. The disposal includes collection, sorting, reduction, recycling, and reuse. The majority of available literature concerns smart dustbins, many of which are in the area. Nevertheless, household waste management is still difficult. There are very few scientific articles on biodegradable and nonbiodegradable waste classification and biodegradable compost management.

Furthermore, household products produce toxic gasses that cause environmental problems. This encourages us to develop a two-tier tool to separate household waste—determining whether it is biodegradable or not—and to send alarm messages using sensor-based machine training approaches. Compost and nonbiodegradable waste were used separately for further isolation. A smart city idea cannot be accomplished without people and climate. For handling waste, the author proposes a green company with advanced technology, such as detecting toxic gas and automatic opening and closing for users gas. Compost is generated using biodegradable waste, so waste is divided into two groups: biological and nonbiodegradable household waste. Google Messenger sends the info to the supervising business and local authorities.

In another example of IoT-based waste management system, Smart Bag, the proposed system has two different compartments for biologically degradable and nonbiodegradable waste. In this model, the green society's waste management program is applied at two levels: Level A is for the domestic level, and Level B is for the community.

Level A: The House Level Smart dustbin has two house-level compartments. One for nonbiodegradable waste and another for biodegradable waste. There are two buttons for nonbiodegradable waste compartment as Green and Red buttons. The sensor for the facility notifies users when it is filled to the predefined level or an alert device used to detect toxic gas [32].

Level B: The second-level isolation conveyor belt is filled with nonbiodegradable waste of different forms. The inductive nearness sensor of the conveyor belt is responsible for metal waste isolation. Metal is sensed when it is moved into the box. The proximity sensor for plastic and wood waste identification is on the conveyor belt. Plastic is sensed and switched to a plastic and wooden enclosure. This model cannot further distinguish the remainder of nonbiodegradable waste. If the nonbiodegradable waste separated reaches the 90% threshold, a warning message will be sent to the municipal corporation for recovery. Level A biodegradable mixture of household waste with dropped leaf green areas, remaining roots, earthworms, etc.

Effective waste management is a significant problem for densely populated urban areas. Day by day, living in clean, healthy urban areas without polluting the environment is challenging. Due to the inadequacy of the waste management strategy, issues such as waste pollution are harmful to the climate. Contaminated conditions help to spread outbreaks of numerous diseases. Developing economies have a long-term waste management sustainability problem. The increasing population, urbanization, and waste management are improved in smart cities. In this modern technology era, technology-based solutions are needed.

A smart IoT architecture introduces a network of identification, automatic doors, control and communication tools, environmental hygiene, and sustainable urban life. Even if the traditional waste management system is effective, an IoT-based integrated waste management system can share details. The utility of the planned waste management system definitely enhances the sustainability of a green city for healthy urban living.

2.7 CONCLUSION

The design of a green city to meet a sustainable, better living environment with harmony needs the adaption of IoT technologies as the essential design element. Technologies enable the green city designers to transform the green city into a "sustainable green city," especially technology's role in controlling CO_2 emissions, waste management, mobility control, and effective utilization of building space. The designer must not be optimistic about the decisive role of IoT in the economy without taking into account the environmental impacts of IoT (e.g., the disposal of e-waste). The direct and indirect impacts must be better understood and taken into account when IoT strategic decisions are taken. Instead of having IoT as an objective, a more pragmatic view of a material future is aimed at meeting a human need (demand) while minimizing the impact on the environment. In order to implement the IoT in a sustainable way, assignment, action, and appropriate implementation are necessary. The efficiency and maturity of technologies include stability in resources and electricity consumption, among the critical aspects of ensuring the sustainable use of IoT technologies.

REFERENCES

[1] Parry, Roberta. "Agricultural phosphorus and water quality: A US environmental protection agency perspective." *Journal of Environmental Quality* 27, no. 2 (1998): 258–261.
[2] Salinero, Kennan Kellaris, Keith Keller, William S. Feil, Helene Feil, Stephan Trong, Genevieve Di Bartolo, and Alla Lapidus. "Metabolic analysis of the soil microbe Dechloromonas aromatica str. RCB: Indications of a surprisingly complex life-style and cryptic anaerobic pathways for aromatic degradation." *BMC Genomics* 10, no. 1 (2009): 351.
[3] Hammi, Badis, Rida Khatoun, Sherali Zeadally, Achraf Fayad, and Lyes Khoukhi. "IoT technologies for smart cities." *IET Networks* 7, no. 1 (2017): 1–13.
[4] Nitoslawski, Sophie A., Nadine J. Galle, Cecil Konijnendijk Van Den Bosch, and James W.N. Steenberg. "Smarter ecosystems for smarter cities? A review of trends, technologies, and turning points for smart urban forestry." *Sustainable Cities and Society* 51 (2019): 101770.
[5] Saravanan, K., Golden Julie E., and Herold Robinson Y. "Smart cities & IoT: Evolution of applications, architectures & technologies, present scenarios & future dream." For

the upcoming book series. *Intel.Syst.Ref.Library*, vol. 154. Valentina E. Balas, et al. (Eds.), *Internet of Things and Big Data Analytics for Smart Generation*. 978-3-030-04202-8, 467407_1_En (7). www.springer.com/us/book/9783030042028.

[6] Brand, Peter. "Green subjection: The politics of neoliberal urban environmental management." *International Journal of Urban and Regional Research* 31, no. 3 (2007): 616–632.

[7] Bateman, Philip W., and Patricia A. Fleming. "Big city life: Carnivores in urban environments." *Journal of Zoology* 287, no. 1 (2012): 1–23.

[8] Kambites, Carol, and Stephen Owen. "Renewed prospects for green infrastructure planning in the UK." *Planning, Practice & Research* 21, no. 4 (2006): 483–496.

[9] Saravanan, K., and P. Srinivasan. "Examining IoT's applications using cloud services." In P. Tomar and G. Kaur (Eds.), *Examining Cloud Computing Technologies Through the Internet of Things* (pp. 147–163). Hershey, PA: IGI Global, 2017. doi:10.4018/978-1-5225-3445-7.ch008.

[10] Darko, Amos, Albert Ping Chuen Chan, Ernest Effah Ameyaw, Bao-Jie He, and Ayokunle Olubunmi Olanipekun. "Examining issues influencing green building technologies adoption: The United States green building experts' perspectives." *Energy and Buildings* 144 (2017): 320–332.

[11] Li, Shancang, Li Da Xu, and Shanshan Zhao. "5G internet of things: A survey." *Journal of Industrial Information Integration* 10 (2018): 1–9.

[12] Magno, Michele, Tommaso Polonelli, Luca Benini, and Emanuel Popovici. "A low cost, highly scalable wireless sensor network solution to achieve smart LED light control for green buildings." *IEEE Sensors Journal* 15, no. 5 (2014): 2963–2973.

[13] Chang, Qing, Xiaowen Liu, Jiansheng Wu, and Pan He. "MSPA-based urban green infrastructure planning and management approach for urban sustainability: Case study of Longgang in China." *Journal of Urban Planning and Development* 141, no. 3 (2015): A5014006.

[14] Mao, Xiaoping, Huimin Lu, and Qiming Li. "A comparison study of mainstream sustainable/green building rating tools in the world." *2009 International Conference on Management and Service Science*, pp. 1–5.IEEE, Beijing, China, 2009.

[15] Premalatha, M., S.M. Tauseef, Tasneem Abbasi, and S.A. Abbasi. "The promise and the performance of the world's first two zero carbon eco-cities." *Renewable and Sustainable Energy Reviews* 25 (2013): 660–669.

[16] Shaikh, Faisal Karim, Sherali Zeadally, and Ernesto Exposito. "Enabling technologies for green internet of things." *IEEE Systems Journal* 11, no. 2 (2015): 983–994.

[17] Norton, Briony A., Andrew M. Coutts, Stephen J. Livesley, Richard J. Harris, Annie M. Hunter, and Nicholas S.G. Williams. "Planning for cooler cities: A framework to prioritise green infrastructure to mitigate high temperatures in urban landscapes." *Landscape and Urban Planning* 134 (2015): 127–138.

[18] oglu Huseynov, Emir Fikret. "Planning of sustainable cities in view of green architecture." *Procedia Engineering* 21 (2011): 534–542.

[19] Patel, Keyur K., and Sunil M. Patel. "Internet of things-IOT: Definition, characteristics, architecture, enabling technologies, application & future challenges." *International Journal of Engineering Science and Computing* 6, no. 5 (2016).

[20] Zhao, Wei, Ester Van Der Voet, Yufeng Zhang, and Gjalt Huppes. "Life cycle assessment of municipal solid waste management with regard to greenhouse gas emissions: Case study of Tianjin, China." *Science of the Total Environment* 407, no. 5 (2009): 1517–1526.

[21] Sodhro, Ali Hassan, Sandeep Pirbhulal, Zongwei Luo, and Victor Hugo C. de Albuquerque. "Towards an optimal resource management for IoT based Green and sustainable smart cities." *Journal of Cleaner Production* 220 (2019): 1167–1179.

[22] Rajkumar, Dr M. Newlin, M. Sruthi, and Dr V. Venkatesa Kumar. "IoT based smart system for controlling Co2 emission." *International Journal of Scientific Research in Computer Science, Engineering, and Information Technology* 2, no. 2 (2017): 284.

[23] Eldrandaly, Khalid A., Mohamed Abdel-Basset, and Laila Abdel-Fatah. "PTZ-surveillance coverage based on artificial intelligence for smart cities." *International Journal of Information Management* 49 (2019): 520–532.

[24] Zhou, Zhenyu, Fei Xiong, Chen Xu, Yejun He, and Shahid Mumtaz. "Energy-efficient vehicular heterogeneous networks for green cities." *IEEE Transactions on Industrial Informatics* 14, no. 4 (2017): 1522–1531.

[25] Oztemel, Ercan, and Samet Gursev. "Literature review of Industry 4.0 and related technologies." *Journal of Intelligent Manufacturing* 31, no. 1 (2020): 127–182.

[26] BS, R. "Automatic smart parking system using Internet of Things (IOT)." *International Journal of Scientific and Research Publications* 5, no. 12 (2015): 629–632.

[27] Bong, D.B.L., K.C. Ting, and K.C. Lai. "Integrated approach in the design of Car Park Occupancy Information System (COINS)." *IAENG International Journal of Computer Science* 35, no. 1 (2008).

[28] Khansari, Nasrin, Barry G. Silverman, Qing Du, John B. Waldt, Willian W. Braham, and Jae Min Lee. "An agent-based decision tool to explore urban climate & smart city possibilities." *2017 Annual IEEE International Systems Conference (SysCon)*, pp. 1–6. IEEE, Montreal, Quebec, Canada, 2017.

[29] Goodman, Anna, Judith Green, and James Woodcock. "The role of bicycle sharing systems in normalising the image of cycling: An observational study of London cyclists." *Journal of Transport & Health* 1, no. 1 (2014): 5–8.

[30] Dubey, Sonali, Pushpa Singh, Piyush Yadav, and Krishna Kant Singh. "Household Waste Management System Using IoT and Machine Learning." *Procedia Computer Science* 167 (2020): 1950–1959.

[31] Anagnostopoulos, Theodoros, Arkady Zaslavsky, Kostas Kolomvatsos, Alexey Medvedev, Pouria Amirian, Jeremy Morley, and Stathes Hadjieftymiades. "Challenges and opportunities of waste management in IoT-enabled smart cities: A survey." *IEEE Transactions on Sustainable Computing* 2, no. 3 (2017): 275–289.

[32] Bhade, Ishwari M., Damodar V. Hegde, Rucha R. Kelkar, and Atharva V. Shastri. "IoT Enabled Toxic Gas Detection and Safety Recommendation System." *Dnyanamay Journal* 3, no. 1 (2017): 43–47.

[33] Anagnostopoulos, Theodoros, Arkady Zaslavsy, Alexey Medvedev, and Sergei Khoruzhnicov. "Top-k query based dynamic scheduling for IoT-enabled smart city waste collection." *2015 16th IEEE International Conference on Mobile Data Management*, vol. 2, pp. 50–55. IEEE, Montreal, Quebec, Canada, 2015.

3 A Green and Sustainable Campus Dwelling

Proposition of a Hybrid Indicator

Yash Shah and Madhusree Kundu

CONTENTS

3.1 INTRODUCTION

'*Smart* city' is sort of a misnomer! Rather, we should aim for a *sustainable city*. In this vein, let us brief the context and necessity of the term *sustainability*. Ecological threats thwarting humanity is simply because of our demands, more than Mother Earth can provide. According to a survey done in 2013, 12 billion hectares of biologically productive land and water (*biocapacity*) on Earth were available; in other words, 1.72 global ha were provided per person (dividing 12 billion ha by the number of people alive in that year, 7 billion). The biocapacity (BC) measures the bio-productive supply, that is, the biological production in an area. It is an aggregate of the production of various ecosystems within the area, for example, arable, pasture, forest, and productive sea. In contrast, the average per-person *ecological footprint* worldwide is 2.8 global hectares. The ecological footprint is the impact of human activities measured in terms of the area of biologically productive land and water required to produce the goods consumed and to assimilate the wastes generated using the prevailing technology [1]. The *ecological footprint*, therefore, is indicative of demand for biocapacity in the form of resources. Sustainability signifies maintaining an overall equilibrium between the available resources (biocapacity) and their utilization (ecological footprint). Sustainability is a constraint of consumption and waste generation, satisfying the human need at present without compromising the needs of the future [2–4]. Hence, one can imagine the propensity for an apparently inevitable commotion concerning sustainability.

The drastic change in lifestyle around the globe is an impending danger toward the sustainability of this planet. Carbon and water footprints are the two major benchmarks in environmental sustainability. The *carbon cycle* plays an important role in biochemical dynamics of Earth [5]. Anthropogenic activities have altered the global carbon cycle, leading to an alarming increase of CO_2 concentration in the atmosphere, which is unprecedented in the sweeping history of many species present on Earth at present [6]. The carbon footprint describes the biocapacity required to mitigate untreated carbon waste and avoid a carbon build-up in the atmosphere. Mere carbon sequestration is not the solution to avoid carbon buildup; rather, it is shifting the burden from one ecosphere to another.

Assessment of sustainability can be of three classes, namely, indicator and indices, product-related assessment, and integrated assessment. A commonly used environmental indicator is the global warming potential (GWP) of greenhouse gases, which reveals the amount of heat trapped by greenhouse gases with respect to carbon dioxide (CO_2) in the atmosphere over a specific time span [IPCC, 1996]. The GWP of CO_2 gas is standardized as 1. A greenhouse effect of increased temperature of Earth's surface is caused by radiations emitted and absorbed by some of the emissions from anthropogenic activities such as carbon dioxide (CO_2), methane (CH_4), fluorinated gases, and water vapor within thermal infrared range. The CO_2 equivalent of a fluorinated gas is quantified by multiplying the mass of the specific fluorinated gas (in metric tonnes) with the gas's GWP. Other important environmental indicators are the *ozone depletion potential, aquatic acidification potential, eutrophication potential, toxic release intensity,* and *hazardous solid waste.* There are energy-related indicators such as *specific energy intensity, energy efficiency,* etc. A product-related assessment refers to various material and energy flows right from the product's inception to its consumption stage, within agreed boundary conditions. Among various gauges of sustainability available, life cycle inventory analysis (LCIA), carbon footprint, and water footprint are significant. Carbon and water footprint may be considered as specific cases of LCIA assessment. The carbon footprint is estimated as carbon dioxide equivalent using the relevant 100-year global worming potential (GWP100). There are various avenues for carbon footprint calculations, including (a) direct GHG emission related to the product's manufacturing, for example, CO_2 emission in the refinery of gases in natural gas processing units; (b) emissions due to electricity that is consumed during the production process—India's average GHG emissions are about 1 kg CO_2 equivalent per 1 kWh electricity production; and (c) the GHG emissions involved in various material production related to the process at its various stages, including transport, material of construction, and so on.

The *water footprint* estimates direct and indirect water consumption due to a product and a process. One should differentiate between groundwater (blue water) and rainwater (green water) while calculating water footprint.

Most of the world-class universities in their vision aim to create a healthy and sustainable society [7–8]. They are the primary educators of future generations and are expected to instill the awareness among students to create a sustainable nexus among technology, knowledge, tools, and life. Sustainability in university campuses has become an active research pursuit in academia [9–13]. Ecological footprint analyses of several world-class universities, including Holme Lacy College, UK, Swansea, University of Wales, University of Newcastle, AU, University of Redlands, USA, University of Toronto at Mississauga, Canada, and Oxford Brookes, UK, have been reported [9, 14–18]. Most of those surveys estimated energy consumption associated with transportation to campus, usage of buildings, food and water consumption, and waste production [14].

The present chapter proposes a new indicator model with the help of a case study based on the NIT Rourkela (NITRKL) Campus at Orissa, India. The study has considered 6000 students and a total of 800 families living on campus, which adds 4000 more to the population. This study also provides a scheme of a survey, EnvoQuant, as a part of practice/ethics/guidelines regarding four-wheeler usage by campus residents. The proposed indicator model can be simulated anywhere in the globe to ascertain a sustainable dwelling.

3.2 CONCEPT OF THE PROPOSED INDICATOR

The proposed indicator is applicable for urban/campus dwelling and based on the lifestyle of people living in it. A *lifestyle indicator* is a summation of allowable ingesting ratios of various privileges related to environmental, social, and economic aspects of life, which should ideally not be exceeding 1.0 for a sustainable dwelling. It is expressed as a weighted summation of prevailing ratios of various environmental, social, and economic footprints (per capita) on the NITRKL campus to Indian standards. The life style indicator is expressed as a weighted sum of the following components in the campus compared to the Indian average: (a) ratio of CO_2 equivalent emission per capita (due to type of food and beverage, personal care products, electricity, liquefied petroleum gas (LPG) cylinder, plastics, paper consumption, and transport usage; (b) ratio of per capita water footprint; (c) ratio of per capita income; (d) ratio of prevailing population density; (e) ratio of per capita expenditure in education; and (f) ratio of per capita health expenditure. The weight factors (termed as *liberty factors*) have been chosen intuitively as significant accountabilities representing disparity/heterogeneity between two standards. The prevailing ingesting ratios can be curbed either of two ways: (a) by increasing the consumption capability of the higher population or (b) using constrained consumption by the smaller population. For any country having heterogeneity between groups of population, the privileged region bears the responsibility of constraining consumption or exercising restraint in consumption to ensure a sustainable life for the entire population. The lifestyle indicator reveals the quantitative measures of restraint in those essential consumptions exercised by the privileged population, provided the larger population is in a continuous effort to uplift their prevailing standards. This subtle indicator is a manifestation of a sustainable life style.

3.3 A BRIEF QUALITY PROFILE OF NIT ROURKELA CAMPUS

The NIT Rourkela Campus, with its air and water quality, its position, and some demographical data are tabulated in (Table 3.1). The campus is surrounded by the Vindya and Satpura mountain ranges and full of vegetation. Naga pond is the vast water body in the campus, ultimately making its way to the river. The whole drainage system of the campus and the rainwater from the hills make their way to the Naga pond. When the water level in the pond reaches a certain height, the water gets channeled into another three storage ponds for retention and preliminary treatment. This water is partly used for organic farming in the campus and finally channeled out of the institute.

TABLE 3.1
Quality Profile of NITRKL Campus

Annual Rainfall	1448 millimeters
Temperature	Maximum 49.70° Celsius (during summers) and minimum 5° Celsius (during winters)
Latitude	22° 15' 11.79" N
Longitude	84° 54' 5.90" E
Altitude	219 meters above sea level
Campus Area	1024 acres (4.144 sq. km)
Density of Population	2414 per sq. km approximately
Literacy	95%
Per Capita Income	Rs. 135000 (approximately)
Population	10000 (including students and family of faculty members and supporting staff)
Water Quality	
Biochemical Oxygen Demand (BOD)	87 mg/l (limit: 3–5 ppm)
Turbidity	25.6 NTU (not more than 5 NTU (for drinking purposes))
Conductivity	385 microsiemens (limit: 1500 microsiemens)
pH	6.5 (limit: 6.5–8.5)
Dissolved Oxygen	5.3 mg/l (4–7 mg/L DO is good for many aquatic animals, low for cold water fish)
TDS	198 ppm (25.44–89.5 ppm)

Air Quality: Within permissible limit for sulfur dioxide (SO_2), nitrogen dioxide (NO_2), particulate matter PM10, particulate matter PM2.5, carbon monoxide (CO).

3.4 DEVELOPMENT OF THE LIFE STYLE INDICATOR MODEL

3.4.1 COMPONENT 1: CO_2 EMISSIONS PER CAPITA IN THE CAMPUS TO THE PER CAPITA INDIAN AVERAGE

3.4.1.1 Food Habit and Consumption in Campus and Consequences in Terms of Carbon Footprint

Thinking about nonvegetarian food is evident worldwide [19]. Food production is responsible for environmental degradation (contributes to 26% of global GHG emissions), and, interestingly, it has been estimated that the total emission from global livestock is 7.1 gigatons (GT) of CO_2 equivalent per year, which constitutes 14.5% of the world's greenhouse gas emissions [20, 21]. The global food system, including production and post-farming processes such as processing and distribution, contributes to GHG emission; 21% of food's emissions (nitrous oxide (N_2O or NO_x) release due to application of fertilizers and manure; methane (CH_4) from rice production; and carbon dioxide from agricultural machinery and conversion of forests, grasslands, and other carbon 'sinks' into cropland or pasture) comes from crop production owing to direct human consumption, and 6% is due to production of animal feed [22]. In India, over a span of 1987 to 2010, per capita poultry consumption has registered an exponential growth of 525% [23]. Livestock products are a major source of CO_2, CH_4, ammonia (NH_3), and N_2O. According to the Intergovernmental Panel on Climate Change (IPCC), the food print of a vegetarian is about half of that of a nonvegetarian [24]. CO_2 equivalent emissions of various nonvegetarian and vegetarian products are taken from literature ([25–27] and Clune et. al [29], respectively). Weekly kilograms of CO_2 equivalent emissions associated with nonvegetarian and vegetarian food in the campus of NIT Rourkela are presented in the Tables 3.2 and 3.3, respectively. A dairy cow produces about 75 kg of CH_4 per year through digestive processes [24]. According to Mishra [28] et. al., it takes more than 9084 liters of water to produce 1 pound of (0.45kg) meat, while to grow 1 pound of wheat, 95 liters of water is required. At NIT Rourkela, consumption of chicken, mutton, and fish is dominant among nonvegetarian products; hence, the present study limits its domain to these three nonvegetarian food products. Quantification has been done according to the following equation:

$$\text{Amount consumed} = (\text{Amount consumed per head/family per day}) \times$$
$$(\% \text{ of populace}/100) \times (\text{Number of persons/family}) \times$$
$$(\text{Number of days of consumption})$$

TABLE 3.2
Weekly kg CO_2 Equivalent Emissions Associated with Nonvegetarian Food in the NITRKL Campus

Food product	Amount consumed by the students (weekly basis)	Amount consumed by NITR (weekly basis)	Kg CO_2 equivalent emissions	Water required (l) (as per Mishra et. al) [26]
Chicken	270g × 0.7 × 6000 × 3	1200g × 0.7 × 800 × 3	14086.8	1.09 × 108
Mutton	270g × 0.4 × 6000 × 1	1200g × 0.4 × 800 × 1	24252	2.1 × 107
Fish	270g × 0.5 × 6000 × 1	1200g × 0.5 × 800 × 1	1677	2.6 × 107
Total			40015.8	15.6 × 107

Amount consumed = (Amount consumed per head/family per day) × (% of populace/100) × (Number of NITRians/families) × (Number of days of consumption)

TABLE 3.3
Weekly kg CO_2 Equivalent Emissions Associated with Vegetarian Food in the NITRKL Campus

Food Product	Amount consumed by the students (weekly basis)	Amount consumed by other NITRians (weekly basis)	Kg CO_2 equivalent emissions (weekly basis)
Wheat	75g × 2 × 6000	75g × 2 × 0.9 × 4000	734.4
Rice	65g × 2 × 6000	65g × 2 × 0.9 × 4000	3182.4
Dal	25g × 2 × 0.8 × 6000	25g × 2 × 0.8 × 4000	204
Vegetables	60g × 2 × 0.6 × 6000	60g × 2 × 0.6 × 4000	266.4
Butter	5g × 0.4 × 6000	5g × 0.3 × 4000	207.36
Milk	100g × 0.4 × 600	100g × 0.9 × 4000	834
Miscellaneous			0.3 × 10000
Total			8428.56

3.4.1.2 Vehicular Emission in Campus and Consequences in Terms of Carbon Footprint

Vehicles commuting (two-, three-, or four-wheelers) all are sources of greenhouse gas emissions in campus and surrounding. The greenhouse gases emitted by the vehicles include CO_2, CH_4, N_2O, CO, NO_x, and NMVOC (non-methane volatile organic carbon). The Honorable Supreme Court of India has banned the sales of BS (Bharat Stage) III vehicles, effective since April 2017. Since April 2017, all the vehicles have been produced as per the BS (Bharat Stage) IV emission norms. Most of the institute vehicles are run by either petrol (gasoline) or diesel oil. An estimated total number of vehicles inside the campus is as follows: petrol passenger cars (225, out of which 200 are BS-III and the rest are BS-IV vehicles), diesel passenger cars (60 BS-III and 15 BS-IV, respectively), two-wheelers (670 BS-III and 130 BS-IV), and three-wheelers (auto rickshaws) (25 BS-III). The estimated daily maximum limit of GHG emissions from the four-wheelers and two- or three-wheelers in the campus are presented in Tables 3.4 and 3.5, respectively. Average distance traveled by any vehicle is considered as 4.5 km per day (for two-wheelers and four-wheelers) and 10 km per day for three-wheelers. All the two-wheelers are petrol vehicles.

TABLE 3.4

Daily Maximum Limit of GHG Emissions from the Four-Wheelers on the NITRKL Campus (as per BS-III and BS-IV Norms Prescribed by the Government of India)

Fuel Type	Distance traveled daily (approximately, in kilometers)	Stage	CO (g)	HC (g)	NOx (g)	HC + NOx (g)	PM (g)
Petrol	200 × 4.5 = 900	BS-III	2070	180	135	–	–
	25 × 4.5 = 112.5	BS-IV	112.5	11.25	09	–	–
Diesel	60 × 4.5 = 270	BS-III	172.8	–	135	151.2	13.5
	15 × 4.5 = 67.5	BS-IV	33.75	–	16.9	20.25	1.69
Total	**1350**	–	**2389.05**	**191.25**	**295.9**	**171.45**	**15.19**

#PM=particulate matter

TABLE 3.5

Daily Maximum Limit of GHG Emissions from the Two- or Three-Wheelers of NITRKL Campus (as per BS-III and BS-IV Norms Prescribed by the Government of India)

Vehicle Type	Distance travelled daily (approximately, in kilometers)	Stage	CO (g)	NOx (g)	HC + NOx (g)
Two-wheeler	670 × 4.5 = 3015	BS-III	3015	–	3015
	130 × 4.5 = 585	BS-IV	820.75	228.15	403.65
Three-wheeler	25 × 10 = 250 (Diesel)	BS-III	156.25	–	156.25
Total	**3850**	–	**3992**	**228.15**	**3575**

3.4.1.3 Personal Care and Hygiene Products Consumption and Consequences in Terms of Carbon Footprint

Consumer products such as soaps, detergents, shampoos, toiletries, etc., require fossil fuel energy, from their raw materials processing, manufacturing, packaging, and transportation to their use by consumers. Koehler *et al.*, [30] have calculated the fossil cumulative energy demand (CED_{fossil}) for different consumer products like bar soap, liquid soap, powder detergent, liquid detergent, bath cleaner, kitchen cleaner, and window cleaner. Energy footprints, in terms of CED_{fossil}, have been calculated by them for three major processes: raw chemical production and supply, packaging production and supply, and finished product manufacturing. Products like bar soap, liquid soap, and powder and liquid detergents are used by all the students, but some products such as bath cleaners and kitchen cleaners are used only by families residing on the NIT Rourkela campus. Table 3.6 furnishes the total fossil cumulative energy demand of the campus due to various consumer product usage on a weekly basis. The water consumed along with the usage of those products and discharged into the Naga pond is also estimated.

TABLE 3.6
CED_{fossil} of Various Consumer Products and the Corresponding Water Drainage into Naga Pond (Inside the Campus)

Product	Functional unit of quantification	Quantity used	Water con- sumed	Used by	Weekly usage (g)	Total CED_{fossil}*	Water drainage to Naga pond #
Bar soap	One-time hand washing	0.35 g	0.91 L	All residents (twice a day)	$(0.35 \times 2 \times 7) \times 10000 = 49000$	539	127400
Liquid soap (hand wash, facewash)	One-time hand washing	2.3 g	0.64 L	All residents (twice a day)	$(2.3 \times 2 \times 7) \times 10000 = 322000$	16229	89600
Powder detergent	Washing one load of laundry (1 kg)	13.6 g	9.8 L	All residents (for students: twice a week, and for 800 families, four times a week)	$(13.6 \times 2 \times 6000)+(13.6 \times 4 \times 800) = 206720$	6535.352	148960
Toilet-care product	One-time toilet flushing	0.16 g	6 L	800 families (twice a week)	$0.16 \times 2 \times 800 = 256$	16.768	9600
Bath cleaner	Cleaning a small wash basin	4.7 g	0.55 L	500 families (twice a week)	$4.7 \times 2 \times 500 = 4700$	62.04	550
Kitchen cleaner	Cleaning a kitchen sink (0.24 m2)	4.7 g	0.55 L	500 families (daily after washing utensils etc.)	$4.7 \times 2 \times 500 = 4700$	72.38	550
Total						23454.54	376660

*(CED_{fossil} measured in MJ equivalents) #(Water drainage measured in liters)

Detergents used on daily basis are composed of anionic and nonionic surfactants. According to the Handbook of Detergents [31], detergents are generally composed of 15% anionic surfactants and 10% nonionic surfactants. The remaining composition consists of foam boosters, enzymes, builders, brighteners, stabilizers, softeners, fragrances, dyes, etc., which instigates the need of a further investigation on Naga pond. Ying et. al., [32] has calculated the toxicity of different types of surfactants against some fish species. There should be a close monitoring of those surfactant sources due to campus consumption, though nothing could be done so far in this regard.

3.4.1.4 Plastic Consumption at NIT Rourkela Campus and Consequences in Terms of Carbon Footprint

Most of the plastic items used in daily life of NIT Rourkela are nonbiodegradable. Plastic consumption comes in the form of food packets (namkeen, Lays, Kurkure etc.), carry bags, plastic spoons available at canteens, use-and-throw water bottles, and cold drink bottles. Plastics not only are a source of greenhouse gases but are also major source of pollution for the Naga pond. The aquatic life is affected by this pollution. Royer et. al. [33] investigated the production of hydrocarbon gases from polyethylene (PET and low-density polyethylene, LDPE) and other plastics at ambient temperature, with an emphasis on methane (CH_4), one of the most potent atmospheric greenhouse gases, and ethylene (C_2H_4), which reacts with moisture in the atmosphere and increases CO concentrations [34–35]. Different types of plastics consumed by campus residents are polypropylene (food packets of chips, namkeen, packaged food snacks of 50 g and plastic cutlery (spoons, forks, plates, bowls, ice cream cups of 3 g): 2.5 g; LDPE (food packets of chips, namkeen, kurkure etc. of 50 g): 0.5 g, and PET (soft drink (Pepsi, Coca-Cola, Sprite, etc.) bottles of 250 ml and water bottles (Kinley, Bisleri, Rail Neer, etc.) of 1000 ml): 25 g. Food packets are mainly made up polypropylene and LDPE, in layers. On the other hand, the soft drink/water bottles are made up of PET. As per the sales data of the canteen located at the Vikram Sarabhai Hall of Residence (which houses ~1000 students), nearly 150 food packets, 100 soft drink bottles, and 300 plastic cutlery items are sold every week. This serves as the basis for the estimation of weekly consumption of various plastic commodities in the campus and the estimated use of various types of plastics (Table 3.7). CH_4 and C_2H_4 emissions caused by various plastic usages are also estimated in this study. Estimated weekly CH_4 and ethylene (C_2H_4, emissions from the NITR campus are 23322500 (pmol per week) and 19727900 (pmol per week), respectively. The CH_4 and C_2H_4 production potential of some commonly used plastics were taken from Royer et. al. [33].

3.4.1.5 Estimated Beverage Consumption Inside Campus and Consequences in Terms of Carbon Footprint

Consuming beer in either a glass bottle or a can has a considerable global warming potential (GWP) that needs to be considered. Amienyo et. al [36] have calculated GWP for 1 liter of beer available in a glass bottle, an aluminum can, and a steel can. Total GWP includes the GWP of various processes such as packaging, transport, beer production, waste management, and raw materials for beer production. According to a WHO report [37] (2016), 30% of the adult Indian population consumes alcohol

TABLE 3.7
Estimated Usage of Different Plastics inside NITR Campus

Product	Number consumed	Polypropylene (g)	Polystyrene (g)	LDPE (g)	PET (g)
Food packets (chips, namkeen, Kurkure, etc.) 50 g packet	1500	2250	–	750	–
Soft drink (Pepsi, Coca-Cola, Sprite, etc.) bottles 250 ml	1000	–	–	–	4950
Water bottles (Kinley, Bisleri, Rail Neer, etc.) 1000 ml	500	–	–	–	9900
Plastic cutlery (spoons, forks, plates, bowls, ice cream cups, etc.) 3 g	3000	3000	6000	–	–
Miscellaneous items, 2 g	900	–	–	1800	–
Total		**6250**	**6000**	**2550**	**14850**

As per the sales data of canteen located at the Vikram Sarabhai Hall of Residence (which houses ~1000 students), nearly 150 food packets, 100 soft drink bottles, 300 plastic cutlery items are sold every week. Miscellaneous items are 2 g of LDPE used in the form of shopping bags, carry bags, packets etc.

regularly. The average Indian consumes 5.7 liters of alcohol per year, according to the same report. Estimated weekly consumption of beer/alcohol and its GWP at the NITR campus are 328.85 L and 224.2 kg CO_2 equivalent, respectively.

3.4.1.6 Electricity Consumption in NIT Rourkela and Consequences in Terms of Carbon Footprint

According to M. Lakra [38] et al., the average monthly consumption of electricity inside NIT Rourkela campus is 9720 kwh, and the average monthly loss is 1296 kwh, summing up to a monthly electricity consumption of 11016 kwh. This is calculated for the academic campus. In this study, the academic campus, student housing and the campus households are considered to be the electricity consumers. Consumption by the campus households and student housing are assumed to be equal to the electricity consumption of the academic campus, that is, 11016 kWh. Hence, the total electricity consumption of NIT Rourkela is: (11016 + 11016) kWh = 22032 kWh.

As per a German report by WINGAS [39], it is estimated that around 1.142 kg of CO_2 is transferred to atmosphere for 1 kWh of electricity production using hard coal. This estimation takes into account the contributions due to transport, production, and processing of materials. Production of 1 kWh electricity using natural gas generates 0.572kg of CO_2. "The global median GHG emission intensity of the hydropower reservoirs is 18.5 gCO$_2$-eq/kWh—this is grams of carbon dioxide equivalent per kilowatt-hour of electricity generated due to hydropower over a life-cycle" [40]. Monthly electricity consumption of NIT Rourkela campus and kg equivalent CO_2 emissions due to monthly consumption of electricity in the campus are 22032 (kWh) and 8700 kg equivalents, respectively. It is important to mention that percent consumption of electricity generated by hydroelectricity plants is 32%, and percent consumption of electricity generated by thermal plants is 68% [41].

3.4.1.7 Paper Usage in NIT Rourkela and Consequences in Terms of Carbon Footprint

Paper is a basic part of academic institutions like NITRKL. In India, people still rely on pen and paper-based education/studies; by contrast, most of the institutes in Western and European countries have gone almost paperless. Paper production involves cutting down trees. Each tree being cut for this purpose diminishes the scope of exclusion of carbon dioxide from the atmosphere. According to an MIT report [42], 0.018 kg of CO_2 is released by a single sheet of paper (both in terms of carbon emitted during processing and carbon that could have been sequestered if the trees had remained alive).

The same report estimated that an average university student consumes around 53.5 sheets of paper every week. We take this as the basis of our study. With the estimated weekly paper consumption by a student at 53.5 and the estimated per head weekly paper consumption by campus families (other than students) at 12, the estimated weekly CO_2 emission due to paper usage at NITRKL is 6642 (kg equivalent).

3.4.1.8 LPG Gas Cylinder Consumption in NIT Rourkela and Consequences in Terms of Carbon Footprint

Most of the foods are cooked using LPG gas cylinders in NITRKL. The KR Patel Cafeteria, feeding 850 boarders of the DBA and MSS halls, typically consumes seven 14.2 kg (33.3 L) LPG gas cylinders every day. A study by the University of Exeter [43] has estimated that 1 liter of LPG gas usage corresponds to 1.51 (0.0018×1.51 kg) liters of CO_2 emissions. Daily consumption of LPG cylinders by 6000 boarders residing in the halls of NIT Rourkela and resulting CO_2 emissions on a daily basis are 1481.85 L and 2237.5935×0.0018 kg equivalents, respectively.

In this study, the consumption of 0.013 of an LPG cylinder was assumed to be used per day per family for 300 families living on the campus (on the basis that for a usual Indian family, one LPG cylinder lasts for 2.5 months). So the total daily LPG consumption for all the campus families is $0.013 \times 33.3 \times 300 = 129.87$ liter of LPG gas usage = 196.28 L CO_2 equivalent emissions.

Total NITR campus consumption (daily) = 1611.717 L of LPG = (2237.5935 + 196.28) Liters of CO_2 emissions = 2433.87×0.0018 kg CO_2 equivalent emissions daily. CO_2 equivalent emissions in seven days due to LPG consumption = $2433.87 \times 0.0018 \times 7 = 30.668$ kg.

3.4.1.9 Total CO_2 Equivalent Emission

The yearly CO_2 equivalent is calculated as follows:

CO_2 equivalent emissions per week = 59727.12 kg. Annual CO_2 equivalent emission: 3105810.552 kg = 3105.810 mt = 0.0031058 MT (0.0000031 Gt/), and it is about 0.00011% of (2.8 Gt,) the CO_2 equivalent emissions of India during 2018 [mt = metric ton, MT = megatonne, Gt = gigatonne].

Actual CO_2 emissions per capita (10000 population) in the campus = 0.31058/0.65 = 0.4778 metric ton (because CO_2 emissions due to buildings, furniture, etc., were not included). In 2018, CO_2 emissions per capita for India was 1.94 metric tons [44].

Ratio of CO_2 emissions per capita in the campus to the CO_2 emissions per capita Indian average is 0.2462. Campus CO_2 emission is only 24.62% of the Indian standard. From a campus lifestyle point of view, a very nominal percentage of the prevailing ratio may matter to account for the disparity between two standards. 0.1% of the prevailing ratio is considered to be significant in ascertaining the disparity between the two standards. The curbing of the Indian per capita CO_2 emissions is recommended.

3.4.2 COMPONENT 2: WATER FOOTPRINT

The standard norm for domestic water usage in India is 135 litres per capita per day (lpcd), prescribed by the Central Public Health and Environmental Engineering Organization [45]. Average water consumption per day per person in the NIT campus is estimated to be180 litres. The ratio of campus consumption per capita to the Indian average per capita consumption is 1.33, which can be curbed toward equivalence by decreasing water consumption in the campus. Indian per capita consumption is 75% of the campus consumption. Nearly 18.75% of the prevailing ratio is significant from the lifestyle point of view of campus residents (they can only enjoy up to 18.75% of the prevailing ratio) in making up the disparity between the campus and the Indian standards. For every 1 inch (2.54 cm) of rainfall over 1000 square feet (0.023 acre) of the campus, around 620 gallons (2346.7 L) water can be harvested. Rourkela has an average annual rainfall of 145 cm (57.1 in.). So, for every 0.023 acre of area, around 134000 liters of water can be harvest every year. If a dedicated area of around 5 acres is planned for rainwater harvesting inside the institute, then 2.9×10^7 liters of water can be harvested every year, resulting in 7.9 lcpd saving per day. This water can be used for car washing, gardening, bathroom cleaning, and many other such purposes and can curb the municipal water usage by campus residents.

3.4.3 COMPONENT 3: INCOME FOOTPRINT

Average per capita income in the campus is estimated as INR 50000 × 12. Per capita income in India in the year 2018–2019 was INR. 92,565 [46]. Indian per capita income is nearly 15% of campus per capita income. Prevailing ratio of campus per capita income to the Indian average is=6.48. The ratio should be curbed towards uniformity not by decreasing the per capita income of the campus people rather by pulling up the national standard. Nearly 2.3 % of the prevailing ratio is significant from the life style point of view of the campus inmates (they can ideally enjoy up to 2.3 % of the prevailing ratio) towards mitigating the disparity between the campus and the Indian standards.

3.4.4 COMPONENT 4: POPULATION FOOTPRINT

Population density in the campus=2414 per sq. km approximately. Ideal population density is 50–100 people per square km and until 150 is bearable but not desirable [47].

The ideal population density is 4 % of NITRKL Campus. Ratio (=2414/150=16.09) can be curbed towards uniformity by reducing the population density of the campus. Hence, nearly 0.4 % of the prevailing population ratio can bring the ideal population density in the campus as 2414 × 0.064=155 per sq. km. 0.4 % of the prevailing ratio is considered to be significant in restoring parity between the campus and the Indian standards.

3.4.5   COMPONENT 5: EDUCATIONAL EXPENDITURE FOOTPRINT

Average per capita annual expenditure in education (primary plus secondary plus technical education) in the campus = INR 48400. Average per capita annual expenditure in education in India was assumed to be INR 18035 (states like Maharashtra and Chhattisgarh spent as high as INR 18035 and INR 17223, respectively, per student, for academic session 2016–17) [48]. Average per capita Indian expenditure is 37.26% of campus expenditure. The per capita expenditure ratio can be curbed towards uniformity (Ratio = 48400/18035 = 2.68) not by compromising campus standards but rather by pulling up the national standard. Nearly 13.9% of the prevailing ratio is considered to be significant from the lifestyle point of view of campus residents (they can only enjoy up to 13.9% of the prevailing ratio) to mitigate the disparity between the campus and the Indian standards.

3.4.6   COMPONENT 6: HEALTH EXPENDITURE FOOTPRINT

Average per capita annual expenditure in health in the campus is INR 25000. The Indian average per capita health care expenditure = INR 4116 [49]. It is alarming that India spends INR 3 per day on each citizen on health care [49]. Indian health expenditure is only 16.4% of the campus expenditure. The ratio of per capita health care expenditure of the campus to the Indian average (6.07) can be curbed toward uniformity by pulling up the Indian average health care standard. Nearly 2.7% of the ratio (6.07 × 0.027 = 0.164) can be considered to be significant from the lifestyle point of view of campus residents (they can only enjoy up to 2.7% of the prevailing ratio) to rectify the disparity between the campus and the Indian standards.

3.4.7   ESTIMATION OF LIFESTYLE INDICATOR

The *lifestyle indicator* is expressed as a weighted summation of prevailing ratios of various environmental, social, and economic footprints (per capita) of the NITRKL campus to the Indian standards. Those chosen weight factors are *liberty factors*. The liberty factors chosen were: w1 = 0.001 or 1% for component 1 (ratio of CO_2 equivalent emission per capita to the Indian average), w2 = 0.1875 (18.75%) for component 2 (ratio of per capita water footprint to the Indian average), w3 = 0.025 (2.5%) for component 3 (ratio of per capita income to the Indian average), w4 = 0.004 (0.4%) for component 4 (ratio of prevailing population density to the Indian average), w5 = 0.139 (13.9%) for component 5 (ratio of per capita expenditure in education to the Indian average), and w6 = 0.027 (2.7%) for component 6 (ratio of per capita health expenditure to the Indian average). The derived lifestyle indicator was (0.00478 +

0.249 + 0.1552 + 0.064 + 0.3726 + 0.164) = 1.0095. Those liberty factors have been chosen intuitively as significant accountabilities responsible for disparity between the two standards. The prevailing ratios of components (1 to 6) can be curbed either by enhancing the consumption/ability of the higher population or using constrained consumption for the smaller population. This subtle indicator is a measure of a sustainable life style. The problem can be posed as an optimization problem, as well:

Minimize OBJ
{(*lifestyle indicator*) − 1} = 0
Subjected to $w1$, $w2$, $w3$, $w4$, $w5$, and $w6$
Constraints: $0 < (w1, w2, w3, w4, w5,$ and $w6) < 1.0$.

In that case, one needs real-life data for a particular country or a region for generating more practical constraints. The optimized value of the weighted factors or liberty factors obtained can be used with greater confidence in the model proposed (see Annexure 3.1 for a sample calculation).

3.5 ENVOQUANT SURVEY SCHEMATIC

In this survey scheme proposed, the user will be provided with the CO_2 equivalent emissions for his/her weekly driving patterns. Along with this, recommendations will be provided as to how these emissions can be reduced at his/her level, in addition to savings in terms of money. The EnvoQuant survey interface with the user is not limited to the data input; nonetheless, it acts as a concern awakener regarding the environment degradation and pollution. Figure 3.1 reveals the schematic of the proposed survey form.

3.5.1 ENVOQUANT APP PLAN AND THE QUESTIONNAIRE

  I. Age group of the user.
 II. Commute details, including route and number of days. This may be done using the Google Maps implementation in the app, so that the user just has to enter two places between which he/she travels, and Google Maps provides the exact distance of the commute. Mileage for suburban areas will be the mean of city and highway mileage.
III. Fuel type (Petrol/Diesel/CNG) of the user's vehicle.
IV. Car details: Here the user will just select the name of the car manufacturer from the list, and in the very next list he/she should be able to get the names of all the cars of that particular car manufacturer. Suppose the user selects Hyundai as car manufacturer; then, in the very next dropdown list, he/she should get the names of all the existing Hyundai cars in India. Based on the car model and fuel type, the proposed app will automatically calculate the car (BS-II, III, IV) mileage from the available database.
 V. Weekend driving (in terms of kilometers) for shopping/tripping.

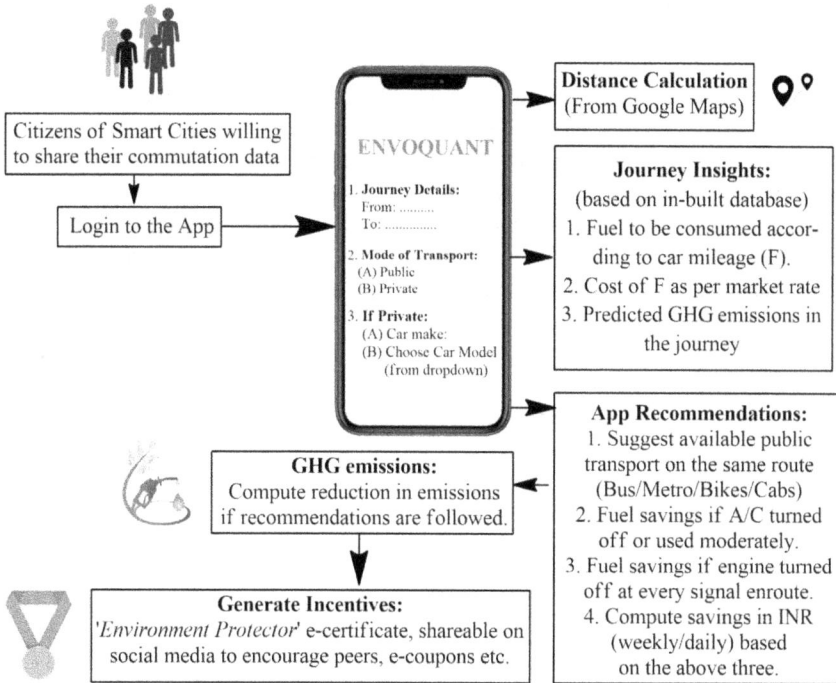

FIGURE 3.1   EnvoQuant survey schematic.

VI. Is the A/C on while driving? If yes, the percentage of driving with the car A/C on?

VII. Does the user turn off the engine while waiting at traffic signals?

3.5.2   RECOMMENDATIONS AND SCOPE OF IMPROVEMENT

I. Suggest available public transport available on the same route.

II. Reduction of emissions (if car engine is off at the signals).

III. Reduction of emissions (if A/C is used within prescribed limits).

IV. Reduction of emissions (if car is driven at particular speed limits).

V. Net reduction in emissions (sum of II through IV).

VI. Weekly savings in INR if the recommended routine is adapted.

The core *EnvoQuant* app principles and values are as follows: A privacy notice has to be given to the user before the user starts to provide personal data. Every survey should end with a quote showing some useful statistics and highlights manifesting the user's contribution in saving the environment by participating in this survey. Finally, the user may share the same survey with his/her community through social media (WhatsApp, Facebook, Instagram, etc.).

3.6   EPILOGUE

This *lifestyle indicator* may be achieved by adapting a smarter approach towards life, for example, by adapting solar power, reducing consumption of paper and plastic articles, adapting a vegetarian diet, harvesting rainwater, banning alcohol consumption, aiming for optimum usage of health/personal care products, and relying more on battery-operated vehicles or cycles instead of two-, three-, or four-wheelers. The kitchen wastes in the student /community housing can be used to generate biogas, which can supplement LPG usage in the housing units. Collecting organic waste in the campus and using it for organic farming is a green effort. Finally, a survey schematic, EnvoQuant, is proposed to reduce the pollution level due to increasing four-wheeler usage.

3.7   ACKNOWLEDGMENT

We are thankful to the NIT populace for helping us with various required data. We also acknowledge the cooperation extended by Professor Kakoli Karar Paul and her PhD students and NIT Rourkela in-house startup Phoenix Robotics.

REFERENCES

[1] Wackernagel, Mathis, Chad Monfreda, Karl-Heinz Erb, Helmut Haberl, and Niels B. Schulz. "Ecological footprint time series of Austria, the Philippines, and South Korea for 1961–1999: Comparing the conventional approach to an 'actual land area' approach." *Land Use Policy* 21, no. 3 (2004): 261–269.

[2] D'Souza, Clare. "Ecolabel programmes: A stakeholder (consumer) perspective." *Corporate Communications: An International Journal* 9, no. 3 (2004): 179–188.

[3] Ruževičius, Juozas, and Eva Waginger. "Eco-labelling in Austria and Lithuania: A comparative study." *Engineering Economics* 54, no. 4 (2007).

[4] Ciegis, Remigijus, Jolita Ramanauskiene, and Bronislovas Martinkus. "The concept of sustainable development and its use for sustainability scenarios." *Engineering Economics* 62, no. 2 (2009).

[5] Schlesinger, William H., and Emily S. Bernhardt. *Biogeochemistry: An analysis of global change*. Academic Press, Cambridge, MA, USA, 2013.

[6] Pearson, Paul N., and Martin R. Palmer. "Atmospheric carbon dioxide concentrations over the past 60 million years." *Nature* 406, no. 6797 (2000): 695–699.

[7] Clugston, Richard M., and Wynn Calder. "Critical dimensions of sustainability in higher education." *Sustainability and University Life* 5, no. 1 (1999): 31–46.

[8] Savage, Mike. *Social class in the 21st century*. Penguin, UK, 2015.

[9] Conway, Tenley M., et al. "Developing ecological footprint scenarios on university campuses: A case study of the University of Toronto at Mississauga." *International Journal of Sustainability in Higher Education* 9, no. 1 (2008): 4–20.

[10] Moffatt, Ian. "Ecological footprints and sustainable development." *Ecological Economics* 32, no. 3 (2000): 359–362.

[11] Savinainen, Antti, and Philip Scott. "Using the force concept inventory to monitor student learning and to plan teaching." *Physics Education* 37, no. 1 (2002): 53.

[12] Rees, William E. "Eco-footprint analysis: Merits and brickbats." *Ecological Economics* 32, no. 3 (2000): 371–374.

[13] Rees, William E. "Impeding sustainability." *Planning for Higher Education* 31, no. 3 (2003): 88.

[14] Dawe, Gerald F.M., Arnie Vetter, and Stephen Martin. "An overview of ecological footprinting and other tools and their application to the development of sustainability process: Audit and methodology at Holme Lacy College, UK." *International Journal of Sustainability in Higher Education* 5, no. 4 (2004): 340–371.

[15] Griffiths, A. "Ecological footprint analysis: Swansea (University of Wales)." Unpublished BSc. dissertation, University of Wales, Swansea, 2002.

[16] Flint, Kate. "Institutional ecological footprint analysis: A case study of the University of Newcastle, Australia." *International Journal of Sustainability in Higher Education* 2, no. 1 (2001): 48–62.

[17] Venetoulis, Jason. "Assessing the ecological impact of a university: The ecological footprint for the University of Redlands." *International Journal of Sustainability in Higher Education* 2, no. 2 (2001): 180–197.

[18] Chambers, Nicky, Craig Simmons, and Mathis Wackernagel. *Sharing nature's interest: Ecological footprints as an indicator of sustainability.* Abingdon, UK: Routledge, 2014.

[19] Brown, Lester Russell. *Who will feed China?: Wake-up call for a small planet.* W.W. Norton & Company, New York, USA, 1995.

[20] Tukker, Arnold, R. Alexandra Goldbohm, Arjan De Koning, Marieke Verheijden, René Kleijn, Oliver Wolf, Ignacio Pérez-Domínguez, and Jose M. Rueda-Cantuche. "Environmental impacts of changes to healthier diets in Europe." *Ecological Economics* 70, no. 10 (2011): 1776–1788.

[21] Food and Agriculture Organization (FAO), "Key facts and findings". www.fao.org/news/story/en/item/197623/icode/ (accessed May 23, 2020).

[22] Choudhary, Ashutosh Kumar, and Nagendra Kumar. "environmental impact of non-vegetarian diet: An overview." *International Journal of Engineering Sciences & Research Technology.* Doi:10.5281/zenodo.843908"

[23] Ritchie, Hannah. "Food production is responsible for one-quarter of the world's greenhouse gas emissions." https://ourworldindata.org/food-ghg-emissions (accessed November 6, 2019).

[24] Watson, R.T., M.C. Zinyowera, R.H. Moss, and D.J. Dokken. *The regional impacts of climate change: An assessment of vulnerability*, p. 53. Cambridge University Press, Cambridge, UK, 1998.

[25] Ibidhi, R., A.Y. Hoekstra, P.W. Gerbens-Leenes, and H. Chouchane. "Water, land and carbon footprints of sheep and chicken meat produced in Tunisia under different farming systems." *Ecological Indicators* 77 (2017): 304–313. https://doi.org/10.1016/j.ecolind.2017.02.022.

[26] Greenhouse gas emissions from ruminant supply chains. "A global life cycle assessment". www.fao.org/3/i3461e/i3461e04.pdf (accessed September 10, 2019).

[27] Eady, S.J., P. Sanguansri, R. Bektash, B. Ridoutt, L. Simons, and P. Swiergon. "Carbon foot-print for Australian agricultural products and downstream food products in the supermarket." *7th Australian Conference on Life Cycle Assessment*, Melbourne, 2011.

[28] Mishra, Mukesh Kumar. "Modern lifestyle, non veg food and its impact on environmental aspects." *Saving Humanity Saving Humanity* (2012): 376.

[29] Clune, Stephen, Enda Crossin, and Karli Verghese. "Systematic review of greenhouse gas emissions for different fresh food categories." *Journal of Cleaner Production* 140 (2017): 766–783.

[30] Koehler, Annette, and Caroline Wildbolz. "Comparing the environmental footprints of home-care and personal-hygiene products: The relevance of different life-cycle phases." *Environmental Science & Technology* 43, no. 22 (2009): 8643–8651.

[31] Zoller, Uri, and Paul Sosis. *Handbook of detergents, part F: Production.* CRC Press, Boca Raton, FL, USA, 2008.

[32] Ying, Guang-Guo. "Fate, behavior and effects of surfactants and their degradation products in the environment." *Environment International* 32, no. 3 (2006): 417–431.

[33] Royer, Sarah-Jeanne, Sara Ferron, Samuel T. Wilson, and David M. Karl. "Production of methane and ethylene from plastic in the environment." *PLoS One* 13, no. 8 (2018): e0200574.

[34] Saunois, Marielle, R.B. Jackson, Phillippe Bousquet, Ben Poulter, and J.G. Canadell. "The growing role of methane in anthropogenic climate change." *Environmental Research Letters* 11, no. 12 (2016): 120207.

[35] Aikin, A.C., J.R. Herman, E.J. Maier, and C.J. McQuillan. "Atmospheric chemistry of ethane and ethylene." *Journal of Geophysical Research: Oceans* 87, no. C4 (1982): 3105–3118.

[36] Amienyo, David, and Adisa Azapagic. "Life cycle environmental impacts and costs of beer production and consumption in the UK." *The International Journal of Life Cycle Assessment* 21, no. 4 (2016): 492–509.

[37] WHO Report. www.who.int/substance_abuse/publications/global_alcohol_report/profiles/ind.pdf?ua=1 (accessed May 23, 2020).

[38] Lakra, M., and S. Chinara. "Smart and energy efficient power saving system based on wireless sensor network." (2015). http://ethesis.nitrkl.ac.in/7067/1/Smart_Lakra_2015.pdf (accessed May 23, 2020).

[39] "Natural gas is the most climate-friendly fossil fuel in electricity production." www.wingas.com/fileadmin/Wingas/WINGAS-Studien/Energieversorgung_und_Energiewende_en.pdf (accessed May 23, 2020).

[40] "Greenhouse gas emissions." www.hydropower.org/greenhouse-gas-emissions (accessed May 23, 2020).

[41] "Executive summary of month of November 2015." (PDF). *Central Electricity Authority, Ministry of Power, Government of India.* Archived from the original (PDF) (2016, March 4) (accessed December 15, 2015).

[42] "Save paper." http://mit.edu/~slanou/www/shared_documents/Daniel/Save%20paper.docx (accessed May 23, 2020).

[43] "Calculation of CO2 emissions." https://people.exeter.ac.uk/TWDavies/energy_conversion/Calculation%20of%20CO2%20emissions%20from%20fuels.htm (accessed May 23, 2020).

[44] "India: CO_2 emissions per capita." https://knoema.com/atlas/India/CO2-emissions-per-capita (accessed May 23, 2020).

[45] Grover, Aruna Ramani. "Evaluation of norms and standards for domestic water use in Indian cities." *American Journal of Sustainable Cities and Society* 1, no. 6 (2017): 44–55.

[46] "GDP per capita of India." http://statisticstimes.com/economy/gdp-capita-of-india.php (source: *Ministry of Statistics and Programme Implementation*: (2004–18), Economic Survey of India 2014–15, World Bank: Nominal, PPP, IMF World Economic Outlook (April-2019) (accessed September 9, 2019).

[47] "Ideal population-density." https://forums.digitalspy.com/discussion/684316/ideal-population-density (accessed May 23, 2020).

[48] "At Rs 9, 167, UP spends least on per child school education, reveals study." *The Times of India Education.* (2017, May 12). https://timesofindia.indiatimes.com/home/education/news/at-rs-9-167-up-spends-least-on-per-child-school-education-reveals-study/articleshow/58646292.cms.

[49] "India's per capita expenditure on healthcare among lowest in the world; govt spends as little as Rs 3 per day on each citizen." www.firstpost.com/india/indias-per-capita-expenditure-on-healthcare-among-lowest-in-the-world-govt-spends-as-little-as-rs-3-per-day-on-each-citizen-4559761.html (accessed May 23, 2020).

ANNEXURE 3.1

Liberty factor sample calculation:

Per capita health expenditure (Indian avg.): INR 4116 . . . [A]
Per capita health expenditure (NIT campus): INR 25000 . . . [B]
Ratio of Indian avg. to NIT campus: A/B . . . [C]
Ratio of NIT campus to Indian avg.: B/A . . . [D]

Liberty factor (in %): {C/D} × 100 = {0.164/6.07} × 100 = 2.7%

4 Electromagnetic Pollution and Its Management
An Overview

Amiya Kumar Mallick

CONTENTS

4.1 INTRODUCTION

Of late, a new emerging concept of smart cities is in the air. All the smartness of the smart cities that inhabitants of the cities want to realize and comprehend depends squarely upon the successful deployment of the Internet of Things (IoT) technologies and other associated electronic devices. As a result, all the weaknesses and liabilities of the electronic devices as a whole need to be taken care of. It is well known that in general, all electrical devices are vulnerable and prone to electromagnetic interference (EMI) or, in other words, electromagnetic pollution. If any such pollution happens to occur, the IoT and its other supportive components either enter into a malfunction state or do not work at all. Under such circumstances, the entire project's mission becomes jeopardized, and it turns into a concrete jungle. In order to defend the needs and aspirations of the city's inhabitants, city planners should be aware of and understand electromagnetic pollution as they do for other types of pollution, such as atmospheric, environmental and sound pollutions. They should take adequate and proper measures in advance to mitigate and alleviate such dangerous and unpleasant consequences or interference of any kind before they crop up. For that reason, this chapter is very significant and beneficial for city planners and engineers taking part in the successful and sustainable smart city mission. It appears that experts in the field of green technology for smart cities may not be fully conscious about the existence of electromagnetic pollution and their hazardous and perilous offences. Now they should take care of these from the very outset of the project in order to make their projects successful, sustainable and really smart in all respects.

What is bliss for you may cause grief for others. Imagine a situation in which you are enjoying music from an FM receiver, possibly at high volume, and not caring for others who are nearby and attentively engaged with their work. Now try to ponder the awkward situation of those people present over there. Certainly, what you are enjoying is creating interference to the attention of others. In other words, under such circumstances, you are the 'culprit' troubling the normal functioning of others, and others are 'victims', as they are getting disturbed.

Now, imagine a similar scenario in the electromagnetic domain in which a number of electronic devices are operating together in a closed compound. It may be observed that an AM/FM transmitter radiating modulated signals may affect the normal operation of an aircraft passing very close to the transmitter. Similarly, a radar station may slip into malfunctioning mode when a mobile set momentarily comes very close to it and interferes in its normal operation. Of late, with the increased proliferation of electronic and electrical equipment in various sectors—household, industrial, telecom, hospital, IT segment, automobile, defence etc., use of high frequency digital circuits has been spiraling up like anything. This has inadvertently created a lot of undue electromagnetic radiation sources—a new kind of invisible giant nuisance-makers in the environment. In this electronic era, this is a new phenomenon generating electromagnetic 'smog' affecting the normal operation of electronic gadgets and even upsetting human health, thus causing annoyance, delay, loss of revenue, significant property loss and even death and serious injuries.

This highly uninvited, unsolicited and disturbing interfering signal producing toxic waste in the electromagnetic environment may be referred to as electromagnetic pollution or electromagnetic interference (EMI). The following paragraphs are going to present the general characteristic features of the unseen demon creating awful nuisance and its adverse effects in electronic technology. Methods to mitigate such challenging mischievous problems will subsequently be discussed.

To start with, one should know the meanings of various terms related to such an abnormal, unique massive phenomenon. This is possibly present in all electronic and electrical systems, that is, electromagnetic compatibility (EMC), EMI, EMC conditions, EMI sources, EMI effects, and associated problems with non-compliance. In addition, various concepts and definitions associated with the event will be brought into light. In order to mitigate the related troubles, the outline of norms and standards formed by various regulatory bodies at national and international levels will be discussed. Some relevant design guidelines, methodology and process control will be presented as well. To start with, the inner significance of some important terminology related to EMI and EMC are presented.

4.1.1 CONNOTATION OF RELEVANT RELATED TERMS

EMI: Modern electronic equipment is designed using complex electronic components and dense assembly techniques. These devices emanate various spurious unwanted signals which interfere with normal function of other sensitive receiving devices or even with themselves, causing havoc or complete failure of operation. This type of signal is normally branded as electromagnetic interferences (EMI). It is also known as radio frequency interference (RFI). This may be referred to as *electromagnetic pollution* as well.

EMC: The ability of an electronic device to function satisfactorily in an electromagnetically polluted environment without inflicting intolerable disturbances on any other electronic devices nearby is designated as its electromagnetic compatibility (EMC). It is, therefore, a matter of coexistence and nothing else. In order to satisfy the compatibility conditions in an EMC environment, electronic devices are to observe certain norms. For example, they should

- Not cause interference with the operation of other electronic devices in close proximity.
- Not be vulnerable themselves to emissions from other electronic systems and should be able to tolerate a specified degree of interference, or they must have immunity to survive it.
- Not have interference with themselves; in other words, they should be self-compatible.

4.1.2 EMI SOURCES

Basically, the prime causes of electromagnetic interference are many, but for the sake of expedience, the EMI sources may be subdivided as follows: (1) man-made and (2) erratic.

4.1.2.1 Man-Made Noise
- Electrostatic charges generated within the human body.
- Noise caused due to ignition in automobiles.
- Noise generated by electric machines such as kitchen blenders, hair dryers, refrigerators, AC, electric shavers, washing machines etc.
- Switched mode power supply generating noise signals at the switching frequency, and its harmonics that conduct through the powerline.
- High-voltage transmission lines.
- Household appliances like microwave ovens, laptops, printers, computer peripherals.
- Fluorescent lamps.
- Industrial control systems, medical devices, communication receivers.
- Electromagnetic pulses (EMP) due to nuclear detonation producing gamma rays.
- Fast-switching digital devices, ICs and many more.

4.1.2.2 Erratic Noise
Atmospheric noise.
Space noise.
Lightning and discharge.

Some of the practical examples of possible effects of electromagnetic interference (EMI) commonly observed in electromagnetic domain are listed here:

- False activation of weapons in warfront causing a failure in interception with target.
- False alarming in security zones in vulnerable installations causing risk and hazard.
- Interference in voice/data/video signals used by radio and television networks causing poor quality of reception.
- Loss of stored data in digital systems.
- Health hazards; destructive effects on biological cells of humans, animals, vegetation.
- Electronically operated automobiles malfunctioning simply when the radiated pulse signal from a piezoelectric cigarette lighter is lit nearby.
- The stoppage of all the sensitive instruments in an aircraft from working when a transmitting antenna starts radiating signals close by.
- Signals emitted from the local security system (LAN) in a super-specialty hospital collapsing the operation of all the sophisticated diagnostic instruments, causing false and misleading diagnoses.
- In green technology for smart cities, mission fails in case of EMI attack on the IoT complex of the smart project.

With the spiralling rise of popularity and production of electronic gadgets specially based on the digital principle, it is realized, with heavy heart, that the electromagnetic pollution created in some critical situations has reached beyond the limit of tolerance, especially in the electromagnetic domain. In order to combat such an unendurable state of affairs, imposition of some effective electromagnetic compatibility (EMC) conditions must be enforced so as to have happy and satisfactory coexistence of electronic devices in a chaotic electromagnetic environment. Hence, the importance of EMC must be appreciated in these circumstances. That is, presence of proper EMC rules means alleviation of EMI effects in this issue.

4.1.2.3 Significance of EMC Issues

Electromagnetic compatibility requires that systems/equipment must be able to tolerate a specified degree of interference and not generate more than a specified amount of interference. Further, EMC is becoming more and more important because there are so many opportunities today for EMC issues. At the same time, there is increased use of electronic gadgets in different application areas; For example,

- Automotive applications.
- Personal computing/entertainment/communication.
- Increased potential for susceptibility/emission.
- Lower supply voltages.
- Increasing clock frequencies, faster slew rates.
- Increasing packaging density to enhance compactness.
- Demand for smaller, lighter, cheaper, lower-power devices.

In the case that EMC conditions are not being properly abided by in practice, inconveniences that may be caused are classified mainly in three categories: minor, major and in-between, as per their seriousness and nature of gravity. Minor effects, for example, may cause annoyance or delay, say, due to AM/FM/XM/TV interferences or cell phone interruptions. This can be tolerated or ignored. Non-compliance of EMC regulations may sometimes create major problems as serious as death or severe injuries that may be due to, say, radar interruption in aircraft landing systems, erroneous ordnance firing or improper deployment of airbags. Neither major nor minor, in-between hassles may still cause loss of revenue, minor property loss and significant property loss. This may be due to critical communication interruptions or interference and automated monetary transactions.

However, it may be mentioned that nowadays, non-compliance of EMC issues does not pose any major problem in industry because, fortunately, industry is well regulated and standards are comprehensive. Hence, major EMC issues are relatively rare. It is to be noted that for cost-effective compliance of EMC, regulations are carefully considered and applied throughout product/system development from the very beginning of planning and design of the gadget, not on a piecemeal basis before final testing of the device.

4.2 MODELING OF INTERFERENCE PROCESS

The mechanism of interference, basically, concerns the leakage of interference emission from an electromagnetic source, which gets coupled to a very sensitive receiver nearby through a suitable routing path, creating serious functional problems to a receiver meant for some other purpose. In other words, EMC deals with the mechanism of interference in the following manner. Basically, it requires

- A source of emission of undesired interference (call it a *culprit*).
- A transmission path between the source (culprit) and the receptor (call it a *victim*).
- A receptor (victim)—the most affected device—which malfunctions upon reception of interfering signal through a coupling path.

Pictorially, the model appears as depicted in Figure 4.1:

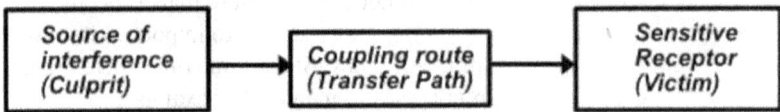

FIGURE 4.1   Schematic block diagram representing the mechanism of an EMC problem.

In Figure 4.2, Susceptibility and immunity are qualities of the victim, but they are antonyms to each other. However, susceptibility represents the sensibility of victim's response to interference and immunity stands for the ability to withstand any interfering signal without being perturbed. Here, coupling path transferring interfering signal can be of two categories, in general: conductive (wired) or radiative (wireless) in nature. The radiative path is again subdivided in two varieties: magnetic or electric. It all depends on how close the victim is with respect to the culprit. If the victim is in the near field of the culprit's emission radiation zone, where the electric field is dominant, the coupling path is electric in character, while if it is the radiative near field region, where the magnetic field is dominant, it is magnetic in nature. On the other hand, when the culprit as well as victim are connected together through a conductive wire, as may happen when they are a common power grid supply or have a common signal database link, the coupling path is always conducted in character. In short, EMI is due to conduction having common impedance or due to capacitive or inductive coupling when the two devices are in the near-field region, and EMI is electromagnetic emission in nature when they are in the far-field region.

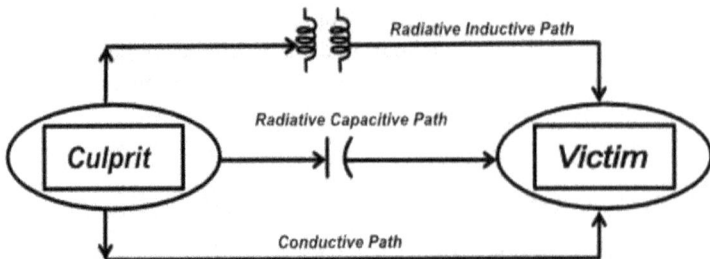

FIGURE 4.2   Detailed schematic view of EMC coupling path between culprit and victim.

Depending on the coupling conditions, EMI entry into a victim instrument that originated elsewhere, in an external culprit device, may be interpreted in the following manner.

i. Presence of impedance mismatch and discontinuity.
ii. Common-mode mismatch in a circuit leading to differential noise signals.
iii. Capacitively coupled unbalanced network is susceptible to voltage variation (dv/dt) in one causing noisy displacement current on the other, through the mutual capacitance in between them, for example, 1 mA/pF due to voltage fluctuation of 1 V/ns.
iv. Mutual inductance sometimes creates a difficult source of interference, if high value of di/dt induces a noise voltage on the other circuit through mutual inductance, for example, 1 V of noise voltage is contributed through a mutual inductance of 1 nH for a change of 1 A/ns of I.

4.3 CONCEPTS AND DEFINITIONS

Effective EMC design has become a critical component in the design of most of modern electronic devices/systems. The EMI environment is becoming more and more cluttered/clumsy with the increasing application of small-sized high-speed wireless components in the marketplace. In order to combat the entire EMC problem, one has to carefully identify types of coupling mechanism of fields between emitter (culprit) and receptor (victim). Logically, it is either wireless or wired. However, based on this categorization of coupling path, specific EMC components now boil down to four major types, as follows:

- Radiative emission (wireless).
- Conductive emission (wired).
- Radiative immunity/susceptibility (wireless).
- Conductive immunity/susceptibility (wired)

In order to explain the nature of EMC problems classified here, help of a schematic view (Figure 4.3) depicted has been employed. Here, DUT refers to device under test.

4.3.1 PREVENTION OF EMC ISSUES

In order to arrest completely or else at least to alleviate the EMC problem within its specified limit, it is realistic to consider following four steps.

1. Suppression of the unwarranted emission at its source point (culprit, Figure 4.2).
2. Making the unintended coupling path as inefficient and ineffective as possible.
3. Making the inappropriate receptor (victim, Figure 4.2) less susceptible or more immune to the unwanted emission, if any. Or else, it should be able to withstand a specific degree of interference from nearby emissions (culprit).
4. Combination of all three.

In other words, the possible steps that one has to comply with in order to mitigate EMC problems are the following (Table 4.1).

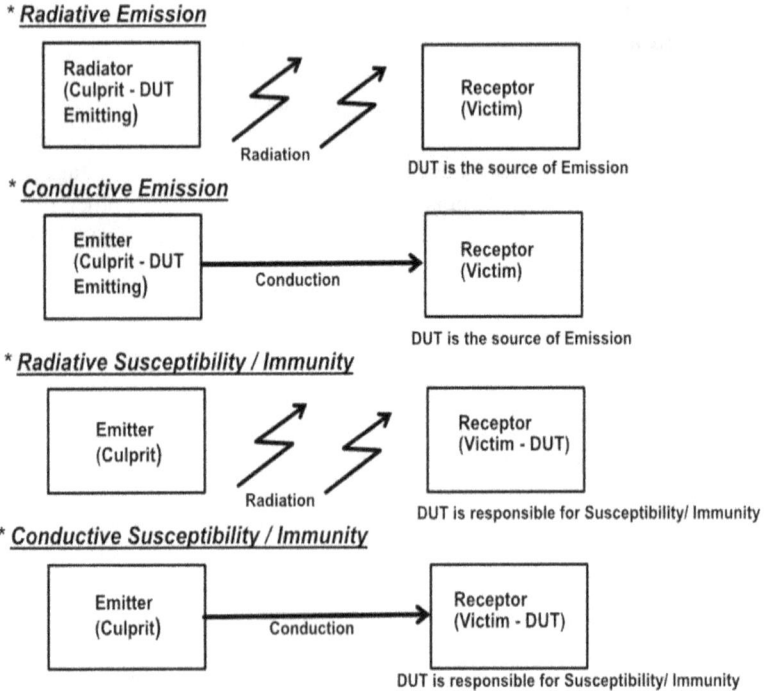

FIGURE 4.3 Pictorial interpretations of the four types of EMC problems.

TABLE 4.1
A Few Steps to Mitigate EMC Problems

Source (Culprit)	Modify Signal Routing	Add Local Filtering	Reduce Signal Level	Spread Spectrum
Coupling Path	Increase Separation	Shielding	Reduce Interconnections	Filter Interconnections
Receptor (Victim)	Modify Signal Routing	Add Local Filtering	Operating Frequency Selection	___

4.3.2 Some Other Aspects of EMC Problems

There are some other aspects of EMC problems that are equally interesting and noteworthy; just to name a few: electrostatic discharge (ESD); electromagnetic pulse (EMP); lightning; and TEMPEST (secure communication and data processing).

4.4 EMC STANDARDS

In order to deal with the emerging and ever-increasing problems of EMI, in 1933, the International Electrotechnical Commission (IEC) arranged a meeting in Paris and recommended the formation of the International Special Committee on Radio Interference (CISPR). The CISPR reconvened another meeting after World War II in London in 1946. Subsequent meetings provided various technical publications, which concerned measurement techniques as well as recommended emission limits. Some European countries adopted versions of CISPR's recommended limits. The Federal Communications Commission (FCC) of the United States of America suggested a rule that was the first regulation for digital systems in the United States, and the limits follow the CISPR recommendations with variations peculiar to the US environment. Most manufacturers of electronic products within the United States already had internal limits and standards imposed on their products in order to prevent "field pollution" associated with EMI. However, the FCC rule made what had been voluntary a matter of legal compliance. With the increasing application of digital devices in electrical systems ranging from satellite communication to radar to television to electronic typewriters, the EMI (both conducted and radiated) became significantly predominant, creating nuisance in electromagnetic domain. The FCC in the United States promulgated a regulation in 1979 fixing the permissible limit of emission of digital devices that are saleable in legitimate markets. This has alerted the manufacturers of the digital device-systems. The FCC defines a digital device as: "Any unintentional radiating device or system that generates and uses timing pulses at a rate in excess of 9000 pulses (cycles) per second and uses digital techniques". This rule includes any electronic system employing digital circuitry using a clock signal in excess of 9 KHz like electronic typewriters, calculators, personal computers, printers, modems etc.

In order to combat this problem of EMI, some standard-making bodies are formed by various government agencies at their respective national level. They try to achieve a goal of creating an atmosphere so as to have a certain degree of compatibility among the electronic gadgets while working together in a congested region. As for example, a list of such regulatory bodies in different countries is provided in section 4.4.1. However, numerous EMC standards exist. But the common fundamental theme is to fix up the limits of emissions (conducted and radiated) and immunity or susceptibility (conducted and radiated) of the electronic gadgets.

4.4.1   REGULATORY INSTITUTIONS FOR EMC STANDARDS

- Federal Communication Commission (FCC) in United States America.
- US Military (DoD).
- European Union (EU).
- Radio Technical Commission for Aeronautics (RTCA).
- Comite Internationale Special des Perturbations Radioelectrotechnique (CISPR).
- National Association of Radio and Telecommunication Engineers (NARTE).
- British Standard Institution (UK).
- Verband Deutscher Elektrotechniker (VDE) in Germany.
- Voluntary Control Council for Interference (VCCI) in Japan.
- China Compulsory Certification (CCC) in China.

It may be carefully noted that all countries do not have exactly the same requirements. It is the responsibility of the government of a country to enforce the EMC

TABLE 4.2
FCC Standards for Conducted Emissions [FCC Part 15]

**********	Frequency [MHz]	Quasi-Peak Limits [dBµV]	Average Limits [dBµV]
Class A	0.15–0.50	79	66
	0.50–30.00	73	60
Class B	0.15–0.50	66 to 56*	56 to 46*
	0.50–5.00	56	46
	5.00–30.00	60	50

*Decreases as logarithm of frequency

TABLE 4.3
FCC Standards for Radiated Emissions [FCC Part 15]

******************	Frequency [MHz]	Field Strength Limits [µV/m]
Class A	30–88	90
[10 meters]	88–216	150
	216–960	210
	Above 960	300
Class B	30–88	100
[3 meters]	88–216	150
	216–960	200
	Above 960	500

regulations as much as is required for that country, matching with its EMC environment. So the EMC settings of different countries are more or less similar but not identical. However, some of the countries prefer to adopt the CISPR-22 standards instead of having their own. As for illustration, the standards promulgated by the FCC for conductive emissions and radiative emissions for digital devices applicable in United States are as shown in Tables 4.2 and 4.3 respectively.

It may be carefully noted that due to obvious reasons, stated earlier, the limits of conducted and radiated emissions are much more stringent for class B than those of class A digital devices. However, in an identical fashion, other regulatory bodies like CISPR have recast the limits of emissions as a function of frequency, which is also useful and valuable.

In exactly the same way, tables for radiated and conducted susceptibility/immunity are also prepared for setting the corresponding limits required and are available in the respective standards.

4.5 DESIGN GUIDELINES AND METHODOLOGY

4.5.1 NEAR-FIELD AND FAR-FIELD CONCEPTS

While propagating in free space, much away from the source of radiation, the propagation supports only the transverse electromagnetic (TEM) mode. In this case, the electric field vector, E (Volt/meter), and magnetic field vector, H (Ampere/meter), are

orthogonal to each other, as also to the direction of propagation. The wave impedance, η (Ω) of free space is given by the ratio of E/H (~377 Ω). However, it will be
worthwhile now to bring in the concept of the near field and far field of an RF emitter
in order to characterise the EMC problems due to EMI. Considering the field expressions of elementary RF emitters, the range between emitter and receptor [Harrington]
may be divided mainly in two fields, that is, near field and far field (Fraunhofer zone).
Near field is further subdivided into two regions, reactive near field and radiative near
field (Fresnel zone). Figure 4.4 shows the details.

Now, it is evident that the far-field (Fraunhofer) region starts from a distance $2D^2/l$
from the emitter and goes beyond this limit to infinity, when the maximum linear
dimension of the emitter is D. Or in other words, for the far field, $R > 2D^2/\lambda$, at the
same time, $R \gg D$ and $R \gg \lambda$. In this region, E-field and H-field die off at the rate
proportional to $1/R$ and the power density as $1/R^2$. The far-field radiation pattern of
the emitter is independent of R.

The reactive near-field extends up to a distant of $0.62\sqrt{(D^3/\lambda)}$ from the emitter and stays
alive in the immediate vicinity of the emitter. This region is highly reactive in nature, and
the E- and H-fields are out of phase by 90^0 with each other and fall off as $1/R^3$.

The radiative near-field (Fresnel) region ranges between $0.62\sqrt{(D^3/\lambda)} < R$
$< 2D^2/\lambda$, and reactive fields do not dominate, yet radiative fields begin to emerge.
The shape of the radiation pattern in this region may change appreciably with distance R. The values of R and the wavelength λ will determine if, at all, this region
will stay alive or not (especially when $D \ll \lambda$). E- and H-fields are dependent on
R as $1/R^2$.

4.5.2 COMMON-MODE AND DIFFERENTIAL-MODE CURRENTS

In the case that EMC standards are not properly and meticulously taken care of at
the circuit-design stage, there is every possibility of the presence of noise (EMI)
currents, $\bar{I}_C$ flowing through two parallel conducting wires or PCB fields that cannot be interpreted and understood by the circuit theory or transmission line theory
alone. Along with the noise currents, $\bar{I}_C$, the normal functional currents, $\bar{I}_D$, will
coexist, making the net overall currents, $\bar{I}_1$ and $\bar{I}_2$, over the two parallel wires a
linear superposition of these two types of currents. Or in other words, the currents

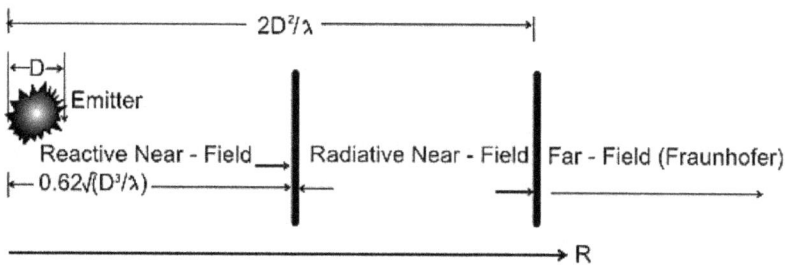

FIGURE 4.4 Depicts the field regions when emitter has a maximum dimension of *D* and *R*
represents the range of the region from the emitter.

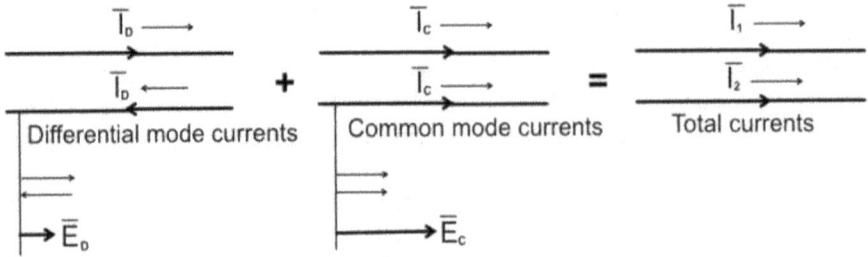

FIGURE 4.5 Illustrates the differential-mode current, $\bar{I}_D$, and the common-mode current, $\bar{I}_C$ form the total net currents as $\bar{I}_1$ and $\bar{I}_2$ on the parallel conducting wires. Correspondingly, they produce radiated emissions all around as $\bar{E}_D$ and $\bar{E}_C$.

$\bar{I}_1$ and $\bar{I}_2$ on the two parallel wires may be resolved into two types of currents, that is, differential-mode currents, $\bar{I}_D$, which on the lines are oppositely directed, and common-mode current, $\bar{I}_C$, flowing in the same direction (Figure 4.5). It may be noted that $\bar{I}_D$ currents are functional and desired. Hence, these are predictable by the circuit theory. On the other hand, $\bar{I}_C$, that is, common-mode currents, represent, mainly, the noise currents in the system and are a non-functional and undesired entity. And these currents are inexplicable by the circuit theory. Therefore, mathematically, the current relations are as follows: $\bar{I}_1 = \bar{I}_C + \bar{I}_D$ and $\bar{I}_2 = \bar{I}_C - \bar{I}_D$. Hence, the expressions of the decomposed current components can be obtained as $\bar{I}_D = \frac{1}{2}(\bar{I}_1 - \bar{I}_2)$ and $\bar{I}_C = \frac{1}{2}(\bar{I}_1 + \bar{I}_2)$. Both these current components, that is, differential-mode currents, $\bar{I}_D$, and common-mode currents, $\bar{I}_C$, by their nature, produce radiated emissions as $\bar{E}_D$ and $\bar{E}_C$. Normally, the magnitudes of $\bar{I}_D$ are much larger than that of the noise currents, $\bar{I}_C$, but due to their nature, the radiated emission $\bar{E}_C$ is much more effective and damaging than $\bar{E}_D$. This is explained in Figure 4.5. Now, it is worthwhile to mention that at the time of designing any electronic product, care should be taken to alleviate or at least mitigate the possibilities of any common-mode currents through conduction paths in the system.

4.5.3 SHIELDING THEORY AND TECHNIQUES

In order to mitigate the radiative EMC problems, mention has been made of making the coupling path between the culprit and victim as inefficient as practicable. Of various methods suggested like grounding, bonding, filtering and the like, effective shielding is used as the first line of defence, at least for mitigating the radiated emission/radiated susceptibility. In case of any applications, if it cannot afford to offer the overall shielding effectively, some other means of controlling EMC may be relied upon. Normally, electric shielding is provided by a solid metal enclosure/chassis. Sometimes, in place of metal housing, the electronic circuit is placed within a cover made up with rubber/polymer impregnated with carbon fibre or carbon powder.

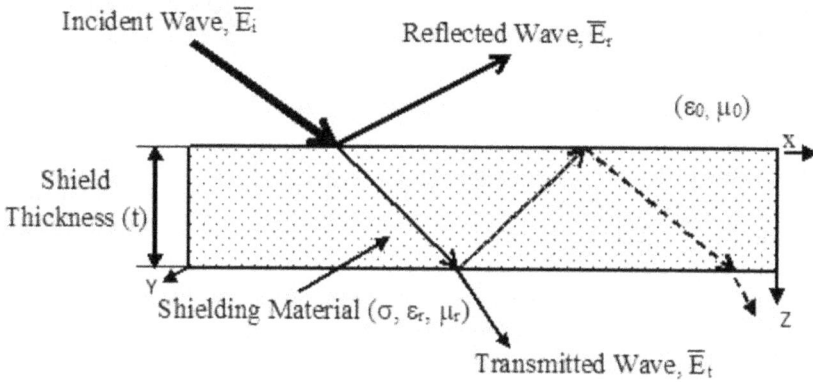

FIGURE 4.6 Mechanism of wave propagation through a shielding material showing how reflected and transmitted waves are generated from an incident wave.

Occasionally, housing may be a plastic cover painted with highly conductive coating or vacuum deposition.

- Reflective losses, R (dB), due to reflection in the plane of incidence.
- Absorption losses, A (dB), due to power absorptions taking place while passing through the material.
- Secondary reflective losses, B (dB), due to secondary reflections within the material along the dotted paths as shown in Figure 4.6. Normally, B is neglected when A > 8 dB.

Under such situation, SE (dB) = R + A + B ≈ R + A.

Some of the useful parameters of the material in controlling the values of SE, are the following: frequency–f, shield material thickness–t, emitter to shield distance–r, permeability of the shield material relative to air–μ_r and conductivity of the shield material relative to copper–σ_r. The shielding materials as used in practice are aluminium, copper, cold-rolled steel and the like, essentially because of their relatively high conductivity and high reflectivity. The SE of those materials theoretically is adequately high and found to be more than 200 dB at 10 KHz and above for aluminium and cold-rolled steel, and for copper at 1 GHz and above.

But all said and done, in reality, the values of SE are very much limited, typically because of the fact that there is inevitably a necessity for apertures and holes on the shield cover, including, for example,

- Removable shield-cover, which may be unavoidable for regular repair and check.
- Holes for ventilation and passage for connectors inside.
- Holes for control of components from outside.

Under such circumstances, some measures of enhancement may be recommended as

- Using the minimum possible number and sizes of holes/apertures/seams, if inevitable.
- In case of metal-to-metal interface, using a gasket/spring-finger for good, well-ensured contacts.
- Having adequate contact surface areas free from paints and debris in the interfaces of seams.
- Ensuring an enclosure with seams has adequate conductive contact areas with as many as possible contact points over adequate overlap areas; gasketing helps ensure good electrical contact between fasteners.

4.6 EMI MITIGATION TECHNIQUES

As far as the conductive interference is concerned, the main power supply line or signal or data links (victim) can be completely isolated from the defective equipment (culprit) causing contamination of the interference in the victim system with the help of the line impedance stabilization network (LISN) having sharp cut-off low-pass filters. This is a good method of alleviating conductive interference causing conduction noise current in a conductive path. In this context, it is to be remembered that the LISN is also used to measure the magnitude of conductive interferences flowing through the phase as well as neutral lines of the defective product. Furthermore, it may be recalled that in the design of a circuit, keeping common mode current produced within the product to its minimal value is another effective way to mitigate conductive interference to a certain extent. In order to combat EMI, it is always advisable to make use of passive components like resistors, inductors and capacitors for design of the LISN and filters.

The coupling path in between the culprit and victim, as discussed earlier, should be designed to be as ineffective as possible so as to produce a minimal amount of radiative interference to be coupled to the victim. One of the suggested methods is the application of good and effective shielding material (with adequate SE value) around the victim as well as culprit or in between them in a suitable manner so as to avoid easy passage of the radiative interference through victim from the culprit. For the purpose of unavoidable ventilation, there is a need for holes or slots required to be drilled on the surface of the shielding material; these are to be properly designed so as to have effectively continuous conductive electrical surface inside the shielding material (based on the fact that every waveguide has a cut-off frequency) without affecting ventilation. Such a type of shielding structure design is standardised and is available in literature.

It is to be kept in mind that an earthing and equipotential bonding implementation is another important area to which one has pay careful attention, not only for human safety alone but for EMC mitigation as well. As a rule, for the sake of human safety, all metal enclosures in offices, commercial and residential buildings where IT installations exist are connected together as a part of an equipotential bonding network to all other parts, such as metal conduits (both water and gas conduits), metal beams, steel girders etc. It will be observed, surprisingly, that such equipotential bonding networks concurrently improve the EMC performance. Hence, it may be inferred that EMI and EMC are heavily dependent on earthing and the equipotential bonding principle, which will ultimately yield both human safety and EMI mitigation.

Besides technological considerations in the product design, as mentioned earlier, strict implementation and imposition of regulations set up by regulatory bodies (like FCC, CISPR, MIL-STD etc.) of different countries are to be properly executed and tested before permitting the manufacturers of digital-based electronic devices to put their products into legal markets for sale. It is the responsibility of the respective governments to look after the proper implementation of this regulatory process with adequate infrastructural facility. Side by side, the violation, if any, of the regulations done by a manufacturer deliberately or otherwise should be subjected to a penalty that may include fines or jail time. Under such circumstances, companies concerned may be afraid of negative publicity for their merchandise.

For the purpose of avoiding last-minute test failure and to achieve cost-effectiveness in respect of mechanized goods, it is always advisable to take suitable steps for necessary EMC considerations before the design, layout and fabrication of the products with EMC experts. EMC certified materials and modules are to be used for manufacturing of the products. Further, for the system installation and project supervision, EMC trained staff and skilled experts are to be employed.

4.7   EMC METROLOGY

The measurement methods of the EMC parameters are as important as the design techniques employed for controlling the EMI in a noisy environment. Yet it should be as reliable as possible—meaning it must be accurate and repeatable irrespective of time, space and person. However, this significant area can be broadly categorized as (1) diagnostics, (2) pre-compliance, (3) full compliance and (4) production. The testing methods should be properly standardized and characterized as per the requirements of a country. However, a brief review of the theories involved in the major EMC measurements is presented here:

Measurable EMC parameters are (1) conductive emission, (2) conductive susceptibility/immunity, (3) radiative emission and (4) radiative susceptibility/immunity.

4.7.1   MEASUREMENT OF RADIATIVE EMISSIONS

The most important requirement for the testing site is to have a noise-free environment, which will ensure a direct link between the equipment under test (EUT) and the test-probe without allowing any other signals, whether reflecting or otherwise, to pollute the receiving signal at the test-probe site. In order to simulate free space conditions, some novel arrangements have been suggested to test and evaluate the radiative emission of EMC measurements. These are as follows: open area test site (OATS), transverse electromagnetic (TEM) cell, gigahertz transverse electromagnetic (GTEM) cell and anechoic chamber. Basic principles of operation of some of these testing concepts are explained and given here.

4.7.2   Open Area Test Site (OATS)

A good account of OATS as regard to its concept and construction has been provided in ANSI C63.4 and CISPR 16.1.1 (Radiative Emissions/Immunity) rules. This is referred to as "CISPR Ellipse" and is as shown in Figure 4.7.

In Figure 4.7, X represents the distance between the foci, A and B, of the ellipse; A is for the position of EUT and B is for antenna test-probe. Preferred value of X is chosen based on the recommendation of the regulation applied as 3 m or 10 m or 30 m. Of the ellipse, the major diameter = 2X and the minor diameter = √3X. The ellipse area must be free of reflecting objects. On the ground plane, a metal mat extends from A to B, as shown by shaded test area within Figure 4.7. Antenna test probe positioning height is adjustable at 1 m to 4 m from ground.

Generally, being an open space, ambient environment prevails over the test area of the OATS, which may swamp the EUT signals with strong broadcast and legitimate radio transmissions of TV, AM, FM and cellular carriers. However, fortunately, there are efficient methods to differentiate these signals. It would have been better if OATS was installed in a remote place surrounded with mountains all around providing natural, nevertheless effective, shielding.

4.7.3   Anechoic Chamber

This is an altogether different approach to rectify the inherent problem linked with the OATS—for example, due to ambient environmental setback, other radio waves may prevail in the testing ground, affecting the authenticity and accuracy of the measured data. An anechoic chamber is well treated, adequately large and perfectly shielded all around, even the roof space. The inner sides of metallic screening walls

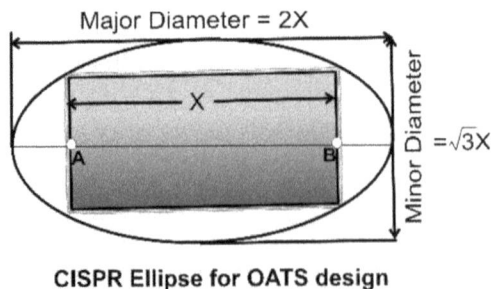

CISPR Ellipse for OATS design

FIGURE 4.7   CISPR Ellipse for OATS design.

of the chamber are covered with RF compatible adhesive, upon which are pasted RF absorbing materials, either in the form of pyramidal carbon-loaded foams or ferrite-loaded flat absorbing tiles available commercially on the market. As such, there will be no risk of infiltration of external radiative emissions at all, if any, within the chamber; neither, internally, will there be any RF reflections. However, a costly ingeniously contrived chamber creates a pure free-space ambience that is ideal for accurate field strength measurement. The measuring set-up is the same as that of an OATS.

4.7.4 ANTENNA FACTOR

The CISPR specified frequency span for radiative emission limits ranges from 30 MHz to 1 GHz. Under such circumstances, the bandwidth requirement for the antenna-probes has to have an adequately large value. For example, during testing the preferred antenna-probes recommended are as follows:

- Bi-conical antennas (30 MHz—300 MHz).
- Log-Periodic antennas (200 MHz—1 GHz).
- Broadband antennas (30 MHz—1 GHz).

Further, it is to be noted that in the measurement, the antenna-probe, on receiving interfering signals from EUT, provides only the magnitude of voltage (volts) at the input terminals of the antenna probe, corresponding to the electric field incident at plane of the antenna-probe. In order to derive the preferred quantity, that is, electric field, E in V/m (say, A) at the antenna plane from the measured quantity, that is, voltage, V in volts, (say, B) at the antenna input terminals, the antenna factor (AF)—a transfer function-like parameter—will be useful to define as the ratio of A/B. Logarithmic unit is extensively used in engineering measurement because of its unique property of data compression and computational advantages of complicated mathematical expressions into summation of data. Hence, AF (dB /m) = E (dB μV /m)—V (dB μV). Or, in other words, E (dB μV /m) = AF (dB /m) + V (dB μV). Therefore, the electric field strength at antenna plane in dB μV /m may now be obtained by adding the measured voltage in dB μV at the antenna terminals with the AF in dB /m. The AF of an antenna probe, being the unique property of an antenna itself, is to be supplied by the manufacturer only. Along with AF, they should supply its calibration chart of its frequency dependence as well, for a correction factor if there is a change in frequency. AF, in general, is a complex quantity. AF for magnetic field may be defined and used in the same way.

4.7.5 MEASUREMENT OF CONDUCTIVE EMISSIONS

The circuit diagram of the line impedance stabilization network (LISN) shown in Figure 4.8 is basically a passive low-pass filter that is extensively and successfully used to measure the conductive emissions, if present in the power cord lines supplying 230 volts, 50 Hz connected with the device under test (DUT) or defective DUT generating interferences to the power cord lines. In fact, LISN isolates the

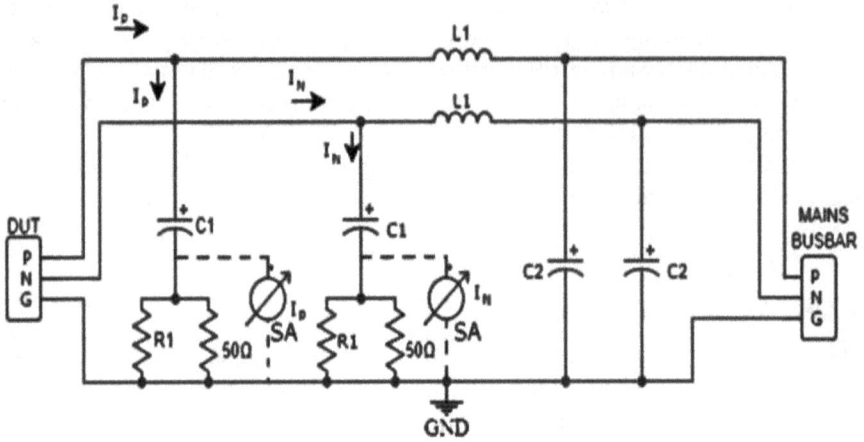

FIGURE 4.8 Depiction of the circuit of the line impedance stabilization network used for the measurement of conducted emissions.

power cord lines from the DUT and vice versa; at the same time, it provides constant impedance to the DUT over the frequency range of the interferences to be blocked. It is to be noted that the conductive emissions limits are regulated by the FCC standards over the frequency range 450 KHz to 30 MHz and by the CISPR 22 from 150 KHz to 30 MHz.

4.8   APPLICATION AREAS

4.8.1   GREEN TECHNOLOGY FOR SMART CITIES

With rapid progress of urbanization all over the world for the last couple of decades, it has been observed that the concept of planning and shaping smart cities with novel and innovative ideas of the citizens concerned is gaining ground in popularity. However, such smart cities are very difficult to conceive in reality, because they do not have a universally accepted definition. It varies from city to city and country to country. It depends very much on the level of expansion desired, willingness to modify, and aspirations of the city inhabitants. The main objective of smart city planning is to endorse cities which can extend the citizen a core infrastructure providing a decent quality of life, a clean and sustainable environment and application of smart solutions.

Being dependent on so many smart and intelligent activities, a smart city might just as well be referred to as an *intelligent city* or *ubiquitous city* or *digital city* or *virtual city*. However, the main thrust of such project may be broadly categorised, as shown here:

- Smart energy.
- Smart transportation.

- Smart data.
- Smart infrastructure.
- Smart mobility.
- Smart Internet of Things (IoT) devices.

It is hoped that these six pillars of aspirations will certainly produce a decent quality of life to city-dwellers. They will certainly provide a pure and clean and at the same time sustainable environment with smart solutions. It is expected that the model will, indeed, work as a lighthouse to other aspiring cities as well.

In order to leverage the collective knowledge and information of the smart city, a seamless networking of the physical infrastructures, ICT infrastructures, business infrastructures and social infrastructures with proper integration of IoT technologies (especially through machine-to-machine communication and human-to-machine communication) is a must for achieving all the smartness of a sustainable, green smart city already defined.

All said and done, it is apparent that the entire smart city mission becomes profoundly dependent on the cutting-edge technology available like digital electronics, IoT, broadband technology, information and communication technology (ICT), renewable energies, smart-phones, social networks and other electronic devices. The IoT mainframe combined with a large number of sensors, actuators and other technological gears collects data from various sensors placed at strategic locations across the cities. All the vital parameters, such as temperature, air pressure, air pollutants, precipitation etc., all over the smart cities are derived from the collected data. On analysis of the data collected. a suitable planned pattern is obtained and used as smart solutions.

Since smart cities as a whole are strengthened by the backbone like electronic systems through the use of IoT devices and other digital technology systems, it is evident that the green technology project is likely to become vulnerable and affected by an unwelcome or unpleasant condition such as electromagnetic interference (EMI) or electromagnetic pollution, if at all they are generated in adequate magnitude. Therefore, the project engineer of the smart city mission should be fully aware of the presence of such pollution, which is invisible, but its effects are hazardous and unsafe to the city planning devices. The engineer, however, should take smart steps, in time, to protect the aspirations of the city dwellers.

4.8.2 DISASTER RESILIENT SMART CITIES

Often fatal in character, catastrophes such as flood, earthquake, tsunami, nuclear detonation, transportation accidents, terrorism etc. cause death and destruction to human life, infrastructure and the environment. Conventional disaster management methods are, in fact, totally obsolete. In reality, they are not adequately equipped and automated to provide an accurate and quick response for such a multifaceted, huge volume of data handling aptitude. It has been observed that with the advent of the IoT technologies along with big data analysis (BDA), smart city project missions have been successfully operational. Now it is proposed to check whether the system could handle disaster management as well in addition to the regular supervision of smart city activities.

The disaster resilient smart cities concept can be implemented with data procurement from various agencies. The main objective remains to provide early warnings, collection of the information in real time and accurate estimation of the damage; quickly work out the evacuation routes; and effectively manage emergency resources. However, the entire work is implemented by processing suitable integration of IoT technologies and BDA system with the aim at minimizing possibilities of casualties and environmental destruction.

4.8.3   HEALTH HAZARDS DUE TO EMC PROBLEMS

With the rapid explosion of digital technology, air space of the urban areas over the globe is filled up with unwanted electromagnetic interference signal components. For example, the use of mobile phones has become an indispensable and fashionable handy handset for easy communication for most of us. In fact, this has practically transformed the scenarios of modern communication technology and pushed its advancement in a forward direction quite a lot. The superfluous high frequency signals thus generated are spread out all over in which human beings are embedded, either conditionally or unconditionally. On the other hand, the signals also act as sources of interference or pollution to all other close by electronic systems. However, the non-ionizing interfering signals at the same time create hazards in the human bodies, damaging their health even to the extent of serious injuries and death. This type of health hazard is commonly observed in all sorts of biological species, from animals to plants, in addition to human beings, all over the advanced countries of the globe.

Scientists are very much worried about the health hazards, especially for human beings. In order to protect human beings from the harmful effects from the non-ionizing electromagnetic exposure, especially in the frequency range from 30 Hz to 300 GHz, the IEEE-SA Standard Board promulgated noteworthy recommendations on the safety levels of maximum permissible exposure (MPE). However, the unit of MPE is mW/cm^2 (milliwatt per square centimetre). These recommendations are strictly applicable to people in controlled and uncontrolled environments. It is not particularly designated for patients who are purposefully exposed to radiation by experts of the healing arts. Typically, for 3 GHz to 300 GHz frequency range, 10 mW/cm^2 has been recommended, with an acceptable safety factor under considerations (IEEE Std C95.1, 1999 Edition).

At this stage, it is necessary to clarify and explain the differences between controlled and uncontrolled atmospheres. In controlled environment, persons concerned are exposed to the microwave radiations because of their occupational obligations. Therefore, it is true that they are certainly aware of the situation and can take necessary preventive measures for their protections in time. MPE levels for them are normally set at a little bit more stringent value. On the other hand, people in uncontrolled surroundings are the general public, who are not aware of the unhealthy condition and not at all conscious of the danger of microwave radiation in which they are all totally immersed. However, the MPE value in terms of power density for such people is set pragmatically at 10 mW/cm² in the frequency of 3 GHz to 300 GHz.

Another dosimetric measurement parameter that has been widely adopted above frequency level of about 100 kHz is the specific energy absorption rate (SAR). In fact, it is the rate at which RF energy is absorbed in human body tissues, and its unit is watts per kilogram (W/kg). SAR may be determined from the following statement that it is the time derivative of the incremental energy (dW) absorbed by an incremental mass (dm) contained in a volume element (dV) of a given tissue density (ρ), that is, SAR = (d/dt) (dW/ρdV). When a human body is exposed to EMI, the associated tissue absorbs electromagnetic energy from space, which in turn increases body temperature inside. It is known as specific absorption (SA) and expressed as (dW/ρdV) with a unit of joule per kilogram (J/kg).

The International Commission on Non-Ionizing Radiation Protection (ICNIRP) published in Health Physics 74 (4): 494–522; 1998, similar guidelines for limiting exposure to time-varying electric, magnetic and electromagnetic fields (up to 300 GHz). Based on the related technical papers published by various scientists, research workers and engineers in different international journals, ICNIRP prepared a comprehensive review plan, which includes:

- Epidemiological study on electromagnetic interferences exposure and cancer, including childhood leukaemia.
- Reproductive outcome study.
- Residential cancer study on the aetiology of childhood cancer. It also includes link between ELF magnetic field exposure and higher risk of cancer.
- Occupational study.
- Cellular and animal study.
- Etc.

Along with the experimental verifications, the epidemiological studies and biological effects within (100 kHz—300 GHz) have been thoroughly discussed, and recommendations on the restrictions on SAR values in respect of electric and magnetic fields were set up up to 10 GHz. However, these guidelines are modified regularly depending upon requirements of the need of time.

In the clinical engineering domain, electromagnetic pollution, if generated, may not allow any sophisticated and sensitive medical equipment such as that for electrocardiography (ECG), electroencephalography (EEG) and electromyography (EMG) to function properly, and thus it may create a most disastrous and unfortunate situation which may lead to serious injury or even to death. Under these circumstances, either this equipment, whose minimum response of physiological signals are always less than a millivolt, may collapse without showing any response or malfunction, giving incorrect results, thus creating confusion in the mind of the doctors. In order to avoid such an awkward state of affairs, there should be a tight collaboration between biomedical workers such as physicians and operators of medical equipment and EMC experts. They should realize the importance of the electromagnetic compatibility between the electronic gadgets working simultaneously along with other medical equipment within a very close proximity. In management of such an uncomfortable and embarrassing situation, the administration should take care of checking and procuring the EMC documents of the medical apparatus involved from the manufacturers and examine whether rules and regulations imposed by the EMC regulation authority of the country are properly maintained and followed or not. The entire establishment must have an appropriate layout of RF and other internal communication cables as per rules with the help of EMC engineers—it is just a simple, effortless coexistence between biomedical units with other information technology (IT) gadgets. However, one should realize that the electromagnetic pollution is not a mystery, neither is it rocket science. Cost-effectiveness of the project of setting up of a super-speciality hospital providing patients with state-of-the-art health care can be improved only if an effective EMC management within the health care facility is properly well coordinated.

4.9 CONCLUSION

With the rapid explosion of the use of digital devices, the generation of high frequency RF components are likely to take place automatically. These signals really are superfluous, unwanted and redundant. Although signal strength is insignificant in magnitude, in such cases, if a number of electronic systems work together in a close proximity, there is every chance of conflict with each other, inadvertently or otherwise. In such a situation, electromagnetic interference or pollution is created and can produce havoc in all the electronic systems by inflicting total failure in their operations. This type of unwanted occurrences, called electromagnetic pollution, is the main focus of this chapter. The nature of the culprits, the sources of electromagnetic pollution, have been systematically characterized, and their behaviours have been analyzed. Usually, they are either conductive (wired) or radiative (wireless) in character. By and large, electromagnetic pollution is man-made. But it can also be created by nature. As instances of such pollution are, most of the time, destructive in nature, the mitigation of such devastating actions has been discussed as well. Framing of rules and regulations has been meticulously prepared and geared up by authorized regulatory institutions for EMC standards. This is strictly followed by the manufacturers of the electronic equipment. It is a punishable offence if there is any deviation in production of the equipment. Some of the regulatory bodies are as follow: the FCC in the United states, CISPR in Europe, VDE in Germany and VCCI in Japan. Another method of alleviating pollution effects has also been discussed in this chapter, which is by technology enhancement. Design guidelines and methodology are proposed in this chapter. Because of the peculiar nature of pollution, some new concepts have been introduced in the chapter, for example, the near-field and far-field concept of radiative component. In the domain of the conductive element, the perception of common-mode current and differential-mode current has been established. Shielding is another region where the chapter has focussed on its design. In order to protect an electronic unit from electromagnetic pollution, the proper shielding design concept has been introduced. Shielding effectiveness (SE in dB), its limitations and improvement for this purpose have been discussed as well.

For the sake of estimation of performance of EMC equipment manufactured in a factory, the method of obtaining accurate results, the chapter has also focussed on EMC metrology. LISN is basically a low-pass filter that not only isolates mains circuit from the defective devices under test but also measures the conductive emissions in dBμV across 50 Ω shunt resistance coming out from the defective device. It is explained with an actual circuit diagram as well. Definition and explanation in regard to antenna factor, CISPR Ellipse and anechoic chamber have been successfully presented and explained.

The defence, telecommunication, e-commerce, hospital and electronic industries, in general, are significantly susceptible and vulnerable to electromagnetic pollution. In case of any attack from electromagnetic pollutions, the entire electronic system will collapse, and there will be complete blackout. Therefore, it is very important to know the methods of winning the battle with electromagnetic pollution in case the need arises.

The chapter will certainly help raise awareness of what electromagnetic pollution is and how to control it in case of emergency and will be useful to those connected to the development of green technology for smart cities.

Health hazard is an equally important area in which electromagnetic pollution plays a great responsibility in creating problems. The chapter indicates that coexistence and simple compatibility of EMC experts and health care providers are essential and indispensable.

In the area of green technology for smart cities, a higher susceptibility to electromagnetic interferences/pollutions is observed where mass increase of low power wireless networks like in the IoT occurs. This happens because of higher concentration of co-located devices. Further, sizeable extension of frequency of emissions of the order of tenths of GHz for IoT applications requires additional improvement in methodology and device designs for standard EMC applications. These are some of the important challenges of EMC to the Internet of Things and hence to the project mission of green technology for smart cities.

4.10 ACKNOWLEDGEMENT

The author would like to express his deep sense of gratitude to Mr. Prashnatit Pal, Research Scholar, NIT-Patna, India, for his umpteen number of technical helps during the progress of manuscript preparation. The author, furthermore, extends his heartfelt thanks and sincere appreciation to Mr. Soumik Dey of Manipal Institute of Technology, Karnataka, India, for preparing computer-Figures (8 nos.) of the manuscript.

BIBLIOGRAPHY

[1] Ryszard Struzak, "Co-Existence: Introduction to Electromagnetic Compatibility", The Abdus Salam International Centre for Theoretical Physics (JCTP), Trieste, Italy, 07 February–04 March, 2005.

[2] William G. Duff, "Technical Notes on an Introduction to Electromagnetic Compatibility (EMC)", Applied Technology Institute, www.aticourses.com/sampler/intro_to_EMI. pdf.

[3] Wolfgag Langguth, "Fundamentals of Electromagnetic Compatibility (EMC)", Hochschule fur Technik und Wirtschaft, University of Applied Sciences, Goebenstrasse, 40, D66 117 Saarbrucken, Germany. Copper Development Association, May 2006.

[4] Clayton R. Paul, *Introduction to Electromagnetic Compatibility*, 2nd Edition, John Wiley & Sons, 2010.

[5] C. Christopoulos, *Principle and Techniques of EMC*, CRC, 1995.

[6] C. Christopoulos, "EMC Fundamentals: Basic Concepts, Theory and Practices", Short course presented at KL Malaysia, August 15, 2003.

[7] Andrew Farrar, "EMC Contributions", IEEE Trans. *Electromagnetic Compatibility*, vol. EMC-25, no. 3, August 1983, pp. 154–156.

[8] James Colloti, "EMC Design Fundamentals", *Telephonic Command Division*, 2005. Internet, www.ieee.li/pdf/viewgraphs/emc_design_fundamentals.pdf.

[9] Karim Loukil and Kais Siala, "EMC Fundamentals", *ITU Training on Conference and Interoperability for ARB Region*, CERT, 02–06 April, 2013.

[10] IET Guidance Document on "Electromagnetic Compatibility and Functional Safety": A Fact-File Provided by the Institution of Engineering and Technology, 2006, www. theiet.org/factfiles.

[11] Stephen E. Lapinsky and Anthony C. Easty, "Electromagnetic Interference in Critical Care", *Journal of Critical Care*, Elsevier, no. 21, October 2006, pp. 267–270.

[12] Mohammad Rouhollah Yazdani, Hosein Farzanehfad, and Jawad Faiz, "Classification and Comparison on EMI Mitigation Techniques in Switching Power Converters: A Review", *Journal of Power Electronics*, vol. 11, no. 5, September 2011, pp. 767–777.

[13] Roger F. Harrington, *Time-Harmonic Electromagnetic Fields*, McGraw-Hill Book Co., 1961.

[14] *Making EMI Compliance Measurements*, Application Note, Agilant Technology.

[15] *EMI, RFI and Shielding Concepts*, Analog Devices Inc., MT-095 Tutorial Internet Address, January, 2009, pp. 1–16, www.analog.com/media/en/training-seminars/tutorials/ MT-095/pdf.

[16] *Module 8: EMC Regulations*, Internet Address, pp. 8–13, www.egr.msu.edu/em/research/ goali/notes/module8_regulations.pdf.

[17] Peter Russer, *EMC Measurements in Time Domain*, Institute for Nano-Electronics, Munich University, Germany. Published in 2011 XXXth URSI General Assembly and Scientific Symposium held in 13–20 August, 2011, Publisher: IEEE. Internet Address, www.ursi.org/proceedings/procGA-11/ursi/ET-1.pdf.

[18] N.A. Omollo and P.G. Wiid, "Toward Time Domain Shielding Effectiveness Investigations Using Reverberation Chamber", *International Conference on Electromagnetic in Advanced Applications (ICEAA)*, 2017, pp. 1444–1447.

[19] P.B. Jana, A.K. Mallick, and S.K. De, "Effects of Sample Thickness and Fibre Aspect Ratio on EMI Shielding Effectiveness of Carbon Fibre Filled Polychloroprene Composites in X-Band Frequency Range", *IEEE Transaction on Electromagnetic Compatibility*, vol. 34, 1992, pp. 478–481.

[20] P.B. Jana, A.K. Mallick, and S.K. De, "Electromagnetic Interference Shielding by Carbon Fibre Filled Polychloroprene Rubber Composites", *Composites* (UK), vol. 22, 1992, pp. 451–455.

[21] P. Annadurai, A.K. Mallick, and D.K. Tripathy, "Studies on Microwave Shielding Materials Based on Ferrite and Carbon Black-Filled EPDM Rubber in the X-Band Frequencies", *Journal of Applied Polymer Science* (USA), vol. 83, no. 01, 2002, pp. 145–150.

[22] V. Kodali, *Engineering Electromagnetic Compatibility: Principles, Measurements, Technologies and Computer Models*, 2ed Edition, IEEE Press, Inc., New York, 2001.

[23] James McLean, Robert Sutton, and Rob Hoffman, TDK RF Solutions Inc., "Interpreting Antenna Performance Parameters for EMC Applications, Part 3: Antenna Factor", Internet Address, http://tdkrfsolutions.com/images/upload/brochures/antenna_paper_part3.pdf published in 2002.

[24] Vito Albino, Umberto Berardi, and Rosa Maria Dangelico, "Smart Cities: Definitions, Dimensions, Performance and Initiatives", *Journal of Urban Technology*, vol. 22, no. 1, February 2015, pp. 3–21.

[25] Teena Maddox, "Smart Cities: 6 Essential Technologies", *Innovation*, August 2016, web page, www.techrepublic.com/article/smart-cities-6-essential-technologis-innovation/.

[26] M. Casini, "Green Technology for Smart Cities", *2nd International Conference on Green Energy Technology (ICGET 2017)*, vol. 83, 2017.

[27] Badis Hammi, Rida Khatoun, Sherali Zeadally, Achraf Fayad, Lyes Khoukhp, "Internet of Things (IoT) Technologies for Smart Cities", *IET Research Journals*, The Institution of Engineering and Technology, vol. 07, no. 01, September 2017.

[28] Saraju P. Mohanty, Uma Choppali, and Elias Kougianos, "Everything You Wanted to Know about Smart Cities: The Internet of Things Is the Backbone", *IEEE Consumer Electronics Magazine*, vol. 05, no. 03, July 2016, pp. 60–70.

[29] Husam Rajab and Tibor Cinkelr, "IoT Based Smart Cities", *2018 International Symposium on Networks, Computers and Communications (ISNCC)*, Rome, Italy, 19 June, 2018.

[30] Syed Attique Shah, et al., "Towards Disaster Resilient Smart Cities: Can Internet of Things and Big Data Analysis Be the Game Changers?", *IEEE Access*, vol. 07, July 25, 2019, pp. 91885–91903.

[31] Yihong Qi, Jiyu Wu, Guang Gong, and Jun Fun, "Review of the EMC Aspects of Internet of Things", *IEEE Transaction of Electromagnetic Compatibility*, December 2017, pp. 1–9.

[32] Megan Ray Nichols, "The Biggest Challenge to Smart Implementation Is Interference", *IoT Magazine*, February 2019.

[33] Kia Wiklundh and Peter Stenumgaad, "EMC Challenges of the Internet of Things (IoT)", The Swedish Defence Research Agency, 2017, Internet Address, www.electronic.se/files/2017/05.

[34] IEEE: SA Standards Board, "IEEE Standard for Safety Levels with Respect to Human Exposures to Radio Frequency Electromagnetic Fields, 3kHz to 300 GHz", *IEEE Std C95.1*, 1999 Edition.

[35] ICNIRP, "Guidelines for Limiting Exposure to Time-Varying Electric, Magnetic, and Electromagnetic Fields (Up to 300 GHz), International Commission on Non-Ionizing Radiation Protection", *Health Phys.*, vol. 74, no. 4, April 1998, pp. 494–522.

[36] ARPANSA, *Radiofrequency Electromagnetic Energy and Health: Research Needs*, Australian Radiation Protection and Nuclear Safety Agency, Technical Report 178, June 2017.

[37] M.T. Ma, *EMC Standards and Regulations: A Brief Review*, NISTIR-3989, NIST (National Institute of Standards and Technology, USA), May 1992.

[38] P.B. Jana, A.K. Mallick and S.K. De, "Electromagnetic Interference Shielding Effectiveness of Short Fibre Polychloroprene Vulcanised by Barium Ferrite", *Journal of Material Science* (UK), vol. 28, 1993, pp. 2097–2194.

[39] P.B. Jana S.K. De and A.K. Mallick, "Electroconductive Polymer Based Composites for EMC Applications – A Review", *Electromagnetic Compatibility Journal*, vol. 7, nos. 1 & 2, April & October 1994, pp. 1–14.

[40] P B Jana, A K Mallick, "Studies on Effectiveness of Electromagnetic Interference Shielding in Carbon Fibre Filled Polychloroprene Composites" – Journal of Elastomers and Plastics, vol. 26, no. 1, January 1994, pp. 58–73.

5 Waste and Material Management in Green Cities

Praveen Kumar and Karan Chandrakar

CONTENTS

5.1 INTRODUCTION

Waste management term is used to manage the biproduct, which is generated by people in variety of forms. This biproduct is not waste if we manage it in proper way. Therefore, the waste management term is coined in this modern scenario to utilize this biproduct. There are multiple practices, which are used to manage these wastes, which are generated from household, industry, and agriculture. While the world's population keeps increasing at a fast rate, sustainable theory will be required to address waste management.

Many Asian countries are facing adverse effect of climate change due to higher generation of waste and its improper utilization. Many studies indicate that world population is projected to increase from 8 billion to 10 billion by 2050, and it is estimated that solid waste generation increased roughly by 70%. Higher population growth means higher demand of products, and it results in higher generation of waste product. This is due to number of factors such as higher population growth, economic growth, urbanization and increasing people purchasing capacity. Therefore, the need of hour is, how to manage this waste and utilized it for generating energy and making other products. Asian cities typically employ a range of schemes for waste management, depending on the technological and financial tools accessible and the existing degree of environmental consciousness in the city concerned. The approach used by low-income urban centres is primarily selective recycling, free-standing waste, partial recovery of recyclables by the informal sector, few landfills and limited composting. Cities in developing countries have expanded their scope for storage, usage of landfills for disposal, use of waste to energy plants and utilization of composting as well as recyclable reuse and recycling, which is shown in Table 5.1. This effective practice guide reflects on crucial strategic elements for implementing an effective solid waste management program, with a review of best practices from across the globe contributing to improved cultural, social and environmental results for communities. These solutions are good practices which include:

- Improving sanitary waste and landfill management.
- Building a waste management system.
- Integrating waste reduction and participation in society.
- Developing creativity in programs to manage waste.
- Supporting business economy growth in the field of waste recycling.
- Using automated imaging for solid waste management.
- Ensuring integrated waste management schemes and their implementation.

Content and solid waste management in green cities involves all activities required for final disposal of waste from its sources, including collection, distribution, monitoring, handling, recycling and final disposal. As such, solid waste management is

TABLE 5.1
Solid Waste Generated in Cities (Quintals per day)

City	Delhi ($\times 10^3$)	Japan ($\times 10^3$)	Singapore ($\times 10^3$)	Kolkata ($\times 10^3$)	Beijing ($\times 10^3$)	Dhaka ($\times 10^3$)	Bangkok ($\times 10^3$)	Jakarta ($\times 10^3$)
Year	2018	2014	2009	2007	2006	2005	2004	2004
Waste production	105	10886.2	151.9	27.2	145.1	4.2	85.2	77.6
Collection	85	10886.2	151.9	16.3	123.5	17.6	0.068	62.1
Recycling	20	5443.1	0.086	2.7	35.5	4.3	–	–
Composting	25	110	–	6.3	6.7	0.13	2.5	0.063
Incineration	30	200	61.6	–	0.27	–	0.18	0.19
Disposal	10	180	3.7	16.3	0.11	17.6	65.9	54.4

an essential part of environmental sanitation, which is one of the municipalities' most resource-intensive services. Many local bodies are responsible for waste management, either through government supervision and distribution or by regulation development and enforcement. In general, structures in emerging Asian cities are still rooted in the processing and disposal of waste while managing much of the waste generated. In these regions, composting and reuse of recyclable materials by way of mechanized or manual processes has not reached adequate rates of use relative to overall waste generation. Open dumping is widespread, although sewage disposal construction is restricted by the minimal approval of such constructed disposal sites by the population. Four mainland landfill sites exceeded capability in Singapore in 1999, during which the country started to use its offshore Semakau landfill [1]. Incineration is used in highly developed cities such as Singapore and Shanghai, which has a high energy consumption. Concerns such as high prices, potential adverse impacts on the atmosphere and resource depletion are major obstacles to using such a technology in most emerging Asian cities [2]. Throughout the countries participating in the group of the Association of Southeast Asian Nations (ASEAN), agricultural and usual solid urban waste is commonly disposed of in large landfills [3]. However, due to the rapid population growth and economic progress of countries, the methods presently used to cope with waste generated in large amounts in Asian cities will not be adequate to manage the projected volume of waste. The introduction of the 3Rs (Reduce, Reuse and Recycle) is an essential method for mitigating the production and efficient management of waste at various levels of suitable sanitation and equipment used to mitigate and recover waste at multiple sites [4]. Green cities are being designed for the future in a particular manner. The following sections include a flow diagram of waste produced from different sources, as well as technology and various approaches that are acceptable as criteria for green city environmental practices as applied in several Asian and European countries. Such criteria provide the essential elements that are required for the management of sustainable operations. The waste-to-energy conversion in green cities is shown in Figure 5.1.

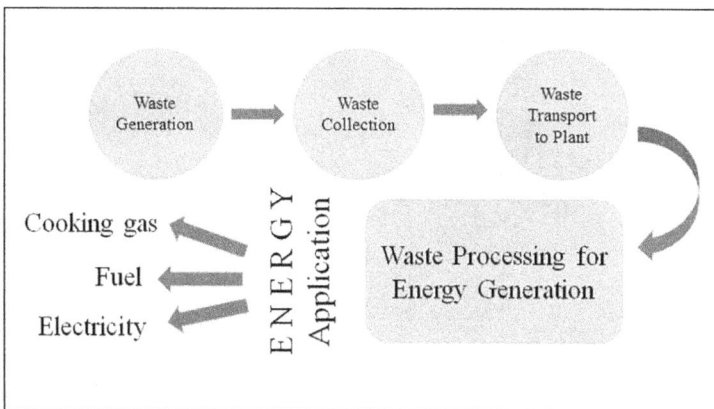

FIGURE 5.1 Waste to energy conversion in green cities.

5.2 MANAGEMENT OF SOLID WASTE IN GREEN CITIES

Green cities' solid waste/garbage disposal is rooted in the prevention of generating waste and the reuse, recycle and reduce concept. Those constitute the upper rank of the system of proper waste/garbage management and act as a general reference to the different waste flow operations, which is shown in Figure 5.2. This depends upon the quality of waste from various sources and the usage, where possible, of new or innovative technology. Open dumps with toxic waste are used for landfills in Asian countries, which can be used as raw material in waste to energy conversion plants. Table 5.2 shows the content of the waste from different generators. The percentage shares of the multiple elements produced in total waste may change with the per capita income level attained by specific cities of different countries.

Waste materials produced in bulk from low-income developing countries are commonly linked to the other biodegradable and waste food materials [5–6]. Figure 5.3 indicates the gross waste or waste movement in the green cities. The reuse of recyclable materials is shown in the figure as a material loop. Once the generated waste is treated and stored, large content chains exist which enables usage by the various generators. In the same area, recycling and eventual reuse of waste materials in specific ways cannot be achieved automatically. The waste flow described by specific waste streams shows the best path to use in a green city solid waste management network through multiple separate elements.

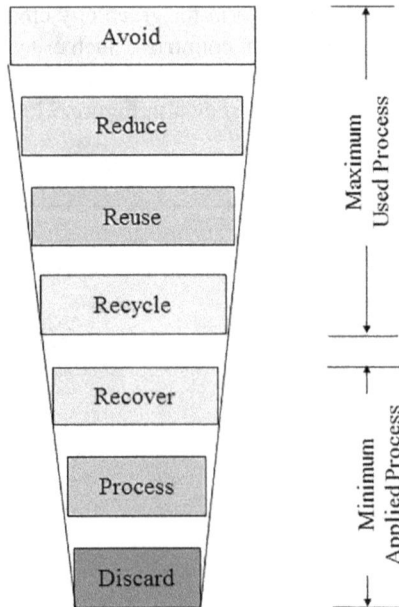

FIGURE 5.2   Solid waste management in green cities.

TABLE 5.2
Types of Solid Wastes in Green Cities

Source	Residential Waste Management	Industrial Waste Management	Commercial Waste Management	Institutional Waste Management	Construction Waste Management	General Services Sector Waste Management
Waste generators	Family units residence	Chemical plants waste	Market and hotel waste	Hospital and educational institutional waste	Building waste	Society and parks waste
Biodegradable waste	Food and plant waste	Food park waste	Paper	Paper	Wood and composite	Tree and plants leafs
Non-Biodegradable waste	Glass plastics and waste gadgets	Glass, metals, plastics, hazardous waste and sludge	Glass, metals, hazardous waste and e-waste	Glass, metals, plastics, hazardous waste and e-waste	Concrete and steel metal scrap	E-waste and sewage sludge

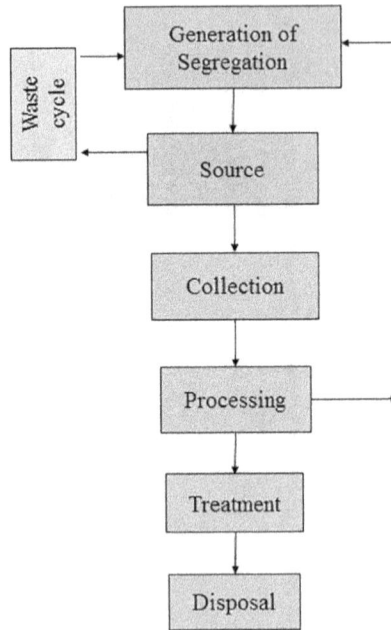

FIGURE 5.3   Waste and material flow chart of a green city.

5.2.1   RESIDENTIAL WASTE MANAGEMENT

The common waste categories in suburban areas comprise food and other biodegradable products, shown in Figure 5.4. The dry, theoretically recyclable materials might be divided into five groups before harvesting. Most of these systems require necessary segregation into components that are both biodegradable and non-biodegradable. Household waste, for example, is divided into four categories: (a) wrapping and packaging paper; (b) inflammable waste; (c) non-inflammable waste and (d) huge bulky waste. The different constituent of waste is divided into the source of origin for potential reuse, storage, sorting and treatment as well as subsequent recycling, based on the waste's composition and the generator size. The food waste should be used specifically for cattle fodder, and in the developing countries, this activity is growing. Other biodegradable components that occurred in the southeast Asian countries [7–9] include the reuse of easily recyclable dry products such as pulp, papers, fibres, metals and glassware waste at sources in the residential areas, as well as discarded appliances or devices. During the trial process, the recovery of related materials occurs well before waste enters transfer stations, resource storage centres or waste areas. Mainly these practices are used in the urban areas in less developed countries as well as the countries where low and medium incomes were generated. Household waste consists of hazardous products, and it should, therefore, be stored directly in the city centre. These materials are then collected and transferred with residual materials from the materials recovery facilities (MRF) or processing plants to final disposal sites or WTE [10].

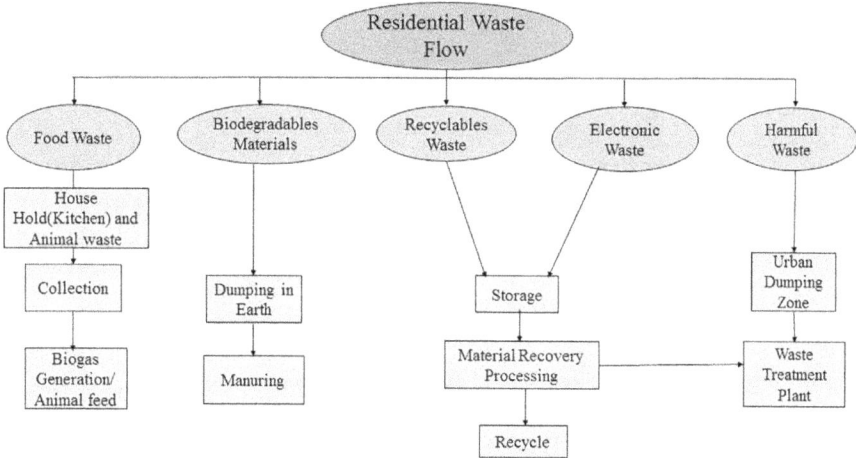

FIGURE 5.4   Residential waste flow areas in a green city.

5.2.2   INDUSTRIAL WASTE MANAGEMENT

In the industrial sector, the waste stored and disposed of is generated mainly from particular, dangerous waste in treatment plants, plants where waste is converting into energy plants or sites of sanitation. Packaged products, as well as other domestic waste, can be reused directly at the source. The majority of waste is collected and then transported to MRFs for sorting such as sorting electronics waste, which is the final step before it is sold at the various recycling centres. Figure 5.5 describes the flow diagram of the segregated food waste which is transferred to the animal feed manufacturing facility units or for composting. Similar to household waste, the incineration waste, which is ash, and MRF remnants and other residuals may be used as a principal component in much ceramic processing and the glass industry [11].

5.2.3   COMMERCIAL WASTE MANAGEMENT

Commercial institutions such as supermarkets, grocery shops, hotels and markets produce mainly food waste, which can be converted in composting plants into soil composting, which is shown in Figure 5.6. Farm waste is processed and treated as animal feed in countries such as Singapore and Japan, which is classified as Ecofeed [12–13]. As well as residential waste, composting plants can manage other biodegradable waste from commercial establishments. Plastic and paper materials are commonly reused in commercial establishments, although the metal waste containers and waste glass items used traditionally often meet with stringent health and protection criteria. All dry waste recyclables and waste that is special, like e-waste, are subject to processing, recovery and storage for distribution in warehouses or the recycling sector.

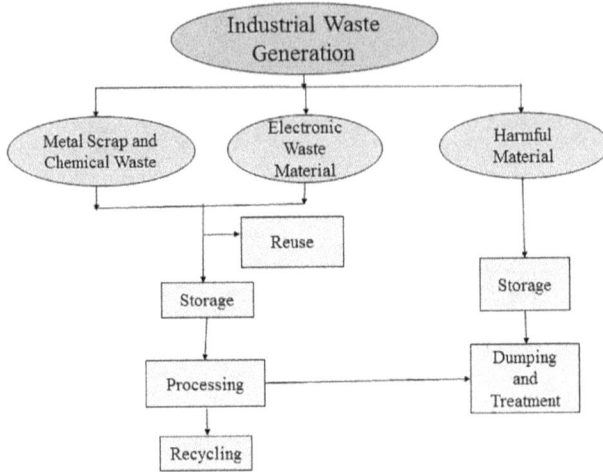

FIGURE 5.5   Industrial waste flow in a green city.

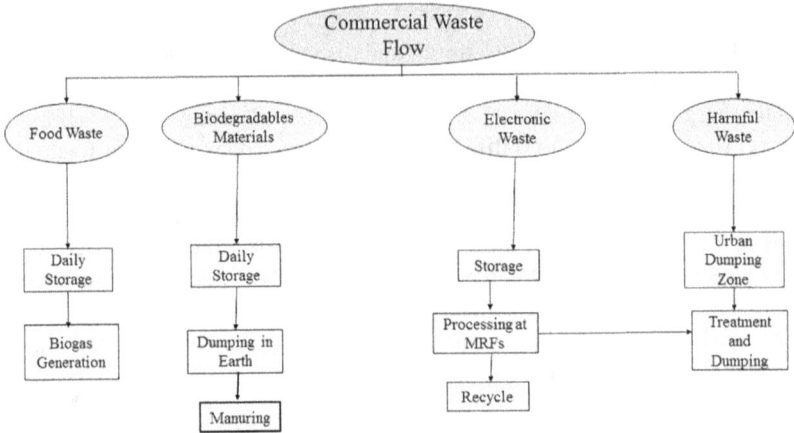

FIGURE 5.6   Commercial waste flow in a green city.

5.2.4   Institutional Waste Management

Institutes and offices usually use paper and the items that are required for office activities, and in this typically blank sheets and printing paper are used for notepads and printing of records. Such paper waste products are processed and often filtered and later recovered. In the case of agricultural waste, the waste is used by animal feed producers and is therefore biodegradable. Dry recyclables such as e-waste are used by many facilities, which are for material recovery and recycling plants, and storage and disposal where the waste is toxic, and this management matches the waste produced by industrial establishments and is defined in the flow chart of Figure 5.7.

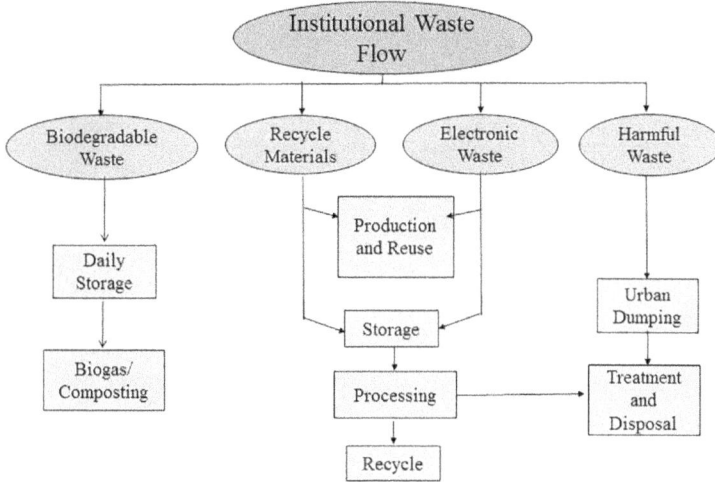

FIGURE 5.7 Institutional waste flow in a green city.

5.2.5 CONSTRUCTION WASTE MANAGEMENT

Building construction waste is typically used for filling low field land holes or as aggregates and checked for suitability. The recovery of reusable materials has increased from factories that build modern machinery in developing countries in eastern Asian countries, with 84% of industrial waste created in 2010 (i.e., 14.2 million tons) [14] and 66% of the industrial and commercial combined waste produced in Hong Kong was recovered in 2010 [15]. Such industrial waste was used in north Asian countries like China as well as in Japan for the creation of wetland parks, using aggregated building construction and demolition (C&D) materials. Remaining waste materials from C&D and facilities for material recovery processing can often be reused as fuel for WTE plants or fuel for other cement plants. As seen in Figure 5.8, wood and other related materials can be used in different construction sites for the production of temporary concrete skeleton structures. In using such C&D waste goods, the conventional practice for disposing of items in sanitary landfills can be stopped.

5.2.6 GENERAL SERVICES SECTOR WASTE MANAGEMENT

Municipal corporations preserve the beauty of a region by offering materials such as tree loppings from roadsides and gardens that are used to produce compost and prepare soil conditioners as fertilizer for the development of productive ground. Road sweeps include biodegradable as well as sandy soil, which provides dust and sand with various forms of biodegradable materials and other non-biodegradable materials which may be directed to MRFs for the recovery of recyclable materials shown in Figure 5.9. Products taken from residual waste are used as a raw material for the recycling plant or in the WTE plant and used as a sludge in the material recovery face phase.

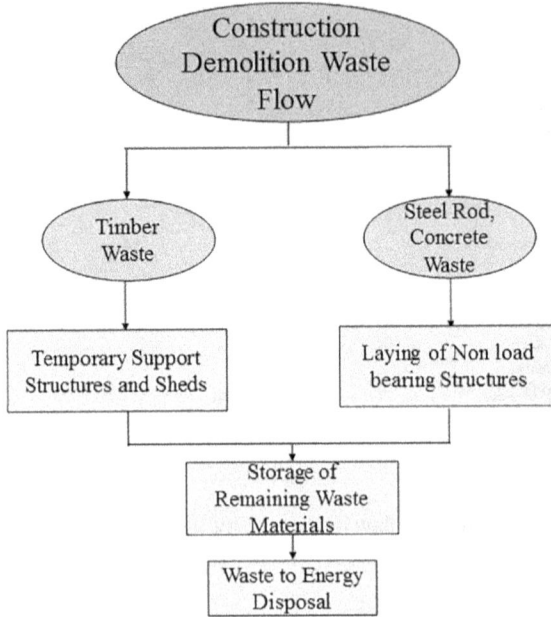

FIGURE 5.8 Constructions and demolition waste flow in a green city.

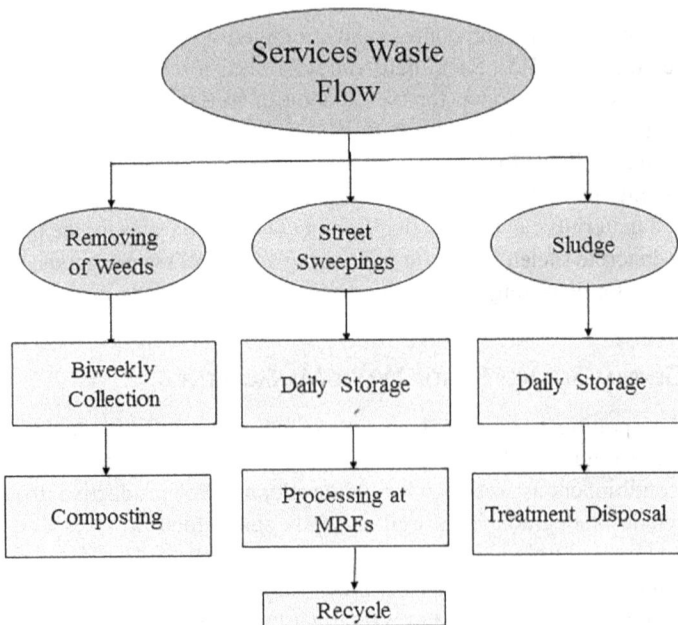

FIGURE 5.9 Services waste flow in a green city.

5.2.7 Integration into the Solid Waste Management System

The waste flows from various fields, and the strategies of processing, disposal and recycling of waste should be incorporated into the solid waste management system in green cities, as seen in Figure 5.10. This method aims to validate the definition of solid waste management and provides further attention to the 3R principle used as a first phase and less concern for free dumping. Outlets (wood-related products, textile rugs, paper, plastics, metal scrap and glass) from which they come for reuse were returned to specifically recycle separated waste from homes, businesses, business enterprises, manufacturing facilities, and building and demolition operations. Most sorted and separated residual waste is processed and directly shipped to the MRFs or composting plants, primarily to transfer centres or stations in the case of cities, where collection units are situated at a significant distance from the waste generation centre. The facilities for recovered materials are designed for agricultural and outside storage of plants. Products provided by MRFs also are used for manufacturing purposes. In cities in low-poverty, medium-wage or semi-developed countries, recyclable materials move from garbage storage to recover facilities centres in order to repurchase system that can be used for the processing of waste goods in recycled plants. Biodegradable agricultural or compost waste may be used as fertilizer or soil conditioners for gardens and farms. The compost amount depends on the sale point of the manure created by waste agro-products. Still, the consistency of the compounds can improve by enhancing the processing skills so it can be marketed at a defined sales price. Residues of non-compostable materials which cannot be processed in composting plants manufacture ash that is used for sanitary landfills, which may be used as aggregates for the development of asphalt, cement and bricks. Upon treatment of waste treatment plants for processing and segregating bulk waste materials, it is crucial to determine the feasibility of landfills, incineration and WTE system operations. RFs are fed into WTE plants or used as disposal to fill sanitary landfills as aggregates.

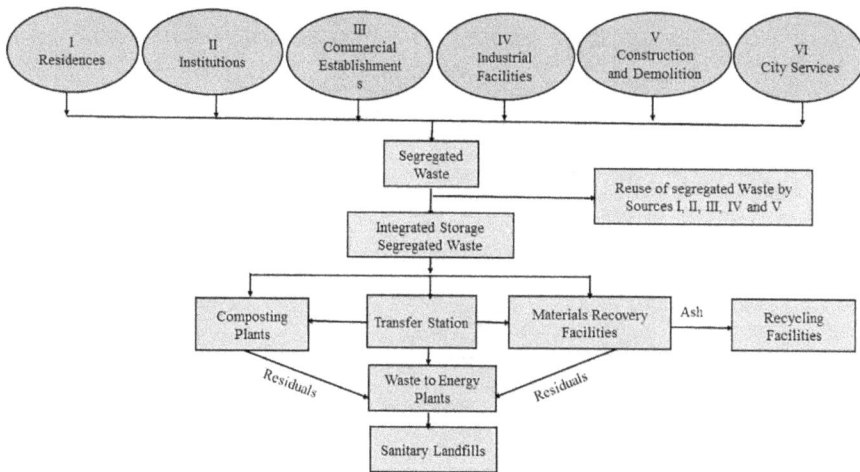

FIGURE 5.10 Integrated waste flow options in a green city.

5.3   METHODS AND TECHNIQUES FOR SOLID
WASTE DISPOSAL IN A GREEN CITY

The methods used for processing and treating biodegradable waste materials, recyclable products, nuclear C&D and nuclear and hazardous waste include basic manual, electrical, thermal and biomechanical procedures. Various techniques for solid waste disposal in a green city are shown in Figure 5.11. The findings of the report focused on the 3R solid waste management net theory of the Asian Institute of Technology, which taught the theory in Bhutan, Myanmar, China, Cambodia, Indonesia, India, the Philippines, Malaysia and Vietnam, are already implementing the 3R concept [17]. The results of studies on 3R solid waste management at the Asian Development Institute indicate that the implementation of the three standards is still underway [17]. The production of compost and the recovery of recyclable items can shift due to varying scales, mainly in the informal sector. Many countries such as Japan, China, South Korea, Singapore and Taiwan have limited the usage of WTE programs.

5.3.1   BIODEGRADABLES, RECYCLABLES, RESIDUALS AND HAZARDOUS WASTE

The majority of all food waste generated in developing countries is fed directly to domesticated animals or other outdoor animals. In order to be commercially feasible,

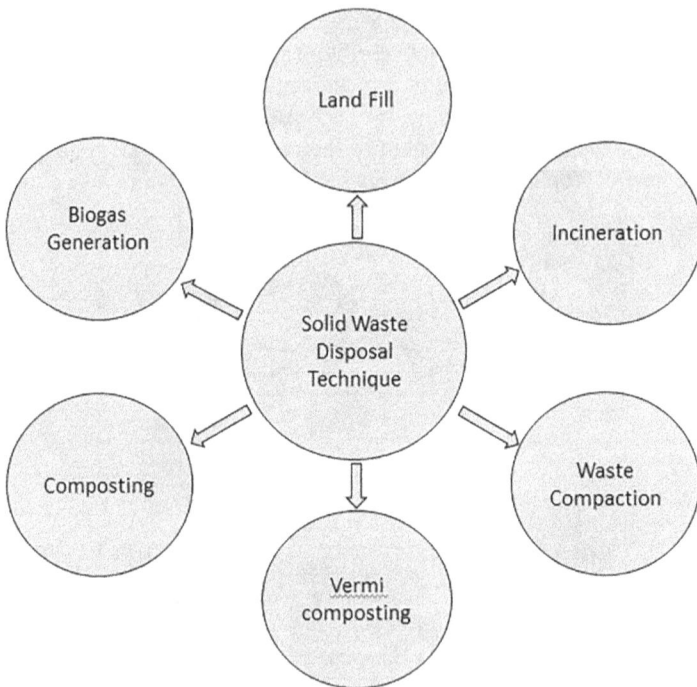

FIGURE 5.11   Various techniques for solid waste disposal in a green city.

the production of food waste from industries to domesticated animal feed is typically carried out by individual businesses, since effective approaches such as synthesis of biogas by food waste require an immense amount of material for their processing. A country like Singapore collected 90.7 million tons of food waste and converted it into animal feed every day in 2005 [18]. Fixed-bin composting and vermicomposting systems along with fixed-bin composting and manual coiling for the processing of manure are typically used for composting operations at this scale. The quality of compost manure provided by worms is high. The material recovery plant must build and operate with six electrically operated rotary drums to treat the biodegradable materials collected. The compost created is used as a fertilizer for growing vegetables and flowers in an urban environment and acts as a soil conditioner, increasing the nutritional value of soil.

The informal market is already in the process of recycling in cities such as Jakarta, Manila, Bangkok, Ho Chi Minh City and Shanghai [6, 19]. Glass, bottles, plastics and metals are valuable recyclable materials obtained from trash bins, waste collection equipment and open dumps. Mechanical machinery required ranges from small to vast recycling operations, which involve hoppers, payloaders, drum panels, conveyor belts, magnetic and pressure insulators, weight scales, balers and large storage areas, all inside big material recovery facilities. The running of such services requires considerable capital expenditure and professionally qualified personnel. Decentralized, community-based composting has been successfully implemented in Dhaka by cooperation between the public and private sectors.

The high calorific content of residual waste from material recovery facilities and industrial waste can be used as a fuel generated from waste fuel in energy generating plants [21]. Split and broken waste residual materials are used in the manufacturing of bricks and hollow blocks that can be used for construction purposes in countries such as Manila [22].

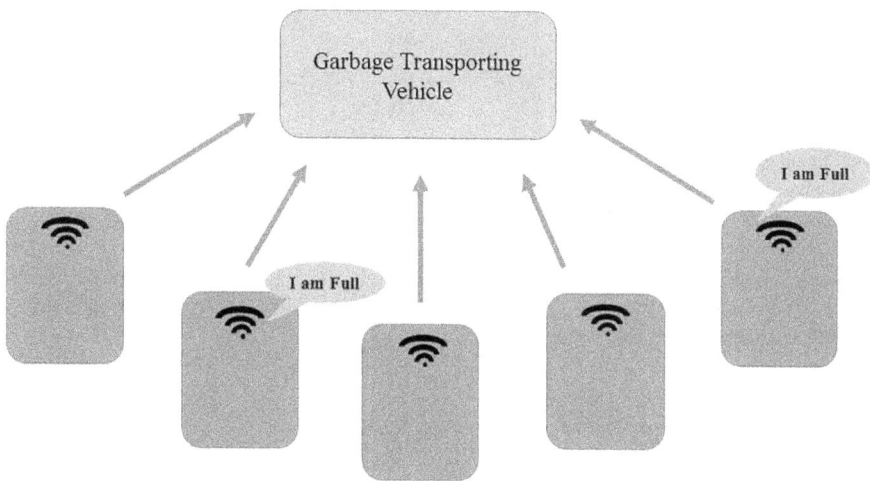

FIGURE 5.12 Various smart bins in green cities.

Many wastes which are hazardous in nature can be burned in incineration plant to minimize their quantity. Ash which can be collected from incinerators can be used to clean concrete bricks, or in the cement industry [23].

5.4　CHARACTERISTICS OF WASTE MANAGEMENT IN GREEN CITIES

The main aspects of waste management in different types of cities (developing, developed and green cities) are shown in Table 5.3, which includes the different types of mechanism of financing facilities require for their development. It also maximizes the chance of changes in many areas such as low-level to high-level production and its processing increasing from rural area to urban area.

TABLE 5.3
Characteristics of Green City Solid Waste Management

Solid Waste Management Components	Developing Country	Developed Country	Green City
Rule and regulations			
Reduce, reuse and recycle programs	X	X	X
Waste collecting sector charges		X	X
Non-sorted fine	X	X	X
Prohibition on open area dumping and burning	X	X	X
Reduction of waste	X	X	X
Sorting at source	X	X	X
Food waste used for cattle	X		
Paper and recyclables regeneration	X	X	X
Construction waste reuse	X	X	X
Methods			
Storage of sorted waste	X	X	X
No sorting no collection	X	X	X
Financial support			
Government financial aid	X		
Initiatives of private sectors		X	X
Equipment and facilities			
Waste bins	X		
High density polyethylene waste bins		X	X
Waste collection vehicles	X	X	X
Waste transfer stations		X	X
Processing plants for food waste		X	X
Types of community composting Plants			
Socially	X		
Privately		X	X
Socially recovery facilities for waste	X		
Privately recovery facilities for waste		X	X
Power plants in land fills	X	X	X
Energy generation from waste		X	X

5.5 POLICIES, REGULATIONS, AND LEGISLATION ON SOLID WASTE MANAGEMENT

National policies and legislation drive the implementation of sound waste management programs. Most Asian countries have devised, with differing degrees of effectiveness, environmental and solid waste management legislation [24]. Generally, the degree to which implementation is successful depends on the nature of the awareness-raising and coordination activities carried out, as well as the degree of non-compliance penalties and this system comprises in South East Asian Countries [25]. The various aspects of green city waste management are shown in Table 5.3, which require the charging of waste collection charges, including warnings and penalties, in the financial elements of solid waste disposal. Past aid has been linked to the elimination of pollution and promotes reuse. This obliges the waste production to conform to the 3R laws and legislation, prohibits littering, free dumping and waste incineration, and collects funds for solid waste collection facilities. Control of waste management payments is usually applied, mainly in high-income countries. Producer responsibilities should be practiced by all major waste-generating companies, which can be used to eliminate, reuse, repurchase or recycle by the producer involved, depending on the product. Both programs promote the financial, commercial and public efficacy of reduce, recycle and reuse procedural operations. Large producing companies may conduct responsible practices or product management at all levels of production. That can be accomplished by economic incentives that inspire the production of environmentally sustainable goods and by holding manufacturers responsible for the expense of maintaining their goods at the end of their useful lives.

5.6 MANAGEMENT OF SOLID WASTE IN GREEN CITIES

Solid waste management systems suitable for a green city could be managed by local communities in the countries with lower incomes. Such facilities typically include informal community participants as part of a broader resettlement plan for dislocated families developing waste collection facilities. The operation of solid waste management plants can gradually be transferred to the private sector as per capita income rises and waste volume increases. Such facilities include transfers, sanitary sites, WTE plants, and even hazardous waste plants and their management.

The remaining waste items, including biodegradable products, must be processed to demonstrate the logistics capacities of the community in question. Funding some investment level is needed if a sound solid waste management system is to be established and implemented. A significant collection method would help source segregation at various per capita income rates that use a broad range of waste collection systems, as long as the separated waste is obtained on a timely basis and transported to waste recycling or disposal installations in house-to-house processing of household biodegradable materials used by green cities in the democratic composting program of the region, which has been introduced by Dhaka [26]. A unique cooperative of waste collectors and urban poor in India was established by the Pune Municipal Corporation, which involves the collection of waste at the door [27].

5.6.1   FACILITIES AND INFRASTRUCTURE

Waste containers are necessary to enable isolation of the source and eventual sepa-rate storage of the waste. There might be water cane, wooden containers, tanks, and concrete or composite bins used for this purpose in the least-industrialized and low-income countries, working countries with limited and poor income. Countries can opt for high-density polyethene (HDPE) bins of different colours and sizes that differ in the volume of waste they contain. The main goal of such waste containers is to hold separated items temporarily, to deter livestock and waste pickers from entering them, and to avoid foul odours or contaminated pollutants from. Plastic and paper bags and cans do not need to be used, because they are less capable of preventing the spill of liquid or odour. Waste generation vehicles allow dedicated waste to be transported to transit points, recycling or storage sites or WTE plants. An essential prerequisite for these vehicles is the non-mixing of source-separated waste materials, a reduction in the emission of foul odours during transport and storage of harmful content that may be produced by the biodegradable or hazardous portion of the waste obtained. Enclosed trucks or small vans are used for the dis-posal of garbage, although open trucks are likely to be used in the least-developed and low-income countries of the world. In this case, sufficient shielding should be given to minimize unwanted odour escape. The smart bins in the green cities are represented in Figure 5.12.

5.6.2   WASTE GENERATED BY FOOD-PROCESSING UNITS

With the usage of vermicomposting for food waste fields, biodegradable materials may be turned into cattle feedstock, biogas and organic manure. Feed may be fer-mented or dried by boiling, baking, curing at lower temperatures or food waste dry-ing at high temperatures [28]. These approaches are used in Japan, India, China and other Asian countries for the production of Ecofeed and pig feed [29]. In Figure 5.13, the conversion of organic waste (food waste, farm waste and other waste) to energy is shown. Both households and the rest of society use biogas from fermenting agri-cultural waste instead of liquefied petroleum gas and for the generation of energy. Earthworms (*Eudrilus eugeniae*), which feed on the shredded and source-grouped food waste, are used for making land fertile and vermicompost, even in low-income communities in cities at every per capita income level, would establish right soil conditions for farms and gardens.

5.6.3   RECYCLING AND COMPOSTING IN GREEN CITIES

The first process of recycling is the collection of waste material from different places and this collection is carried out from different residential societies, ware-house, workshop and industries. After storage, it will be treated in the recycling plant. These treatment plants can have semi-automated machine for sorting and treatment of waste, private sector facilities using conveyors and packaging equip-ment facilities in green cities. Large, fully automated and usually privately owned material recovery facilities are used to sort different waste components according to

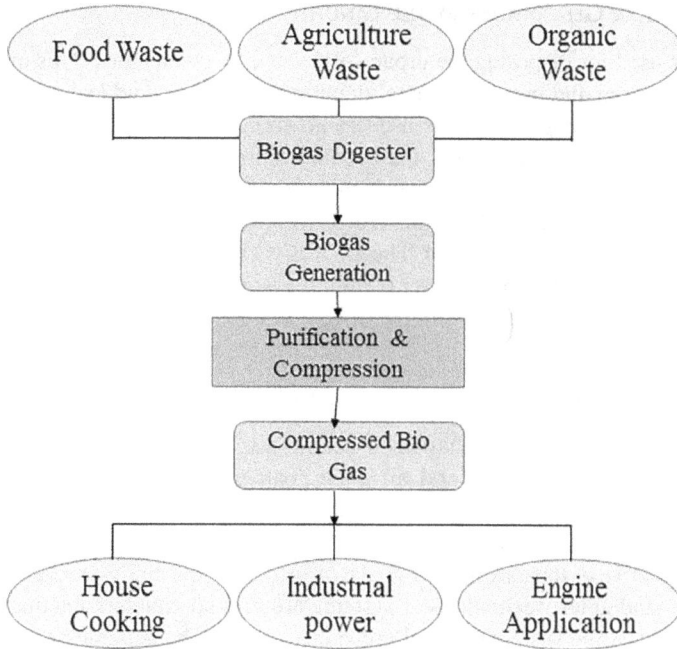

FIGURE 5.13   Food waste to energy conversion.

the density and characteristics of metallic solid waste utilizing conveyors and magnetic separator systems, drums and compressed air system. Biodegradable material composting can be used as a biologically controlled field conditioning agent. The biogas from kitchen food waste and cow dung, obtained by many parts of the rural India, are used for generation of electricity for the rural people. The primary composting input consists of shredded biodegradable waste stream components. The composting method that may be utilized relies on the quantity of waste, the amount of land access, and the quality of the waste itself. This is often used in the fields of food production and unprocessed waste from factories, households and trade establishments. Households and small communities use composting, excavated pits and fixed bins that can completely handle food waste within three months. Waste volumes of more than 5 tons are usually processed by windscreens or spinning drums. Inoculants are typically applied to reduce loading cycles and to allow the conversion of higher efficient compost from the food waste. The efficiency of waste management strategy is determined by waste characterization and mass balance. MRFs and community composting plants will also be built and maintained at or above the community level based on the findings and will rely on waste characterization studies and mass balance studies. This calculates the efficiency of waste management strategies and therefore gives the viability of the expected operations of these installations.

5.6.4 Power Generations by the Landfill Gas

Landfill is used for disposing the urban solid waste in green cities. This urban solid waste suppressed and packed in metal dome with liner to avoid leakage. After few days, it generates biogas which is used as an energy source for power generation. Tapping methane generated from the decomposition of biodegradable components in landfill sites also enables cities using this technology to achieve the agreed-upon emission reductions offered by the Executive Board of the Sustainable Growth Program under the Kyoto Protocol. The 2.1×10^{-3} kW Jana Deposit Facility in Kuala Lumpur and the 100 kW pilot power plant in Payatas, Quezón Region, are tiny LFG installations in Asian cities [30].

5.6.5 Smart Waste Collection System in Green Cities

The concept of a smart city has been commonly adopted in the urban scenario for the implementation of several different goals and developments proposed for the creation of the most productive communities. Such purposes are multiplied to render cities greener and more effective, which enhances people's standard of life [31]. Irrespective of the task field or goals, information and communication technology (ICT) and smart technological systems are crucial enablers for the incorrect advancement of the sustainable and green city model. The geographic information system (GIS) is the most reliable method that can be utilized to play a vital role in decision making and aims to facilitate the study of information [32]. The advancement of the Internet of Things (IoT) and its implementations in smart cities offer appropriate ways to solve challenges. In short, information gathered by sensors can be sent to remote repositories where it is stored, analyzed and used to map, control and eventually make wise decisions on network creation or service management [33]. The network summary of smart waste management in green cities is shown in Figure 5.14, which helps to achieve the goals of the goal. As there is a growth in the usage of sensors in broad fields, the gathering of data obtained by analysis and storing may be explicitly connected to big data, which is necessary to evaluate items and retrieve knowledge efficiently and is commonly used to open up new market strategies and explain a large-scale set of data.

Free and structured information is also known to be a champion of big data technology such as IoT. As a consequence, the incorporation of all cyber-physical structures increases the ability of a wide variety of creative smart and green city solutions [34]. The data which is based on the cloud contributes significantly in the implementation of an IoT centric framework for smart cities to collect and manage data and further work accordingly [35]. Many pieces of information such as types of garbage, garbage levels and garbage vehicles are detected by different sensors, which significantly reduces the labour, time and money [36]. In the background of the implementation of this research, waste-disposal technologies and approaches have been thoroughly researched for stewardship and thus have a significant effect on the maintenance costs of the smart community. However, owing to the decentralizing aspect of decomposition, these issues with high control costs are complicated to address.

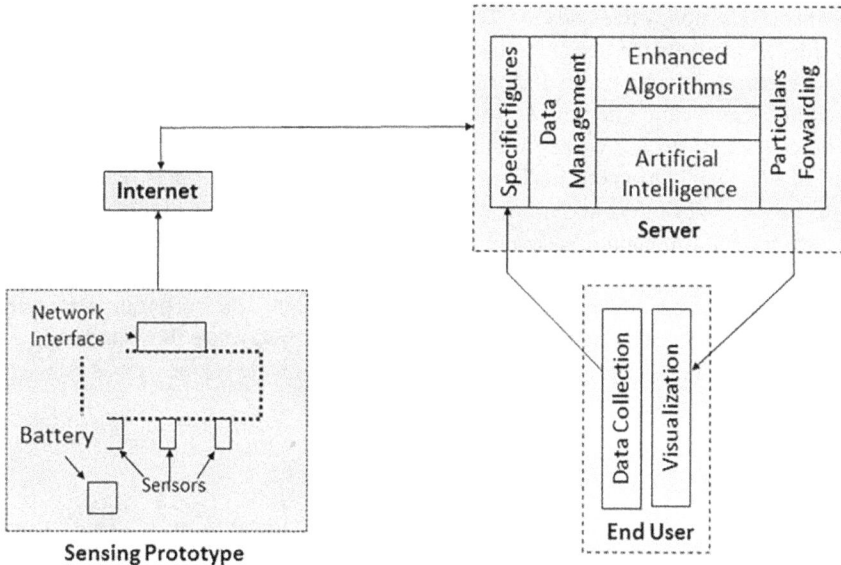

FIGURE 5.14 Smart waste collection system in a green city.

5.7 CONCLUSION

The selection of technology that can be utilized in solid waste management schemes in green cities is diverse and has been adopted in part by most Asian populations. In the beginning of the 21st century, the regulatory framework for the efficient disposal of solid waste using the 3R principle was successfully established in most, if not all, Asian countries. Knowledge and communication systems have undoubtedly been introduced in most nations to increase environmental consciousness across various classes of waste generators. The recycling of recyclables from dumps, truck storage and landfill sites is the prevailing characteristic of the informal sector. This is also not necessary to implement independent waste disposal or set up or operate MRF and green city solid waste treatment combined-waste composting facilities. The design and maintenance of these facilities rely on the reality that the green city region has been given solid waste management status as a consequence of the waste created by the green city waste program. However, these plants have the opportunity to eventually utilize useful recyclable products from all sorts of waste streams. Mixed waste obtained shall be disposed of in open dumps or in landfills or, in the absence of these facilities in individual low- and high-income semi-developed nations, in WTE sites. Despite the apparent lack of political will needed to implement environmentally sustainable waste management approaches, they have been found in a variety of instances in several cities that are compliant with, though on a small scale, green city solid waste management practices. Those have been done in several groups of populations, and these societies are entirely served by solid waste management in a particular manner, in cooperation with various private sector organizations, policy agencies, NGOs etc. This is worth noting because, aside from thermal or mechanized systems, both of these best practices utilized only conventional methods of solid waste disposal. Present laws and rules, including restrictions including fines, would be strictly followed if a solid waste management program is to be developed for the waste producers in all classes. This requires the introduction of non-segregation–no exclusion laws arising from the effects of inadequate solid waste management activities, with a general understanding of the finite ability of both the natural capital base and the tools available to reduce the rate of use of recycled waste materials. Degrading health and natural environment will further lead to effective partnerships between states, institutions, the corporate sector and local communities. The steady accomplishment of desired waste flows and expectations in a green city is at the same rate, irrespective of the region or country's per capita income point.

REFERENCES

[1] Oh, G., 2008. *Integrated solid waste management of Singapore.* Asian Network of Major Cities, Japan. www.asianhumannet.org/db/datas/Integrated_en.pdf.

[2] Visvanathan, C., Tränkler, J., Kuruparan, P., Basnayake, B.F.A., Chiemchaisri, C., Kurian, J. and Gonming, Z., 2005. Asian regional research programme on sustainable solid waste landfill management in Asia. In *Proceeding Sardinia.*

[3] Chamie, J. Mr., 2004. *World Urbanization Prospects: The 2003 Revision.* United Nations Publication Sales No. E.04.XIII.6 ISBN 92-1-141396-0.

[4] Ackerman, F., 2005. Material flows for a sustainable city. *International Review for Environmental Strategies,* 5(2), p. 499.

[5] Zerbock, O., 2003. *Urban solid waste management, waste reduction in developing countries.* School of Forest Resource and Environmental Science, Michigan Technological University, USA.

[6] Visvanathan, C., Adhikari, R. and Ananth, A.P., 2007, November. 3R practices for municipal solid waste management in Asia. *Linnaeus Eco-Tech,* pp. 11–22.

[7] Sapuay, G.P., 2013. Ecological Approach to Solid Waste Management in the Philippines. In S.K. Ghosh (Ed.), *Waste Management & Resource Utilization* (p. 33). Kolkata: Oxford Publishing House.

[8] Shekdar, A.V., 2009. Sustainable solid waste management: An integrated approach for Asian countries. *Waste Management,* 29(4), pp. 1438–1448.

[9] Enayetullah, I. and Sinha, A.H.M., 1999. Urban Management Programme for Asia & the Pacific. In *Community Based Decentralized Composting Experience of Waste Concern in Dhaka.* New Delhi: All India Institute of Local Self Government.

[10] Lin, C., 2009. Hybrid input-output analysis of wastewater treatment and environmental impacts: A case study for the Tokyo Metropolis. *Ecological Economics,* 68(7), pp. 2096–2105.

[11] Cheng, T.W., Ueng, T.H., Chen, Y.S. and Chiu, J.P., 2002. Production of glass-ceramic from incinerator fly ash. *Ceramics International,* 28(7), pp. 779–783.

[12] Sugiura, K., Yamatani, S., Watahara, M. and Onodera, T., 2009. Ecofeed, animal feed produced from recycled food waste. *Veterinaria Italiana,* 45(3), pp. 397–404.

[13] Khoo, H.H., Lim, T.Z. and Tan, R.B., 2010. Food waste conversion options in Singapore: Environmental impacts based on an LCA perspective. *Science of the Total Environment,* 408(6), pp. 1367–1373.

[14] Lu, W. and Tam, V.W., 2013. Construction waste management policies and their effectiveness in Hong Kong: A longitudinal review. *Renewable and Sustainable Energy Reviews*, 23, pp. 214–223.

[15] Jao, A.H., 2016. *Mitigating health and environmental risks from e-waste dismantling via plasma arc gasification WtE and sustainable materials management modeling.* (Master's thesis, Harvard Extension School).

[16] Fong, W.F., Yeung, J.S. and Poon, C.S., 2004, May. Hong Kong experience of using recycled aggregates from construction and demolition materials in ready mix concrete. *Proceedings of the International Workshop on Sustainable Development and Concrete Technology,* May 20, pp. 267–275.

[17] Khoo, H.H., 2015. Review of bio-conversion pathways of lignocellulose-to-ethanol: Sustainability assessment based on land footprint projections. *Renewable and Sustainable Energy Reviews*, 46, pp. 100–119. Rete Technology, pp. 267–275.

[18] Chiu, A.S. and Yong, G., 2004. On the industrial ecology potential in Asian developing countries. *Journal of Cleaner Production*, 12(8–10), pp. 1037–1045.

[19] Edmonds, S., 2008. *Shanghai's municipal solid waste and water sectors, and their respective management.* Consulate General of Switzerland in Shanghai. www.sinoptic.ch/shanghaiflash/texts/pdf/2008/200805_Shanghai. Flash.pdf.

[20] Gendebien A., 2003. *Refuse derived fuel, current practice and perspectives.* Wrc Ref Co5087-4. Accessed: Apr. 08, 2021. [Online]. Available: https://Ci.Nii.Ac.Jp/Naid/10018523483.

[21] Paul, J.G., Acosta, V.L. and Lange, U., 2015. Experiences and lessons learnt from supporting waste sector development in the Philippines. *International Conference on Solid Waste Management ICSWHK,* Hong Kong, May 20–23, 2015.

[22] Badur, S. and Chaudhary, R., 2008. Utilization of hazardous wastes and by-products as a green concrete material through S/S process: A review. *Reviews on Advanced Materials Science.*, 17(1–2), pp. 42–61.

[23] Akenji, L., Bengtsson, M. and IGES, S.O., 2012. Asia-Pacific. In *Global Outlook on Sustainable Consumption and Production Policies: Taking Action together.* UNEP.

[24] Zurbrügg, C., Drescher, S., Rytz, I., Sinha, M. and Enayetullah, I., 2002, July. Decentralized composting in Dhaka, Bangladesh production of compost and its marketing. *ISWA 2002 Annual Congress*, Istanbul, pp. 8–12.

[25] Shekdar, A.V., 2009. Sustainable solid waste management: An integrated approach for Asian countries. *Waste Management*, 29(4), pp. 1438–1448.

[26] Sugiura, K., Yamatani, S., Watahara, M. and Onodera, T., 2009. Ecofeed, animal feed produced from recycled food waste. *Veterinaria Italiana*, 45(3), pp. 397–404.

[27] Yang, S.Y., Ji, K.S., Baik, Y.H., Kwak, W.S. and McCaskey, T.A., 2006. Lactic acid fermentation of food waste for swine feed. *Bioresource Technology*, 97(15), pp. 1858–1864.

[28] Loper, J. and Parr, S., 2007. Energy efficiency in data centers: A new policy frontier. *Environmental Quality Management*, 16(4), pp. 83–97.

[29] Kaewphan, N. and Gheewala, S.H., 2013. Greenhouse gas evaluation and market opportunity of bioplastic bags from Cassava in Thailand. *Journal of Sustainable Energy & Environment*, 4, pp. 15–19.

[30] TCG on behalf of the Global eSustainability Initiative, 2020. *The ICT behind cities of the future.* SMART.

[31] Piro, G., Cianci, I., Grieco, L.A., Boggia, G. and Camarda, P., 2014. Information centric services in smart cities. *Journal of Systems and Software*, 88, pp. 169–188.

[32] Kostakos, V., Ojala, T. and Juntunen, T., 2013. Traffic in the smart city: Exploring city-wide sensing for traffic control center augmentation. *IEEE Internet Computing*, 17(6), pp. 22–29.

[33] Mitton, N., Papavassiliou, S., Puliafito, A. and Trivedi, K.S., 2012. Combining cloud and sensors in a smart city environment. *EURASIP Journal on Wireless Communications and Networking*, 247(1), pp. 1–10.

[34] Gutierrez, J.M., Jensen, M., Henius, M. and Riaz, T., 2015. Smart waste collection system based on location intelligence. *Procedia Computer Science*, 61, pp. 120–127.

[35] Saravanan, K. and Srinivasan, P. (2017). Examining IoT's applications using cloud services. In P. Tomar & G. Kaur (Eds.), *Examining Cloud Computing Technologies Through the Internet of Things* (pp. 147–163). Hershey, PA: IGI Global. doi:10.4018/978-1-5225-3445-7.ch008.

[36] Saravanan, K., Golden Julie, E. and Herold Robinson, Y., Smart cities & IoT: Evolution of applications, architectures & technologies, present scenarios & future dream." For the upcoming book series. *Intel.Syst.Ref.Library*, Vol. 154. Valentina E. Balas, et al. (Eds.), *Internet of Things and Big Data Analytics for Smart Generation*, 978-3-030-04202-8, 467407_1_En (7). www.springer.com/us/book/9783030042028.

6 Smart Energy Management in Green Cities

Manju Lata and Vikas Kumar

CONTENTS

6.1 INTRODUCTION

Global population is currently increasingly living in cities, and this population is increasing yearly at the rate of around 1.05% [1]. In 2018, an estimated 55.3 per cent of the world's population lived in urban settlements. On the other hand, cities are responsible for around 75% of overall energy and overall trade-related greenhouse gas emissions [2]. Generally, large-scale climate change mitigation and environmental sustainability significantly rely on sustainability practices by urban citizens. Current climate change procedures are not very effective, and emissions keep on increasing annually [3]. Overall, with increasing population and urbanization cities place ecological demands beneath social economic demands [4]. Innovative planning is increasing as cities are developing, having a great impact on a world that has urbanized. In the 20th century, the increasing popularity of a discrete and smart city model advanced, and in the 19th century cities became denser, producing a drain on resources and infrastructure. As a result, development resolutions with a smaller amount of ecological impact are required. A variety of thoughts and theories associated with smart cities based on sustainability and environment has been offered [4]. Currently, "smart city" as a catchword and observable fact has drawn better consideration and increased traction amongst the institutions of higher education, research foundations, government, legislators, trades, business and commerce, and specialists across the world. Despite this popularity universally, the concept of a smart city is at a standstill as far as having a collectively approved definition. In other words, a common definition of smart city is still not obtainable or proposed. It is difficult to identify the universal style of smart cities at a comprehensible level [5]. Therefore, several suggestions have been given by various groups to characterize the different definitions. Overall, a wide range of definitions has been recommended in different directions, including:

> A city becomes a smart city when it effectively uses the information communication technology (ICT) to improve the governance, participatory procedures, civic services as well as the transport reserves and is able to ensure the cost-effective growth, improved quality of life, and smart management of natural resources.
>
> [6]

> Smart city is a vast model that includes the physical framework as well as a social and human aspect.
>
> [7]

> Involving the physical, social, business, and informational technology (IT) framework to leverage the combined intellect, a city is designated as being smarter when it is efficient, livable, sustainable, and reasonable.
>
> [8]

> A city becomes a smart city when information communication technology is unified with the conventional frameworks, organized and optimized by the use of innovative digital technology.
>
> [9]

> A smart city is a city that carries the advanced techniques throughout the financial systems, public mobility, governance, surroundings and schedules. It may lead to smart arrangement of behavior and actions of a self-significant, sovereign and responsive society.
>
> [10]

Smart cities growing smarter should be able to solve ecological problems and take action toward socio-economic requirements [11]. Consequently, smarter cities are labeled as the class of cities with an opportunity for a forward-looking vision. This is characterized by constantly increasing connectedness, centrality, permeation, prevalence, and embeddedness of information communication technology into the fundamental underpinnings of the city [12]. In support of these cities, the Internet of Energy (IoE) and big data analytics have been perceived as a powerful director and significant enabler toward the revolution of environment sustainability on a large scale that is able to be implicit, matter-of-fact, intended, and planned. A smart city built upon the latest developments is characterized as a knowledge-based city, a digital and green city, and a low-carbon city [2]. City development will be derived from the comprehensive investigation of large-scale applications of information communication technology (ICT) and IoE based on the Internet of Things (IoT), cyber-physical space, widespread sensing, cloudlet, edge and fog infrastructure, etc. [13]. Applications of these tools and techniques make city life green, with low carbon emissions. Conversely, despite the numerous developments of smart city plans, several energy-related concerns remain while it moves toward deployment. Substantial environmental and green city requirements for smart cities require facilitating urbanization and planning the newest method to manage uncertainty, to enhance efficiency, and to reduce carbon footprint and greenhouse gas emissions.

While the concept of development in smart city planning makes some efforts in resolving the issues of energy-based city development, effectiveness of resource, and the green city concept (GCC), this concept is most relevant as a reply toward the concern of producing greener, denser, and more highly livable cities [14]. An energy-based smart city development model has been proposed by Zygiaris [14]. This model defines smart city-based ecological concepts together with innovative components. Different layers of this model include the city layer, green city layer, interconnection layer, instrumentation layer, open integration layer, application layer, and innovation layer. This model presents a very comprehensive look at the smart city concept and how the different concepts need to be integrated. The vision of energy management in a smart city is to be facilitated by the innovative technologies for energy-efficient digital conversion of essential services, development of quality of life and management of natural resources. The smartness of a city is derived from the plan of a green city with smart city development. Consequently, a green city is defined as a city that endorses renewable energy and energy efficiency during all its actions and comprehensively supports resolutions related to green development systems. The green city is expected to secure its growth at the local level within the rules of green development, equality, and accessibility.

6.2   ENERGY REQUIREMENTS OF GREEN CITIES

Energy requirement is increasing globally, together with growing populations, leading to the continuous utilization of fossil fuel-based energy sources (such as coal, oil, and gas). It further gives rise to a number of challenges such as reduction of fossil fuel reserves, increased greenhouse gas emissions and constant fuel price variations [15]. These challenges make the unsustainable situation that ultimately leads to a severe risk to society, although renewable energy sources are an excellent substitute and the single resolution toward increasing challenges [16]. Renewable energy provisions can significantly diminish the greenhouse gas emissions. The renewable energy distribute logically from constants sources of energy and very much sustainable in nature. It is unlimited and supplies the safe delivery of ecological-based services and production for the city. Dealing with renewable energy, climate change, natural resources, health, food, and water specification needs a very synchronized large-scale supervision and modeling of various aspects with a high ecological orientation [15].

It is obvious, according to some research work, that cities should transform from using fossil fuel energy into renewable energy sources (such as direct solar energy, bio-energy, geothermal energy, wind power, ocean energy, and hydropower) that regularly facilitate the production of sustainable energy [15, 17]. Administrators, legislative organizations, civil society organizations, governments and individuals of the globe now seem to promote the sustainable energy-related approach and opportunities. Therefore, worldwide, the sustainable development goal is ensuring that climate change will be mitigated in the 21st, as well as making sure to leave a legacy in support of future generations [18]. Common approaches and plans are proposed to reduce climate change and related impacts:

- Worldwide cities and related areas have the chance to supply power with renewable energy sources and facilitate mitigate climate change and related impacts.
- Revolutionizing behavior and lifestyle forms is able to reduce the carbon footprint and provide a good arrangement to reduce it.
- Exploring modernizations, tools, and techniques can decrease utilization of land and reduce any devastation from renewable energy sources and their accompanying risks, such as food utilization being challenged by the production of bio-energy.
- Increasing worldwide associations related to government, industries, and institutions help sustain the developing smart cities through infrastructure growth as well as advancement of tools in favor of providing sustainable energy services to reduce the impacts and ecological risk.

In the 21st century, ever-growing cities have driven the revolution of the overall energy system. Cities already comprise almost two-thirds of overall energy utilization and still make up most of the distribution of energy-based CO_2 emissions [19]. Currently, cities have various opportunities to make available the quality of life and sustainable services to the general public. Consequently, increasing requirements for as well as speeding up the of deployment of renewable and efficient energy is proposed as the most excellent way to achieve global and sustainable development goals. That calls for cities' resolutions to be comprehensive, secure, flexible, and sustainable. Using the best possible energy approach provides a global opportunity to control climate change within convenient situations. Developed cities determined the renewable energy objectives of 100% renewable energy with zero net carbon emissions [20]. However, this is not the only way to reorganize the entire urban energy for sustainable green city into renewable means. Various procedures are also being proposed at the universal level [20, 21, 22], for example:

Intentional planning: Cities are able to plan for their renewable energy intentions by supporting stake-holders in bringing up wide-ranging solutions. A small but increasing number of cities are scheduling the move to 100% renewable energy, with zero net carbon emissions.

Regulation: Depending on their authorized ability, cities have a great assertive responsibility to open or release renewable energy prospects by making codes, grid relations and regulations, and technological principles, by scheduling the utilization of land and open or communal housing programs, and particularly by determining the solar regulations.

Process: All processes need to be focused on the renewable energy. For example, the cities own or operate public conveniences via the energy utilities. Thus, they can persuade the power utilities to increase spending on renewable energy sources, on electrifying the network of cooling or heating, and also on creating a sustainable infrastructure of transport.

Utilization: Cities need to control the energy utilization. The big energy users always live with their personal choice in the utilization of energy as well in selecting the energy sources. However, the energy utilization in workplaces, buildings of education and medical, lighting of streets and communal transportation should move towards the renewable sources. This can actually reduce the growing struggle for energy resources and can also decrease the costs and risk [19].

Economics: Cities can behave like investors, creating dues and taxes or offering small-interest mortgages that are inducements to spur action toward renewable energy resolutions. In a number of smart cities, public energy corporations are among the significant shareholders in renewable energy schemes.

Citizen encouragement: Cities can act as authoritative promoters, capable of persuading the behavioral options of residents and industries through increasing consciousness regarding the remuneration of renewable energy. Community establishments are able to build stronger abilities and proficiencies through offering training programs and guidelines for scheduled renewable energy.

The development procedure for cities includes increasing requirements on renewable energy for their citizens and sustaining a peaceful, well living, and green environment and staving off devastating climate change. Renewable energy sources include the possible comprehensive objectives to manage the energy of the city. Renewable resolutions are able to renovate and transform life, society, and the economy. This transformation is producing jobs, and most recent cost-effective activities normalize the energy resources and make sure of energy sovereignty in favor of the nation and economy. Renewable energy can lead to efficient energy and provide power in support of the feasible development of cities [19]. There can be a number of opportunities related to renewable energy sources that consist of energy-related safety, the right to use of energy, socio-economic development, mitigation of climate change, and diminution of ecological and well-being impacts [15].

Many cities in the world have implemented the energy-efficient logic of renewable energy into the sustainable development of smart city to be considered a green city [23]. The aim of a green city is to ensure safe survival, develop the quality of life, secure the environment, and compose comprehensive and shared resolutions [24]. Green cities are evaluated in accordance with achievement of environmental intentions, and in favor of the capability to promote socio-economic growth derived from the green city concept and development. Green city development is implemented with the emerging concept of green cities that transform the urban consumption pattern [25]. Green cities decrease ecological impact and take full advantage of opportunities to develop and sustain their natural surroundings with efficient energy. They become more energy efficient and minimize dependency on fossil fuel. Green cities energetically promote management of waste reduction, creating a flexible and green infrastructure, the cycling of water management as well as low-carbon transfer, and distributing better quality of life results in support of citizens [26]. Additionally, they embrace urban civilization as a responsible holder of sustainability to be an essential aspect in favor of effective implementation of green city development.

6.3   THE INTERNET OF ENERGY AND GREEN CITIES

In daily life, people use different types of energy such as gas, electricity, lighting, heating, and cooling. The universal trend is to renovate the smart city into a green city by using smart tools and techniques in the form of technology [21]. With the help of smart energy technology, the ecological impact can be diminished and the development of a green city life cycle can be sustained with zero ecological impact [22]. Furthermore, the development of green cities is economically and environmental feasible by the utilization of smart energy systems and technologies. Renewable energy became only one component in favor of accomplishing green city infrastructure. For example, smart lights, smart meters, smart IoE, the smart grid, renewable energy, LEDs, smart net meters, HVAC, etc., become the most important technologies have been used to develop green cities [27, 28]. A number of countries have intentions of reducing their carbon footprint nearby 10 to 30% through the deployment of renewable and smart energy technologies [27]. For many, the upcoming vision of smart technologies is very intense and passionate. Therefore, IoE is an upcoming technology that is revolutionary in the direction of providing sustainable energy management toward a green city. IoE technology is also providing a means to solve the complex obstacles offering consumers better insight and management over energy consumption [29]. Consumers across the world are now getting acquainted with the smart meters conception. Tools and techniques related to smart meters are considered to measure energy within green city's homes and energy suppliers [30]. With direct connection of smart meters, consumers and suppliers in a green city are able to acquire a well-versed solution to compute effective energy costs and current accurate billing, dependable for instantaneous consumption of energy, and along with this, demand and supply can be mapped. Clearly, the energy grid is also an essential part of IoE that can help control and manage the safe infrastructure of green cities fundamentally. Consequently, connecting smart grids requires robust safety to protect any type of inconvenience in the process of the Internet of Things (IoT) [31]. Even though smart energy management systems for the green city are still growing, the current stake-holders are moving forward and acting simultaneously to ensure its potential growth, power, and consistency with safety. Some of the potential small applications of IoE will be reviewed in the following sections.

6.3.1 IoE-Based Smart Power Systems

The recent tendency of using fossil fuel in conservative transport schemes is eroding the opportunity of existence on this globe. Some specialists observed that maximum utilization of renewable energy sources is essential toward renewing the globe. The necessity to move to renewable energy sources becomes one of the reasons for complexity of power system procedures because of the changing environment [31]. Additionally, the great diffusion of renewable energy sources becomes an issue of the duck curve, where thermal power plants are not able to manage the steep change in the tariff. To overcome this complexity, there should be maximum use of energy storage capabilities with technology [32]. In particular, these technologies contain the three foremost functions of timely changing of bulk energy, rate of recurrence within low range, regularity solidity within high significance, along with energy trustworthiness [33]. To this point, a variety of energy storage technologies have been used in support of different applications such as pumped-hydro energy storage (PHES) [34], underwater compressed air energy storage (UWCAES) [35], liquid air energy storage (LAES) [36], ocean renewable energy storage (ORES) [37], and blue battery (fly wheel and fuel cell) in green power islands [38], as well as compressed air energy storage (CAES) [39] and advanced rail energy storage (ARES) [40]. Furthermore, these technologies have been tested since their prototype and within pilot balance to work with different applications. Fly-wheel, fuel cell, and different batteries are usually used in small-scale applications such as power quality and frequency regulation [31]. The power system operators have to deal with the transportation, businesses as well as all types of other users to maintain the safe operation of the grid. Consequently, the deployment of IoE with IoT on behalf of real-time data alternates among the system operators, renewable energy sources, and energy storage that is necessary. Thus, IoE within the energy storage sector develops the renewable energy usage along with support, getting good real-time monitoring to restore the inequities with energy storage techniques.

6.3.2 Industrial and Ecological System

Observing scale-based resolutions, an associations-based requirement has been developed among the private and public industries. In the past, this predictable association became the main provider and dealer for the energy requirements [7, 41]. This association is not able to offer suitable resolutions and do business instantaneously. To cope with the city's problem, fostering development with a robust environment, means using a variety of dissimilar providers with parallel proficiencies as well as services that are able to provide resources, pioneering equally with the green city [42]. Private industry is not able to supply the whole thing; in addition, cities are not able to trust everything to one supplier. The industrial and ecological system of IoE formation is able to build the estimation, preparation, scale, and pilot program of a smart city for the purpose of a green city [41]. IoE is able to develop these associations and can make the ecological systems within observable constraints in support of modern industry toward the development of green city.

6.3.3   AIR POLLUTION REDUCTION PLAN

As IoE incorporates the features of the IoT and artificial intelligence, it is able to classify and monitor at the granular level and instantaneously identify the biggest problems of air pollution for green city preservation [7, 43]. With this instantaneous approach of IoE, smart city administrators can acquire the instructive decisions relating to how to deal with such contingencies with a prioritization outlay. At the present time, several air quality sensors are available that can be located in communal transportation, smart tools, for instance on work surfaces, smart lights, etc. [42]. In fact, data from sensors can evaluate the quality of air that is distributed within definite data from cellular phones and a set of connected networks [7]. Alternatively, a number of tools also exist to modify achieving high efficiency from making decisions derived from simultaneous computerized data or processes at specific events. Furthermore, existing smart cities are adapting IoE related tools and procedures that perform regular actions in contiguous regions rather than substitute correspondingly in support of sustainable energy environment.

6.3.4   CENTRALIZED AND SMART GRID

The existing central grid is under duress to sustain the development of the energy system. This challenging concern lessens along with other decentralized systems as solar panels are integrated with the structure [44]. Hence, the German power grid has become, in regard to developed green cities, approximately 100% user-friendly [45]. However, the energy evolution during the trend of renewables so far has exceeded it drastically. Recently, a number of developments within the different ways networks function are getting involved in becoming smart grid [42]. As energy consumption devices are increasing day by day, fueling requirements for electricity are also increasing directly. Apparently, in alignment with the objective, renewable energy sources supply the power involving a relatively innovative mode of managing the energy system that reduces climate change and greenhouse gas emissions [46]. The foremost requirement to manage energy systems becomes smart grids with the use of digital technology. Upcoming smart grids attached the people to the diversity of the IoE, ensuring a reliable, secure, and safe power supply.

6.3.5   PEAK DATA MANAGEMENT

This is a complex issue, as the volume of data is increasing gradually as compared to distributed energy. To overcome this situation, there must be a move toward the large scale of networked devices, while nearly 500 million devices were connected to the

Internet in 2003 [2]. Since then, quite a lot of connected devices have developed quickly and are comprehensively estimated to make a rapid transition to green cities [42]. It is expected that the number of devices involved have moved to beyond 20.4 billion in 2019, from 8.4 billion in 2017 [2]. The industry expected such devices to increase in 2020 to more than 25 billion, and in 2022 it will be 29 billion [47]. The volume of data is growing in velocity. Worldwide, experts on data estimate that data will increase to nearly 175 zeta-bytes in 2025, 10 times more than the 16.1 zeta-bytes generated in 2016. This estimate is backed by market intelligence firm IDC (1 zeta-byte is equivalent to 1 trillion gigabytes) [47, 48]. It is a complex task to know how to utilize this data. Nowadays, developing systems of a green city are resolving this complexity with the support of the latest IoE technology, and others. Particularly, regarding the city data, according to Eric Schmidt, the observable fact of massive information is often referred to as IoE with big data. To connect the requirement of filtering and getting the most recent benefit of available information by 2020, more than 5 billion consumers were connected via mobile devices, approximately 70% of worldwide residents [49]. More citizens were connected via mobile phone, nearly 5.4 billion, than those with electrical energy, about 5.3 billion [49]. The fourth-generation development is led with the web, and mostly through videos that signify mobile traffic. The consumers increase from 2015 to 2020 has been twice as rapid as the increase in worldwide residents. Traffic related to the Internet from mobile devices has reached a volume of 367 exabytes yearly [49]. This large amount of data is not simple to predict; this is equivalent to 7000 billion video clips on YouTube.

The development connects with the network concept of 500 sensors located at strategic points in the green city toward evaluation of vital parameters as well as making it more energy efficient, secure, and livable. Therefore, the sensors include the capability to evaluate and provide, in concurrent data, for instance, scalable temperature as well as air pressure, precipitation, wind, lightening altitude, pollutants (NO_2, CO, SO_2, and ozone), noise level, surface temperature, and pedestrian and vehicular traffic [49]. At the same time, there is also a focus on electric vehicles in the form of dealing with the transport system to get data. These are suitable for commonplace commuting, although the charges and upkeep may be rather higher than in traditional vehicles. Evaluating electric vehicles along with internal combustion engine vehicles, in 2019, there were about 5.6 million electric vehicles around the world, including commuter cars as well as light marketable vehicles with battery electric drive and connected hybrids [50]. There has been a steady increase in the number of electric vehicles, from 3.2 million in 2018 and 1.9 million in 2017 [51]. However, IoE has been an essential concept to get efficient data in the field of transport, in order to organize, sustain, and manage the system and equally to increase the fleet of electric vehicles. Currently, there are globally nearly 160000 charging stations that are reliable, and this is expected to be 200,000 in 2020 [52].

In accordance with the IoE, every new addition to the network enhances energy-related opportunities for the complete system [53]. Sustainable energy management systems enhance the green city infrastructure; as a result, nimble power grids build up within the comprehensive energy-based network. The energy-based network can use the surplus electrical energy from one source to supply new varieties of energy in another sector. Electrical energy is converted into high temperature or exploited toward extracting heat, where it can provide the essential support to generate methane. The entire network-based technology develops most recent consequence during this revolutionary process, and it become most enthusiastic than only transmitting the electrical energy [46].

The IoE is improving the new breed of green city with the amalgamation of energy systems and ICT, where smart cities use the concept of ICT to step up the trend toward a green city infrastructure. The IoE concept is organized as a subpart of IoT, utilizing the smart sensors, smart digital controllers and actuators, and smart meters to sustain the green city environment by being able to switch information by using the information technology infrastructures. Therefore, being a subpart of IoT technology, IoE is able to develop health, quality of life, and education, along with the routine process of urban services, in favor of general public and environmental sustainability. With IoE in the energy sector, A green city is able to consider as a type of universal umbrella over a number of projects such as smart grids, smart vehicles, traffic management systems, energy efficient smart building, etc. [29, 31]. Additionally, other smart objects and features of IoE become required as well as paramount in support of global environment sustainability and energy management systems of green cities.

6.4   ENERGY MANAGEMENT

As the consequence of continuous proliferation along with urbanization, the city is facing global issues concerning infrastructure and greenhouse gas emissions. Smart and sustainable energy management also has become a global issue particularly during the last few years, following the disastrous impacts of global climate change [54]. As a result, smart cities are taking on an essential role using tools and techniques to set up a modern approach to develop sustainable energy management in green cities [23, 55]. The modern approach can reduce the greenhouse gas emissions considerably. Consequently, a number of cities of the world successfully have implemented the energy-efficient-based logistics, plans, and policies into practice for the sustainable development of a green city into specific criteria [24]. The aim of green city is to ensure safe survival, develop the quality of life, secure the environment, and write comprehensive and shared goals.

Green cities are evaluated in accordance with the achievement of environmental goals and the capability to promote socio-economic growth derived from the concept of green city development [56]. Consequently, there is also some research that has analyzed green city development planning to reduce global climate change and greenhouse gas emissions. Researchers classify the monitoring, implementation, and efficient strategies that are able to adapt to the specific characteristics of particular regions' immediate requirements. These strategies apply a unique time frame, a policies-based approach, and planning in city transportation in a balanced attempt to significantly diminish greenhouse gas emissions [57, 58, 59]. Green city strategies recognized and sponsored the essential actions to constrain the external negative environment and the impact of natural adversity. In fact, various cities have had a considerable effect on newly increasing metropolitan areas, traffic obstruction, contamination, impact on the ecological system, and on natural resources. Some strategic energy management recommendations focusing on energy-related issues of green city development [21, 22, 60, 61, 62] will be discussed in the following sections.

6.4.1 STRATEGIC PLAN OF ENERGY MANAGEMENT

Planning policies offer development procedures related to monitor progress in terms of budgetary, performance, and estimation systems. In addition, this strategy includes the determination of quality-based aspects that denote the planning policy of energy management. This strategy mainly defines the planning focusing on stakeholder requirements and management, as well as concerns of human resources. It also is essential to classify unreliability and variations or gap studies addressing comprehensive concerns. Planning policy becomes a standard that signifies the identification of organization growth and puts into operation long-term and short-term objectives and endeavors with the form of regular development of the energy-based environment systems.

6.4.2 LOW-CARBON TRANSPORTATION

Transportation systems are available that are secure, environmentally friendly and reasonable. Low-carbon transport provides continued reduction of the reliance on petroleum-related approaches of transportation as well as improved prominence and support of non-motorized and low-carbon transfer. The main low-carbon transportation proposal is based on transit-oriented development (TOD). Developing city regions through excellent planning produces places that are available, walkable, and contain a green city with green infrastructure. Additionally, cities that provide service through well-organized communal transportation develop into smart spaces in support of public living and serve to decrease specified levels of contamination from greenhouse gas emissions along with developing the living environment.

6.4.3 INDUSTRIAL WASTE REDUCTION

Using recycling, a city is able to salvage 75% of domestic waste. Production industry can make it much larger than the domestic processes [26]. Green Industry should include the complete life cycle of their different yields and the associated side-effects. Also, the consequences of one industrialized method may develop into the other industry. Instituting a circle economy (CE) advances opportunity and requires both a contribution from government as well as from efficient institutional organizations to facilitate satisfactory control and management, with inducements in support of embracing the circle economy. Even though this approach has a few extra charges at the beginning, there will eventually be economic reimbursement. Finally, the development of the environment is considered to outweigh the preliminary opening expenses.

6.4.4 ENERGY-EFFICIENT BUILDINGS

Energy-efficient buildings include the personalized energy essential to take out, process, and transfer and establish new building resources for energy-providing services such as cooling, heating, and providing power. The purpose of the energy-efficient urban infrastructure is to use sustainable construction design principles, including the suitable implementation rights to use, manage, and recycle the resources. Recycling improves the operating efficiency with a decreasing dependence on non-renewable energy as well as consolidation of optional energy sources. Authentication with an evaluation method makes it possible to use green building conventions along with these methods, for instance the Leadership in Energy and Environmental Design (LEED), in addition to the Building Research Establishment Environmental Assessment Method (BREEAM) along with offering profitable aspects [26].

6.4.5 GREEN AND FLEXIBLE INFRASTRUCTURE

Green and flexible infrastructure describes balancing these fundamentals (green and flexible) to improve the infrastructure. Green infrastructure is measured as being a multifunctional network consisting of the semi natural regions, associated aspects, along with the green locations. A flexible infrastructure defines the capability of transportation to endure particular procedures, for example, during flooding, or a natural disaster like earthquakes, and also climate change. Taking action toward climate change as well as dropping the exposure of populations become the key deliberations in support of the provision of a flexible infrastructure, along with integrating in the perception of green infrastructure. Green and flexible infrastructure can be included together with hard and soft engineering proposals.

6.4.6 INTELLIGENT SYSTEM DEVELOPMENT

Intelligent systems include the utilization of ICT, which facilitates the development of the provision of data; delivery of service also effectively encourages residents to contribute. The utilization of this system is not constrained to a smart city or any urban area; it is also able to control changeable conditions during every phase of green city development. Intelligent systems make it possible to help with decision making and develop transport networks by real-time observation as well as manage the progress or action stream of transportation networks with the community into the appropriate locations. This system allows the smart city to measure the decrease of environmental disaster threats along with the utilization of advising in advance of a plan intended to defend society from possible risky conditions.

6.5 CONCLUSION

Smart cities are growing rapidly in both developed and developing countries. However, in order to have a focus on the green city, the initiative must be taken to reduce energy consumption, greenhouse gas emissions, and the carbon foot-print. All of these initiatives together can improve the urban environment significantly. The success of the initiatives depends upon the development of strategic plans, policies, approaches, and methods to follow the energy-based smart city development model. To actually realize the concept of energy savings in green cities, renewable energy must be introduced. Even if renewable energy is not fully introduced, its fundamentals must be introduced to the traditional energy systems. Consequently, to achieve the best possible energy management in green cities, new plans, tools, and techniques need to be introduced for sustainability. In conclusion, the particular findings in this chapter signify that sustainable energy-based technologies become the most important tools in favor of sustainable cities with unique development. Introducing key concepts of green city development with specific cost-cutting mechanisms can facilitate the designing of built-up plans intended for sustainable green city development.

REFERENCES

[1] Worldometer. (2019), "World population prospects: The 2019 revision", Elaboration of data by United Nations, Department of Economic and Social Affairs, Population Division [Available at:] www.worldmeters.info/world-population/

[2] Wang, S. J., & Moriarty, P. (2019), "Energy savings from smart cities: A critical analysis", *Energy Procedia*, *158*, pp. 3271–3276.

[3] Mohanty, S. P., Choppali, U., & Kougianos, E. (2016), "Everything you wanted to know about smart cities: The internet of things is the backbone", *IEEE Consumer Electronics Magazine*, *5*(3), pp. 60–70.

[4] Bibri, S. E. (2019), "On the sustainability of smart and smarter cities in the era of big data: An interdisciplinary and trans-disciplinary literature review", *Journal of Big Data*, *6*(1), p. 25.

[5] Bibri, S. E., & Krogstie, J. (2017), "Smart sustainable cities of the future: An extensive interdisciplinary literature review", *Sustainable Cities and Society*, *31*, pp. 183–212.

[6] Al Nuaimi, E., Al Neyadi, H., Mohamed, N., & Al-Jaroodi, J. (2015), "Applications of big data to smart cities", *Journal of Internet Services and Applications*, *6*(1), pp. 1–15, 25.

[7] Neirotti, P., De Marco, A., Cagliano, A. C., Mangano, G., & Scorrano, F. (2014), "Current trends in smart city initiatives: Some stylised facts", *Cities*, *38*, pp. 25–36.

[8] Chourabi, H., Nam, T., Walker, S., Gil-Garcia, J. R., Mellouli, S., Nahon, K., . . . Scholl, H. J. (2012, January), "Understanding smart cities: An integrative framework", In *2012 45th Hawaii International Conference on System Sciences*, pp. 2289–2297, IEEE, Maui, HI, USA.

[9] Batty, M., Axhausen, K. W., Giannotti, F., Pozdnoukhov, A., Bazzani, A., Wachowicz, M., . . . Portugali, Y. (2012), "Smart cities of the future", *The European Physical Journal Special Topics*, *214*(1), pp. 481–518.

[10] Giffinger, R., Fertner, C., Kramar, H., & Meijers, E. (2007), "City-ranking of European medium-sized cities", *Centre of Regional Science at the Vienna University of Technology*, pp. 1–12.

[11] Kitchin, R. (2014), "The real-time city? Big data and smart urbanism", *GeoJournal*, *79*(1), pp. 1–14.

[12] Bibri, S. E. (2018), "The IoT for smart sustainable cities of the future: An analytical framework for sensor-based big data applications for environmental sustainability", *Sustainable Cities and Society*, *38*, pp. 230–253.

[13] Kuru, K., & Yetgin, H. (2019), "Transformation to advanced mechatronics systems within new industrial revolution: A novel framework in Automation of Everything (AoE)", *IEEE Access*, *7*, pp. 41395–41415.

[14] Zygiaris, S. (2013), "Smart city reference model: Assisting planners to conceptualize the building of smart city innovation ecosystems", *Journal of the Knowledge Economy*, *4*(2), pp. 217–231.

[15] Owusu, P. A., & Asumadu-Sarkodie, S. (2016), "A review of renewable energy sources, sustainability issues and climate change mitigation", *Cogent Engineering*, *3*(1), 1167990, pp. 1–14.

[16] Tiwari, G. N., & Mishra, R. K. (2012), *Advanced Renewable Energy Sources*, p. 483. Royal Society of Chemistry, Cambridge, UK.

[17] Hak, T., Janouskova, S., & Moldan, B. (2016), "Sustainable development goals: A need for relevant indicators", *Ecological Indicators*, *60*, pp. 565–573.

[18] Lu, Y., Nakicenovic, N., Visbeck, M., & Stevance, A.-S., (2015), "Policy: Five priorities for the UN sustainable development goals." *Nature*, *520*, pp. 432–433.

[19] Elliott, D. (2019), "10 cities and renewable energy", In *Sustainable Cities Reimagined: Multidimensional Assessment and Smart Solutions*, p 289. Routledge, London.

[20] IRENA, R. E. S. (2016), "International renewable energy agency", *Renewable Energy Target Setting, Abu Dhabi, UAE*, pp. 1–64.

[21] Guo, L., Qu, Y., Wu, C., & Gui, S. (2018), "Evaluating green growth practices: Empirical evidence from China", *Sustainable Development*, *26*(3), pp. 302–319.

[22] Dlani, A., Ijeoma, E. O. C., & Zhou, L. (2015), "Implementing the green city policy in municipal spatial planning: The case of buffalo city metropolitan municipality", *Africa's Public Service Delivery & Performance Review*, *3*(2), pp. 149–182.

[23] Roseland, M. (1997), "Dimensions of the eco-city", *Cities*, *14*(4), pp. 197–202.

[24] Freytag, T., Gossling, S., & Mossner, S. (2014), "Living the green city: Freiburg's Solarsiedlung between narratives and practices of sustainable urban development", *Local Environment*, *19*(6), pp. 644–659.

[25] Hopwood, D. (2007), "Blueprint for sustainability?", *Refocus*, *3*(8), pp. 54–57.

[26] Lewis, E. (2015), "Green city development tool kit", In *Asian Development Bank Institute,* pp. 1–136. Mandaluyong City, Philippines: Asian Development Bank. [Available at]: www.adb.org; openaccess.adb.org

[27] Bhutta, F. M. (2017, November), "Application of smart energy technologies in building sector: Future prospects", In *2017 International Conference on Energy Conservation and Efficiency (ICECE)*, pp. 7–10, IEEE, Lahore, Pakistan.

[28] Bowen, A., Duffy, C., & Fankhauser, S. (2016), *Green Growth and the New Industrial Revolution*, pp. 1–24. Policy Brief. Grantham Research Institute on Climate Change and the Environment and Global Green Growth Institute, London.

[29] Vu, T. L., Le, N. T., & Jang, Y. M. (2018, October), "An overview of Internet of Energy (IoE) based building energy management system", In *2018 International Conference on Information and Communication Technology Convergence (ICTC)*, pp. 852–855, IEEE, Jeju, South Korea.

[30] Kunold, I., Kuller, M., Bauer, J., & Karaoglan, N. (2011, September), "A system concept of an energy information system in flats using wireless technologies and smart metering devices", In *Proceedings of the 6th IEEE International Conference on Intelligent Data Acquisition and Advanced Computing Systems*, Vol. 2, pp. 812–816, IEEE, Prague, Czech Republic.

[31] Shahinzadeh, H., Moradi, J., Gharehpetian, G. B., Nafisi, H., & Abedi, M. (2019), "Internet of Energy (IoE) in smart power systems", In *2019 5th Conference on Knowledge Based Engineering and Innovation (KBEI)*, pp. 627–636, IEEE, Tehran, Iran.

[32] Arbzadeh, M., Johnson, J. X., & Keoleian, G. A. (2015, July), "Design principles for green energy storage systems", In *Meeting Abstracts* (1), p. 125. The Electrochemical Society.

[33] Arbzadeh, M. (2018), "Green principles, parametric analysis, and optimization for guiding environmental and economic performance of grid-scale energy storage systems", *Natural Resources and Environment in the University of Michigan* (Doctoral dissertation), pp. 1–157, Ann Arbor, MI, USA. [Available at]: http://hdl.handle.net/2027.42/143942

[34] Moradi, J., Shahinzadeh, H., & Khandan, A. (2017), "A cooperative dispatch model for the coordination of the wind and pumped-storage generating companies in the day-ahead electricity market", *International Journal of Renewable Energy Research (IJRER)*, 7(4), pp. 2057–2067.

[35] Moradi, J., Shahinzadeh, H., Khandan, A., & Moazzami, M. (2017), "A profitability investigation into the collaborative operation of wind and underwater compressed air energy storage units in the spot market", *Energy*, *141*, pp. 1779–1794.

[36] Ding, Y., Tong, L., Zhang, P., Li, Y., Radcliffe, J., & Wang, L. (2016), "Liquid air energy storage", In *Storing Energy*, pp. 167–181. Elsevier, Amsterdam, the Netherlands

[37] Shahinzadeh, H., Gheiratmand, A., Fathi, S. H., & Moradi, J. (2016), "Optimal design and management of isolated hybrid renewable energy system (WT/PV/ORES)", In *Electrical Power Distribution Networks Conference (EPDC)*, 2016 21st Conference on, pp. 208–215, IEEE, Karaj, Iran.

[38] Shahinzadeh, H., Gheiratmand, A., Moradi, J., & Fathi, S. H. (2016), "Simultaneous operation of near-to-sea and off-shore wind farms with ocean renewable energy storage", In *Renewable Energy & Distributed Generation (ICREDG)*, 2016 Iranian Conference on, pp. 38–44, IEEE, Mashhad, Iran.

[39] Moazzami, M., Ghanbari, M., Moradi, J., Shahinzadeh, H., & Gharehpetian, G.B. (2018), "Probabilistic SCUC considering implication of compressed air energy storage on redressing intermittent load and stochastic wind generation", *International Journal of Renewable Energy Research (IJRER)*, 8(2), pp. 767–783.

[40] Moazzami, M., Moradi, J., Shahinzadeh, H., Gharehpetian, G. B., & Mogoei, H. (2018), "Optimal economic operation of microgrids integrating wind farms and advanced rail energy storage system", *International Journal of Renewable Energy Research (IJRER)*, 8(2), pp. 1155–1164.

[41] Schaffers, H., Komninos, N., Pallot, M., Trousse, B., Nilsson, M., & Oliveira, A. (2011, May), "Smart cities and the future internet: Towards cooperation frameworks for open innovation", In *The Future Internet Assembly*, pp. 431–446. Springer, Berlin, Heidelberg.

[42] Samad, W. A., & Azar, E. (2019), "Smart cities in the Gulf: An overview", In *Smart Cities in the Gulf*, pp. 3–6. Palgrave Macmillan, Singapore.

[43] Saravanan, K., & Srinivasan, P. (2017), "Examining IoT's applications using cloud services", In P. Tomar & G. Kaur (Eds.), *Examining Cloud Computing Technologies Through the Internet of Things*, pp. 147–163. IGI Global, Hershey, PA. doi:10.4018/978-1-5225-3445-7.ch008

[44] Roscia, M., Longo, M., & Lazaroiu, G. C. (2013, October), "Smart city by multi-agent systems", In *2013 International Conference on Renewable Energy Research and Applications (ICRERA)*, pp. 371–376, IEEE, Madrid, Spain.

[45] Neike, C. (2018, February), "Here are 5 reasons why we need an 'Internet of Energy'", *World Economic Forum*, *13*.

[46] Pieroni, A., Scarpato, N., Di Nunzio, L., Fallucchi, F., & Raso, M. (2018), "Smarter city: Smart energy grid based on blockchain technology", *International Journal on Advanced Science, Engineering and Information Technology (IJASEIT)*, 8(1), pp. 298–306.

[47] Reinsel, D., Gantz, J., & Rydning, J. (2018), "The digitization of the world from edge to core", *IDC White Paper*. [Available at:] www.seagate.com/files/www-content/our-story/trends/files/idc-seagate-dataage-whitepaper.pdf

[48] Patrizio, A. (2018), "IDC: Expect 175 zettabytes of data worldwide by 2025", *Network World*. [Available at:] www.networkworld.com/article/3325397/idc-expect-175-zettabytes-of-data-worldwide-by-2025.html

[49] Casini, M. (2017, August), "Green technology for smart cities", *IOP Conference Series: Earth and Environmental Science*, *83*(1), pp. 1–7.

[50]   Electrive. (2019), "Number of plug-in cars climbs to 5.6M worldwide 2019", [Available at:] www.electrive.com/2019/02/11/the-number-of-evs-climbs-to-5-6-million-worldwide

[51]   Statista. (2019), "Number of fast-charging stations for electric vehicles world-wide in 2013 and 2020.2019." [Available at:] www.statista.com/statistics/283531/electric-vehicles-global-number-offast-charging-stations/

[52]   Statista. (2019), "Worldwide number of battery electric vehicles in use from 2012 to 2018. 2019." [Available at:] www.statista.com/statistics/270603/worldwide-number-of-hybrid-and-electric-vehicles-since-2009.

[53]   Vermesan, O., Blystad, L. C., Zafalon, R., Moscatelli, A., Kriegel, K., Mock, R., . . . Perlo, P. (2011), "Internet of energy: Connecting energy anywhere anytime", In *Advanced Microsystems for Automotive Applications 2011*, pp. 33–48. Springer, Berlin, Heidelberg.

[54]   Merrifield, A. (2013), "The urban question under planetary urbanization", *International Journal of Urban and Regional Research*", *37*(3), pp. 909–922.

[55]   Newman, P., Beatley, T., & Boyer, H. (2009), *Resilient Cities: Responding to Peak Oil and Climate Change*. Island Press, Washington, DC. [Avaliable at]: www.researchgate.net/publication/37717398_Resilient_Cities_Responding_to_Peak_Oil_and_Climate_Change

[56]   Park, J., & Page, G. W. (2017), "Innovative green economy, urban economic per-formance and urban environments: An empirical analysis of US cities", *European Planning Studies*, *25*(5), pp. 772–789.

[57]   Ewing, R., Bartholomew, K., Winkelman, S., Walters, J., & Chen, D. (2008), *Growing Cooler: The Evidence on Urban Development and Climate Change*. Urban Land Institute, Washington, DC.

[58]   Wheeler, S. M. (2008), "State and municipal climate change plans: The first genera-tion", *Journal of the American Planning Association*, *74*(4), pp. 481–496.

[59]   Foley, J., DeFries, R., Asner, G., Barford, C., Bonan, G., Carpenter, S., . . . Snyder, P. (2005), "Global consequences of land use", *Science*, *309*(5734), pp. 570–574.

[60]   Lundqvist, L. (2016), "Planning for climate change adaptation in a multi-level context: The Gothenburg metropolitan area", *European Planning Studies*, *24*(1), pp. 1–20.

[61]   Albers, M., & Deppisch, S. (2013), "Resilience in the light of climate change: Useful approach or empty phrase for spatial planning?", *European Planning Studies*, *21*(10), pp. 1598–1610.

[62]   Andrews, C. (2008), "Greenhouse gas emissions along the rural-urban gradient", *Journal of Environmental Planning & Management*, *51*(6), pp. 847–870.

7 Energy Management in Smart Cities by Novel Wind Turbine Configurations

J. Bruce Ralphin Rose and J.V. Bibal Benifa

CONTENT

7.1 INTRODUCTION

The environmental concerns and globally deteriorating rate of fossil fuels are continuously indicating the mandatory requirement of renewable energy (RE) sources to meet energy demands. In 2014, the United Nations formed a proposal to stop the dependence on fossil fuels by 2050, and regrettably, it is extremely difficult for the developing countries to achieve this goal. It is a well-known fact that many green energy sources haven't demonstrated to be as fruitful as expected. One of the clean and everlasting sustainable energy sources is known as wind energy, which can be acquired through several ways in a cost-effective manner. The plentiful source of wind energy helps nations to add wind power systems in both onshore and offshore environments despite the techno-commercial issues [1]. As an illustrative example, India has the wind energy potential of about 300 GW at the 100 m hub height, according to the estimated report issued by National Institute of Wind Energy (NIWE). However, the present wind energy contribution is limited to 35 to 40 GW only because of the issues associated with land acquisition and environmental impacts. In addition, it is forecasted that India needs more than 900 GW of total power capacity by 2025 to meet all the existing energy requirements [2].

Installation of onshore wind energy systems is an established technology which is undergoing plenty of innovative methods to maximize energy output. It has a comparatively better technology readiness level (TRL) to address various potential challenges such as transportation, precipitation and debris, inconsistent loading because of turbulent wind, and lightning strikes [3]. However, countries like China and India have tremendous urbanization trends that would increase CO_2 emissions as well as energy consumption. The highly populated human activity offers only limited space on land for the establishment of onshore wind farms and solar systems, which are classified as the primary RE sources for smart cities [4]. Hence, offshore wind energy became the preferred choice due to the competing site usage and reduced environmental impacts, with significantly higher power output than the onshore turbines. As stated in the NIWE report, the total available wind energy potential, about 300 GW, is a realistic one to achieve through the effective utilization of seashore of nearly 7600 km with moderately shallow waters.

The conventional and outdated power systems (with fossil fuels) that exist in the developing countries is a strong barrier to the establishment of smart cities [5]. Those power plants are not scalable for the present energy demands, and the expansion of units would initiate climate change and air pollution associated challenges. Hence, the vision of green smart cities can be accomplished effectively through offshore wind energy and smart grid-oriented innovative techniques. The present chapter is devoted to investigating the potential that exists in India towards RE management in green smart cities. The coastline of Gujarat and Tamilnadu alone has the wind energy potential of about 35 GW, according to the NIWE study report [2]. The Indian and global offshore wind energy project developers are devoted to achieve at least 50% (about 15–17 GW) of the potential by 2025 adjacent to the Exclusive Economic Zones (EEZ) of the country. The smart energy management within 12 nm of the coast would minimize the technical hitches involved in the operation of offshore turbines, with less cabling and mooring costs [6].

As the cities become primary habitat for most of the global population, the choice of RE is the predestined one for the electricity needs of smart cities. The goal of a green smart city should be people-centered in terms of sustainable development and excellence of lifestyles that require smart renewable energies sources as an integrated part of it [7]. A typical comparison of wind energy potential between the onshore and offshore turbines is highlighted in Figure 7.1. Here, at 40 m depth, the offshore wind energy potential is three times higher than the onshore energy potential. Hence, the green smart city's energy demands can be fulfilled through the offshore turbines efficiently, as compared to the other energy generation techniques. AI and Internet of Things (IoT) assist with the trouble-free condition-monitoring maintenance of offshore turbines to achieve the maximized RE yield in a cost-effective manner. The turbine information is updated continuously with weather forecast data, and the primitive variables are validated with the incoming signals to achieve the live fault diagnostics, as illustrated in Figure 7.2.

The energy potential increases further in deep waters (60–300 m), but it requires high mooring costs and unique designs for foundations to achieve better stability in the presence of ocean currents. Hence, the vital economic factors involved in the erection of offshore wind turbines, including the installation, stable foundations in

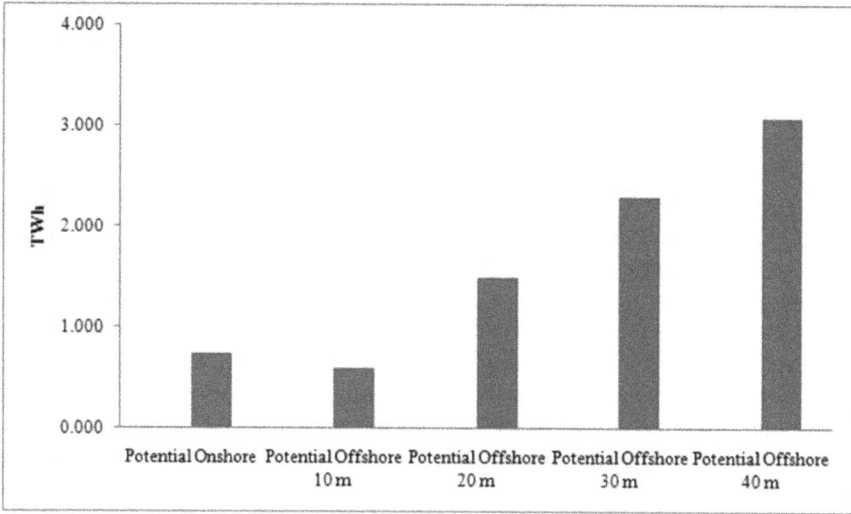

FIGURE 7.1 Comparison of wind energy potential between the onshore and offshore turbines.

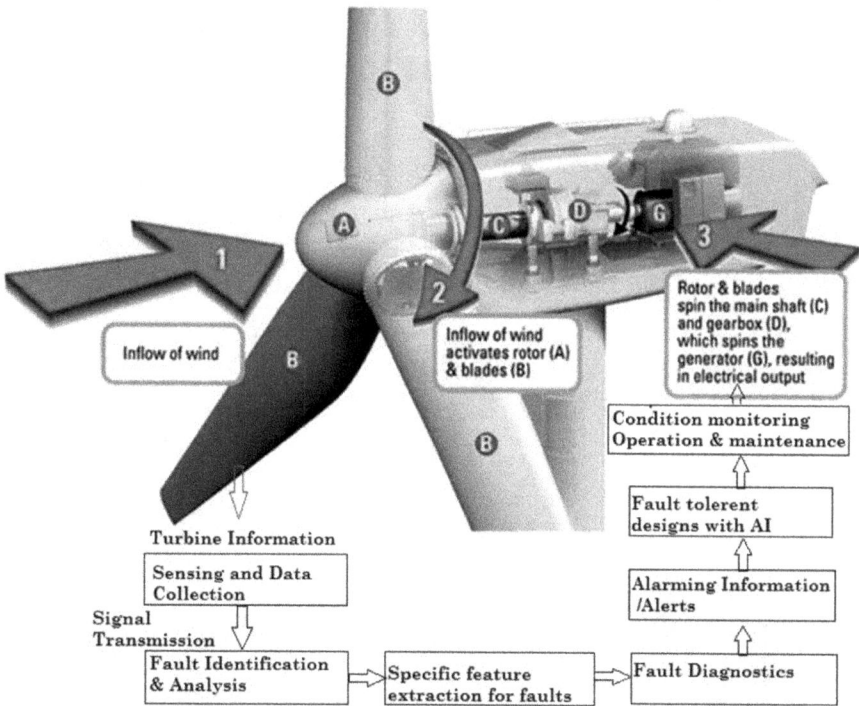

FIGURE 7.2 Automated wind energy systems with AI and IoT for offshore applications.

a hydrodynamic environment, grid infrastructure, and operational costs, should be customized prior to final bidding [8]. The typical offshore wind farm in a shallow water environment is displayed in Figure 7.3.

Hornsea One is the world's largest wind farms, with 1.2 GW capacity, that is located off the Yorkshire coast of United Kingdom to power more than 1 million homes. Originally, the Hornsea project was planned to have a total capacity of up to 6 GW, and its second phase is planned to be completed by 2022 for supplying energy for more than 1.3 million homes. It is apparent that there will be a significant impact on the capacity being generated from renewable sources in the successive phases of offshore wind power projects. Fixed bottom offshore and floating offshore are the two types of foundations used for wind turbines installed at medium depth (5 m to 40 m) and deep waters (> 40 m) respectively. In the fixed bottom configuration, the foundations are firmly installed on the seabed, with minimal operation-maintenance under regular marine winds. Hywind Scotland is the first ever floating wind farm configuration and is located at 29 km off Peterhead, where the rated wind speed is 10.1 m/sec. At this range of rated wind speeds, the new offshore wind farms installed in the United Kingdom have helped to push the use of renewables to move faster towards its zero-carbon emission goal. The onshore and offshore wind velocities in the Aberdeenshire region of Scotland are compared in Figure 7.4, and it is observed that the offshore wind speed is two times its onshore counterpart. Hence, it is evident that the offshore wind turbines would be the key player for the establishment of green smart cities wherever the windiest location exists around the globe. By

FIGURE 7.3 Offshore wind farm in the shallow water environment.

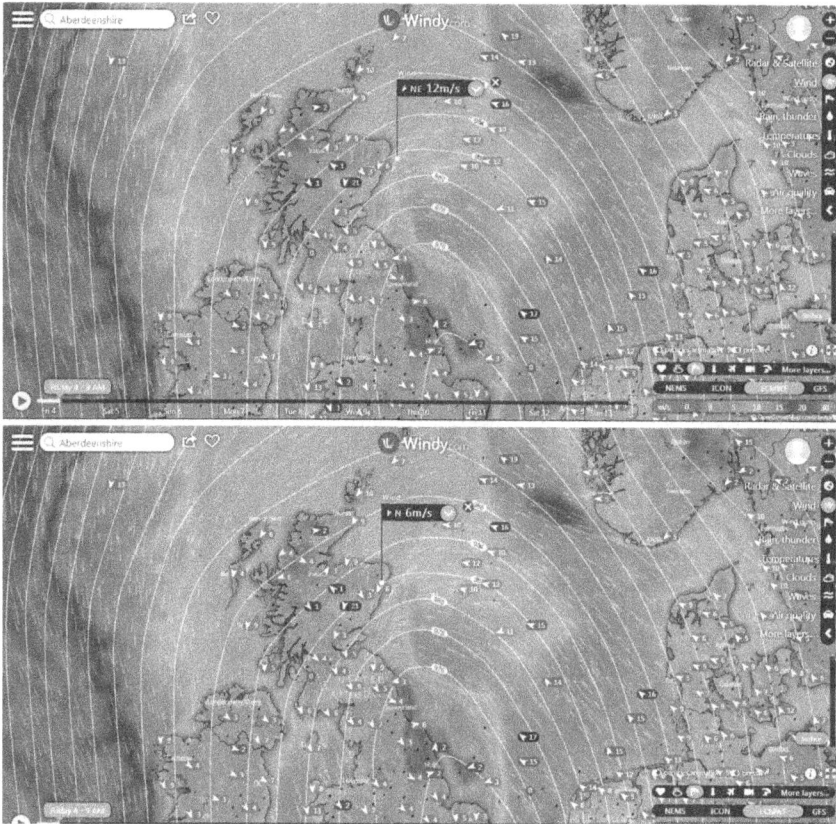

FIGURE 7.4 Comparative wind speeds at onshore and offshore locations in the Aberdeenshire region, Scotland on 4 December 2020.

Source: Courtesy: www.windy.com

2025, the UK government plans to close the coal plants by harnessing the great wind energy potential through several offshore wind farms similar to the Hornsea project.

The comparative merits of offshore turbines with the available state-of-the-art technologies have changed problems into opportunities in the past two decades. The freedom from land acquisition issues and the availability of high wind speeds are the primary elements to consider offshore turbines for smart green energy production. Recently, offshore wind turbines with 12 MW capacity have been installed in Europe, and the testing phase is in progress for 15 MW turbines. Nevertheless, such designs are extremely cost effective, reliable and robust compared to the conventional versions, which offers a push towards the reduced levelized cost of energy (LCOE) [9]. Hence, the present chapter provides a detailed comparative analysis about a 2 MW horizontal axis wind turbine (HAWT) which is assumed to be installed at the onshore and offshore locations of the Indian coastal boundary. Although the performance investigation is done based on the prevailing wind speeds in Indian offshore conditions, the characteristics can be very well matched to any offshore environment across different continents. The simulations are done in the Bladed educational program with the actual wind speed data prevailing in the south Asian environment (Indian Ocean). A detailed statistical wind-built data has been studied in the selected region to make this study as realistic as possible. Presently, around three smart city programs are in place adjacent to the selected geographic location, and this investigation helps to realize the energy security needed for the fulfilment of smart green cities.

7.2 DATA FORECASTING AND MANAGEMENT

Meteorological information relating to the past few years has been considered for the computation of power output for the selected 2 MW wind turbine. Except for the foundation and cabling requirements, the offshore turbine is also identical to the onshore turbines in the 2 MW segment. However, the structural dynamic characteristics and ocean currents should be taken into account with an adequate factor of safety because of the high wind speed applications in the offshore environment. Thanks are due to the AI-based machine control strategies, which help to forecast the adverse weather conditions and control the operational parameters of the turbine instantly to prevent major structural or material failures [10]. Here, the wind power forecast information and local wind speed data (big data) combination is used to maintain the optimal blade pitch and yawing angles to customize the aerodynamic loading on the rotors. A typical hourly wind power generation forecast distribution is presented in Figures 7.5

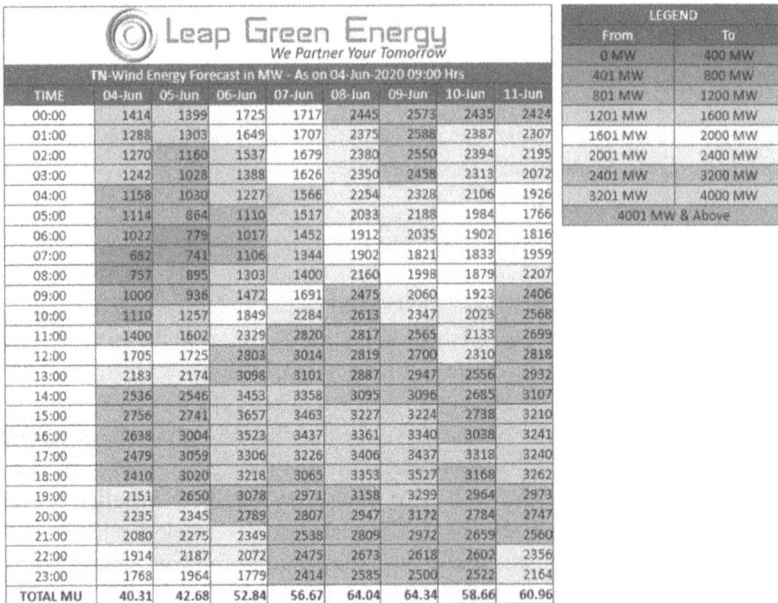

Leap Green Energy — *We Partner Your Tomorrow*

LEGEND	
From	To
0 MW	400 MW
401 MW	800 MW
801 MW	1200 MW
1201 MW	1600 MW
1601 MW	2000 MW
2001 MW	2400 MW
2401 MW	3200 MW
3201 MW	4000 MW
4001 MW & Above	

TN-Wind Energy Forecast in MW - As on 04-Jun-2020 09:00 Hrs

TIME	04-Jun	05-Jun	06-Jun	07-Jun	08-Jun	09-Jun	10-Jun	11-Jun
00:00	1414	1399	1725	1717	2445	2573	2435	2424
01:00	1288	1303	1649	1707	2375	2588	2387	2307
02:00	1270	1160	1537	1679	2380	2550	2394	2195
03:00	1242	1028	1388	1626	2350	2458	2313	2072
04:00	1158	1030	1227	1566	2254	2328	2106	1926
05:00	1114	864	1110	1517	2033	2188	1984	1766
06:00	1022	779	1017	1452	1912	2035	1902	1816
07:00	682	741	1106	1344	1902	1821	1833	1959
08:00	757	895	1303	1400	2160	1998	1879	2207
09:00	1000	936	1472	1691	2475	2060	1923	2406
10:00	1110	1257	1849	2284	2613	2347	2023	2568
11:00	1400	1602	2329	2820	2817	2565	2133	2699
12:00	1705	1725	2803	3014	2819	2700	2310	2818
13:00	2183	2174	3098	3101	2887	2947	2556	2932
14:00	2536	2546	3453	3358	3095	3096	2685	3107
15:00	2756	2741	3657	3463	3227	3224	2738	3210
16:00	2638	3004	3523	3437	3361	3340	3038	3241
17:00	2479	3059	3306	3226	3406	3437	3318	3240
18:00	2410	3020	3218	3065	3353	3527	3168	3262
19:00	2151	2650	3078	2971	3158	3299	2964	2973
20:00	2235	2345	2789	2807	2947	3172	2784	2747
21:00	2080	2275	2349	2538	2809	2972	2659	2560
22:00	1914	2187	2072	2475	2673	2618	2602	2356
23:00	1768	1964	1779	2414	2585	2500	2522	2164
TOTAL MU	40.31	42.68	52.84	56.67	64.04	64.34	58.66	60.96

FIGURE 7.5 Hourly wind energy forecast distribution on 4 June 2020 in the state of Tamilnadu (India).

Source: Courtesy: Leap Green Energy

TIME	05-Jun	06-Jun	07-Jun	08-Jun	09-Jun	10-Jun	11-Jun	12-Jun
00:00	1433	1538	2062	2287	2481	2423	2824	2784
01:00	1333	1343	1969	2191	2306	2396	2641	2720
02:00	1311	1174	1926	2089	1898	2382	2560	2782
03:00	1168	1060	1862	1944	1830	2304	2432	2631
04:00	1063	979	1742	1867	1816	2229	2345	2264
05:00	999	930	1554	1816	1652	2154	2274	2037
06:00	893	905	1379	1709	1719	2070	2129	1911
07:00	913	889	1171	1711	1803	1900	2131	1781
08:00	982	1000	1277	1775	1844	1987	2206	1812
09:00	1117	1218	1532	1959	1963	2029	2415	2000
10:00	1351	1489	1904	2176	2103	2173	2543	2403
11:00	1682	1805	2490	2743	2256	2362	2676	2632
12:00	1987	2202	2962	2953	2458	2589	2790	2819
13:00	2355	2567	3131	3102	2780	2840	2957	2975
14:00	2765	2981	3450	3237	3170	3101	3265	3250
15:00	3129	3159	3334	3356	3434	3323	3337	3394
16:00	3111	3078	3235	3414	3402	3417	3305	3434
17:00	2782	2840	3042	3380	3287	3464	3334	3483
18:00	2560	2754	2854	3346	3188	3536	3218	3553
19:00	2492	2480	2724	3069	2972	3282	3043	3219
20:00	2381	2202	2538	2804	2796	3044	2865	3064
21:00	2175	2037	2464	2674	2618	2900	2809	2858
22:00	1989	1887	2396	2507	2568	2874	2803	2741
23:00	1778	1816	2316	2433	2522	2859	2803	2615
TOTAL MU	43.75	44.33	55.31	60.54	58.87	63.64	65.71	65.16

Leap Green Energy — We Partner Your Tomorrow
TN-Wind Energy Forecast in MW - As on 05-Jun-2020 09:00 Hrs

LEGEND

From	To
0 MW	400 MW
401 MW	800 MW
801 MW	1200 MW
1201 MW	1600 MW
1601 MW	2000 MW
2001 MW	2400 MW
2401 MW	3200 MW
3201 MW	4000 MW
4001 MW & Above	

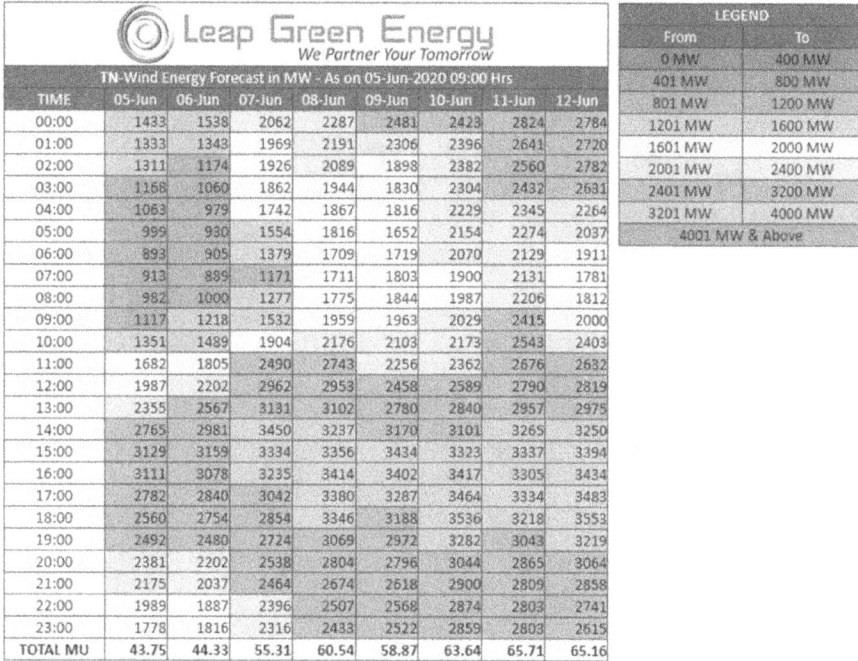

FIGURE 7.6 Hourly wind energy forecast distribution on 5 June 2020 in the state of Tamilnadu (India).

Source: Courtesy: Leap Green Energy

and 7.6. It shows the wind energy forecast done on 4 and 5 June 2020 at 9 AM across the state of Tamilnadu (India), where the average wind power generation per day is about 3200 MW from June to September.

The European Centre for Medium-Range Weather Forecasts (ECMWF) is an independent organization that shares reliable wind speed and weather forecast data through different stakeholders [11]. A comparative analysis of hourly wind speed forecast on 5 June 2020 at the onshore location and the proposed offshore wind farm analysis are presented in Figures 7.7 and 7.8 respectively. From the comparative illustration, it is observed that the offshore wind speed in shallow waters is nearly two times higher than the onshore counterpart. In addition, the wind speed forecast data obtained after three weeks at the same offshore location is presented in Figure 7.9. As the wind season has started, a mean wind speed of about 9 m/sec can be attained constantly at the pinned location shown in Figure 7.5. Hence, a fewer number of turbines are required to achieve the same amount of power output as compared with onshore turbines.

The date-wise wind speed forecast observed on 25 June 2020 in the Madurai and Erode regions of Tamilnadu state are highlighted in Figures 7.10 and 7.11 respectively. Presently, more than 600 turbines with 1.25 MW to 2 MW capacity

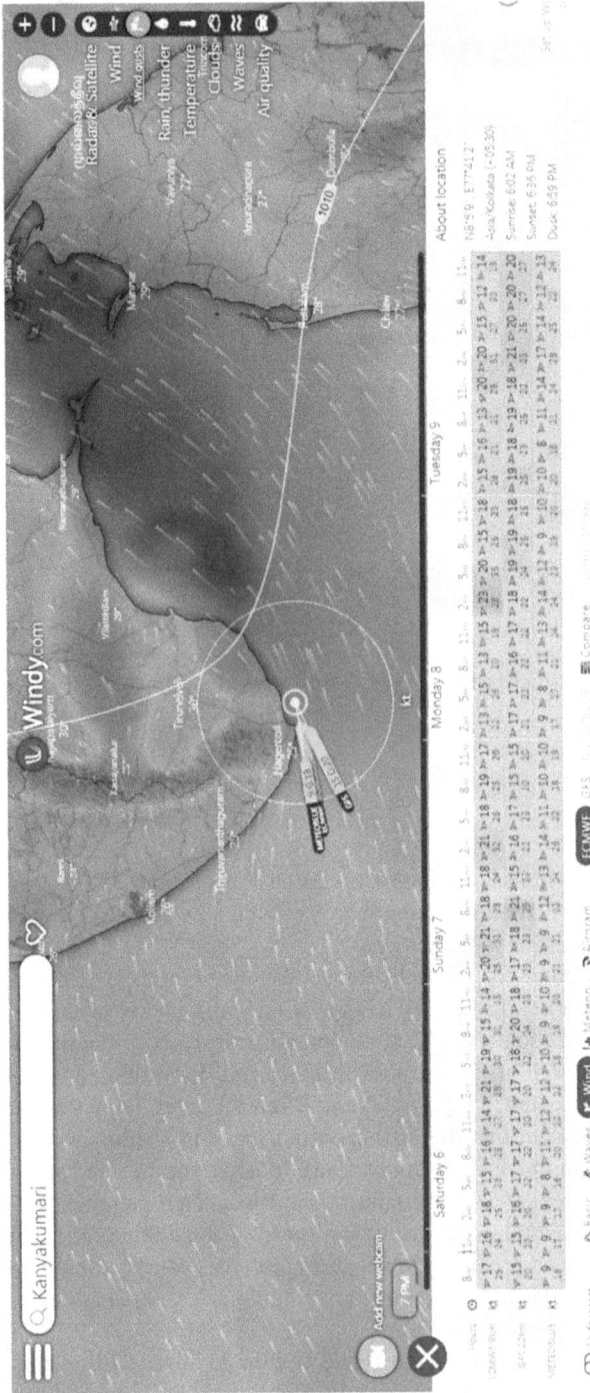

FIGURE 7.7   Hourly wind speed forecast on 5 June 2020 at the onshore wind farm.

Source: Courtesy: www.windy.com

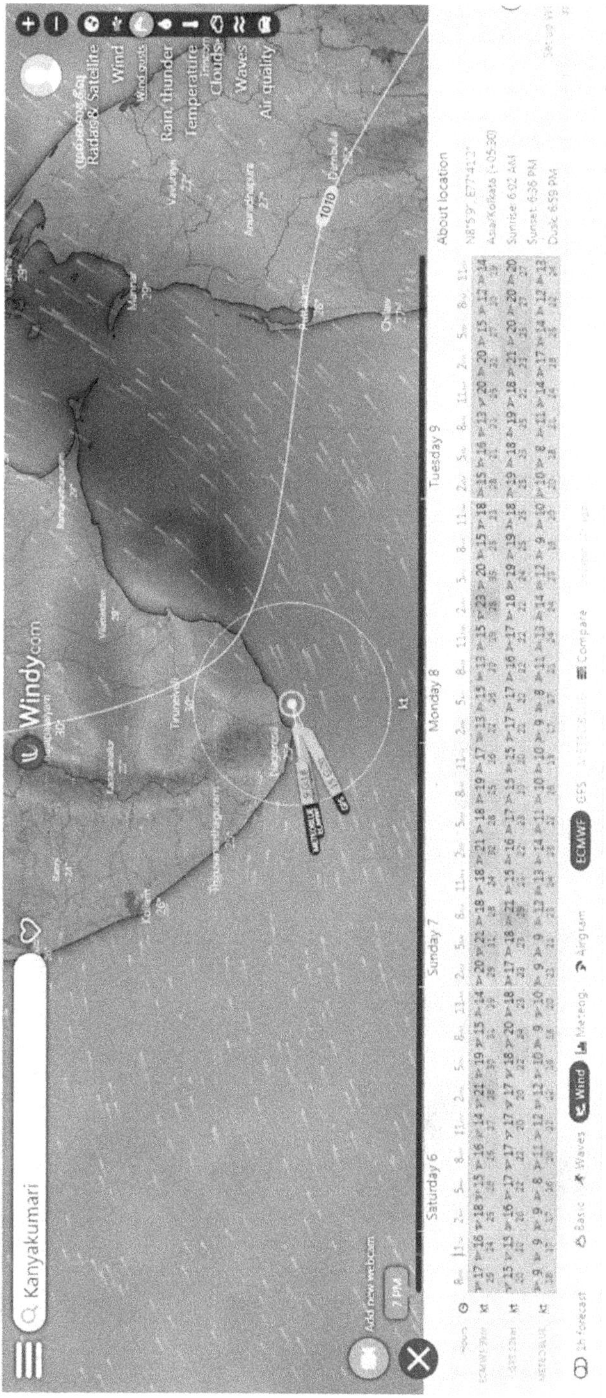

FIGURE 7.8 Hourly wind speed forecast on 5 June 2020 at the offshore location.

Source: Courtesy: www.windy.com

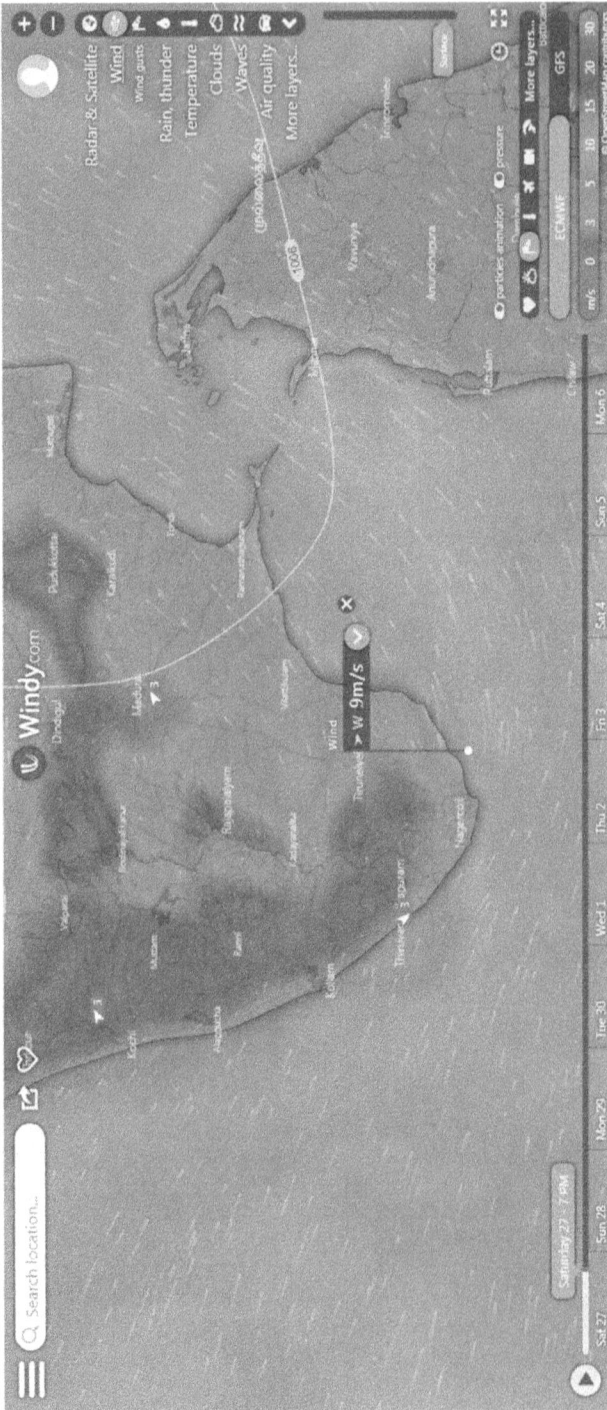

FIGURE 7.9 Hourly wind speed forecast on 27 June 2020 at the offshore location.

Source: Courtesy: www.windy.com

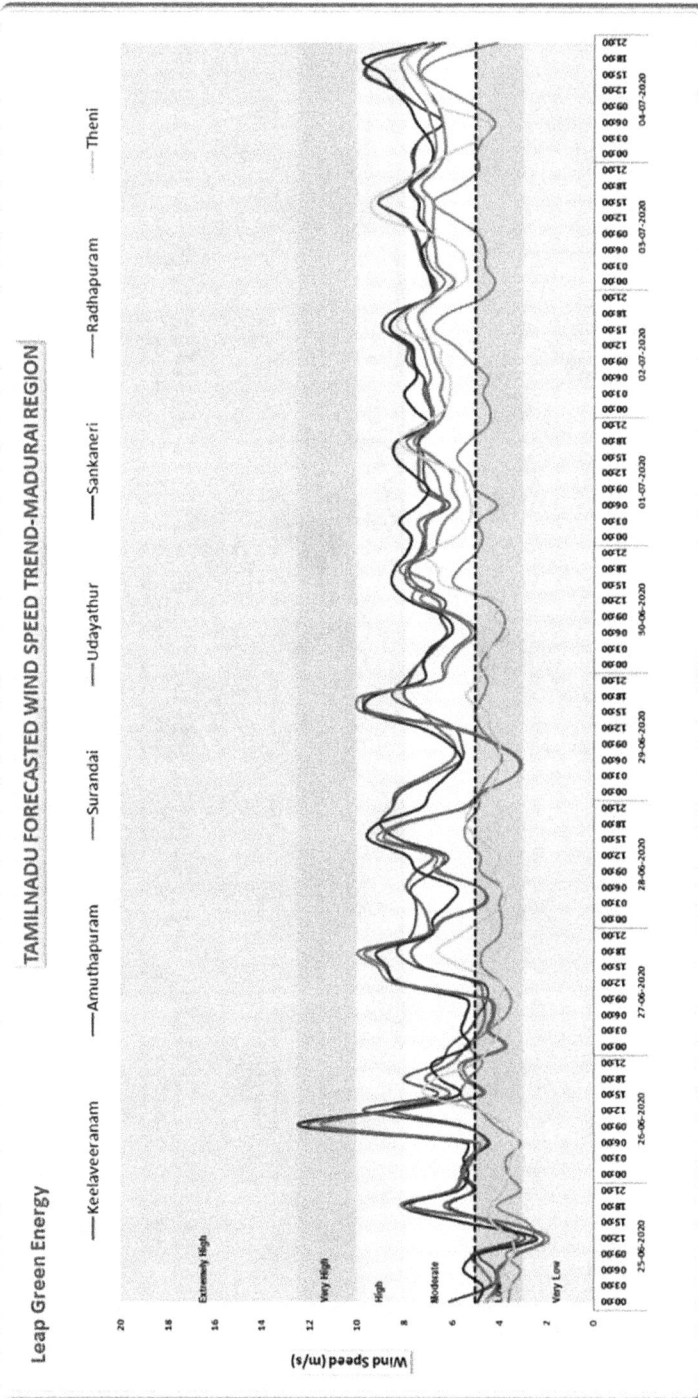

FIGURE 7.10 Date-wise wind speed forecast observed on 25 June 2020 in the Madurai region of Tamilnadu (India).

Source: Courtesy: Leap Green Energy

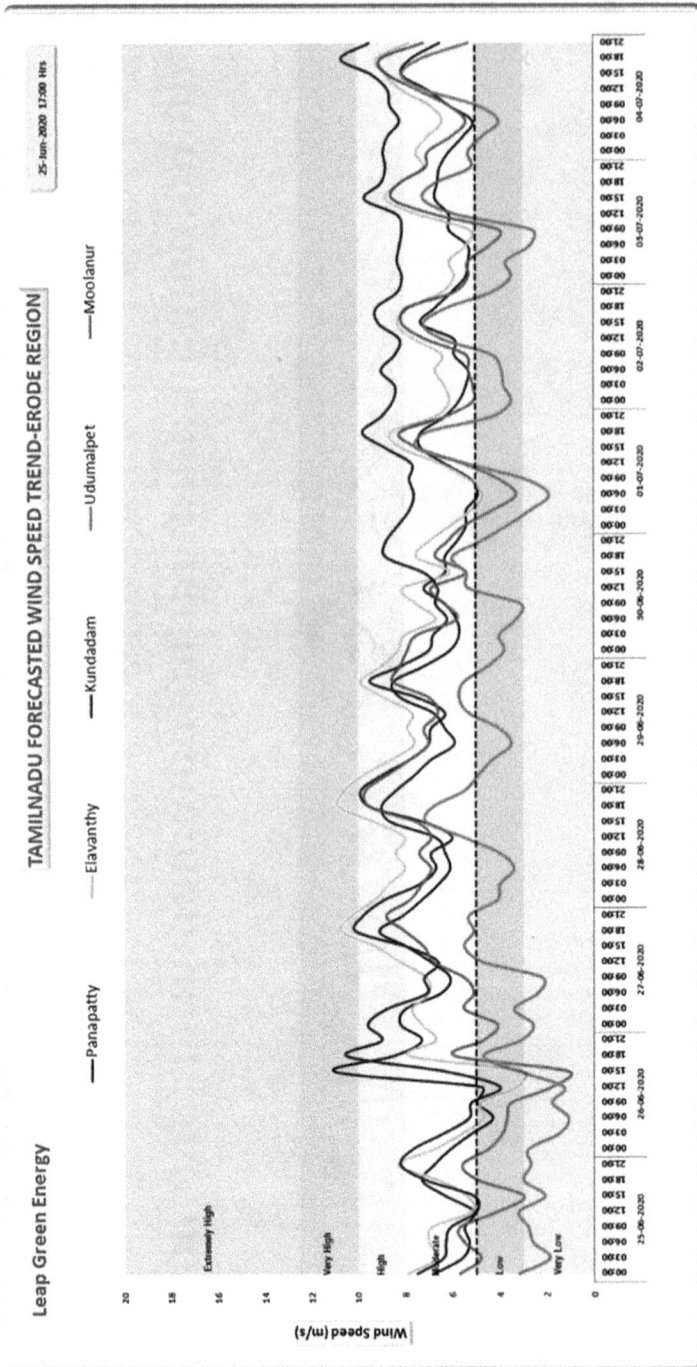

FIGURE 7.11 Date-wise wind speed forecast observed on 25 June 2020 in the Erode region of Tamilnadu (India).

Source: Courtesy: Leap Green Energy

are installed in the two regions, and the corresponding power generation potential has already been discussed (viz., Figure 7.5 and 7.6). At these onshore locations, only moderate to high wind speeds exist, and receiving very high wind speeds is a rare phenomenon. However, at the offshore location the mean wind speed is very high, and it becomes extremely high during the peak wind seasons across the coastal region. Apart from the other unconventional power generation mechanisms, wind energy is an integral part of the green smart cities of the future generation. WINERCOST (WINd Energy technology Reconsideration to enhance the COncept of Smart ciTies) clearly stated the purpose of wind energy incorporation into an urban environment by conquering any technological or economic barriers [12].

The wind power generation potential for the time span of 24 hours according to the ECMWF forecast on 16 June 2020 across Tamilnadu state is presented in Figure 7.12. Subsequently, the complete weekly forecast is presented in Figure 7.13. Though the current scenario is presented in this report, the power output trends concerning each turbine were almost identical, with less than 5% deviations during the past three years. Floating offshore turbines installed in the Hywind Scotland wind farm with the capacity of 6 MW have proved the enhanced efficiency of turbines installed in deep waters. However, the cable and mooring costs should also be considered in addition to the overall advantages of floating offshore configuration in the countries like Japan, where the shallow water environment is limited. Hence, the actual and forecasted wind data have been extensively analyzed before finalizing the conceptual prototype for the proposed offshore turbines to enable green smart cities. From Figures 7.9 and 7.10, we see that the maximum average wind power generation forecast in the state of Tamilnadu is about 4000 MW. Hence, the offshore turbines would roughly double the power output with the present wind energy potential to fulfil the power requirements of smart cities.

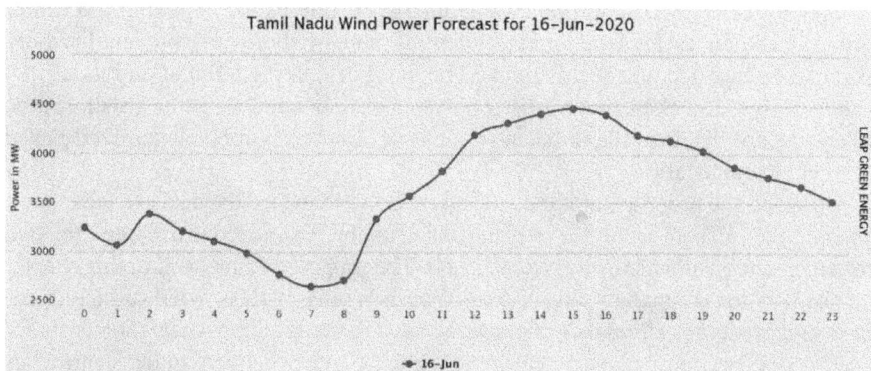

FIGURE 7.12 Hourly wind power generation on 16 June 2020 across the state of Tamilnadu (India).

Source: Courtesy: Leap Green Energy

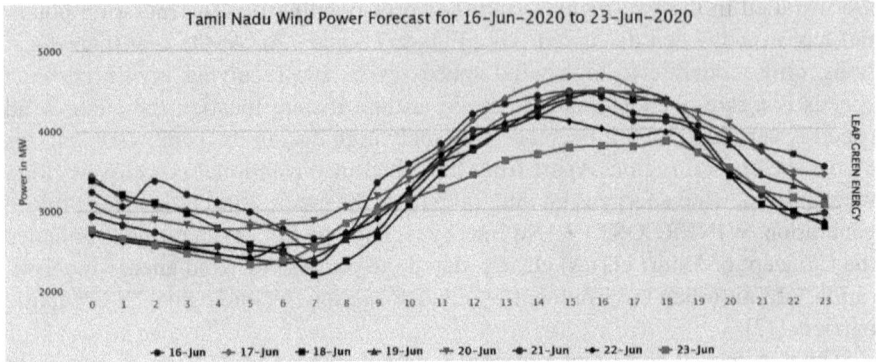

FIGURE 7.13 Hourly wind power generation forecast from 16 June 2020 to 23 June 2020 across the state of Tamilnadu (India).

Source: Courtesy: Leap Green Energy

7.3 WIND TURBINE PARAMETERS

The NACA 0015 and NACA 64–210 airfoil profiles are considered in the design of untwisted/twisted blades for wind turbines. The symmetric configuration known as NACA 0015 is used up to 10 m distance along the pitch axis, and then the supercritical configuration (NACA 64–210) is used to deliver the maximum power output at low wind speeds with hybrid composite epoxy material. Detailed numerical simulations are provided with fixed and variable pitch angles at different possible wind speeds prevailing in the coastal environment, as specified in Figure 7.8. For the maximum wind energy harvesting from the available wind potential, the usual wind turbine blade profiles should be substantially modified with better aerodynamic and structural characteristics [13]. As the present chapter is fully focused towards the renewable energy resources for powering the future green smart cities, the wind energy potential in an offshore environment has only been taken into account at this point. The fixed or floating foundation modelling, tower design and power transmission systems associated with the offshore turbines are not discussed here because these parameters can be customized based on the marine location.

However, the proposed modifications should not influence the decisive factors relevant to the offshore wind power plants, and suitable justifications must be provided for all operation and maintenance costs [14]. The proposed framework in this chapter emphasizes the comparative evaluation of onshore and offshore wind turbine (similar configuration) performance characteristics with optimal implementation costs for energy-efficient smart cities. To ensure sustainable energy delivery in the smart cities which are located adjacent to coastal environments, installation of offshore turbines is the need of the hour, specifically for the developing countries. However, the optimization of the wind farm layout is a convoluted process because of the influence of wake induced flows. As a wind turbine extracts power from the moving wind, the

TABLE 7.1
Wind Turbine Parameters for Bladed Analysis

Sl.No	Parameters	Values
1.	Distance along pitch axis	36 m
2.	Chord (max, min)	3.22 m, 0.030 m
3.	Aerodynamic twist (max, min)	13 deg, 0.3 deg
4.	Thickness (max, min)	100%, 10%
5.	Neutral axis (y') (max, min)	50%, 30%
6.	Reynolds number	2×10^6
7.	Tower height	56 m
8.	Tilt Angle	5 deg
9.	Overhang	3.8 m
10.	Mean water depth	7 m
11.	Inertia of foundation	3.6×10^5 kgm^2
12.	Gearbox ratio	78.55

downstream portion of wind leaving the turbine has more turbulence with reduced velocity. Hence, it is crucial to include the wake losses while locating the turbines to make the most of energy production to acquire the additional costs for infrastructure. The wind turbine parameters considered for the present comparative evaluation are listed in Table 7.1.

The steady speed and undisturbed winds present at the offshore environments are more reliable as compared to onshore winds. Hence, multiple considerations exist for the wind farm layout to increase profitability by using modelling and simulation as a tool to arrive at managerial and technical decisions. Bladed wind turbine software is an efficient tool for optimizing the turbine at various phases of the design and analysis. Wind turbine companies (OEMs) and certification agencies rely on Bladed to a greater extent because it provides critical insights into wind turbine dynamic characteristics to compute the aerodynamic loads and performance.

7.4 AERODYNAMIC PERFORMANCE

The energy management in green smart cities with novel cost effective wind turbine configurations at the offshore environment requires a multidisciplinary investigation. The aerodynamic characteristics of wind turbine blades and the resulting aeroelastic effects are the primary concerns because of the metocean conditions prevailing in the offshore environment. However, the available wind load forecasting strategies and the historical weather data (temperature, pressure and rainfall etc.,) helps to identify the anomalies with condition monitoring maintenance mechanisms that are integrated with AI [15]. Therefore, it is obviously advantageous to use offshore wind turbines as sustainable RE sources for future large-scale power requirements of green smart cities. Hence, a detailed comparison of aerodynamic and performance coefficients of offshore turbines against their onshore counterparts is presented in this chapter for the available wind energy potential existing at the offshore location is shown in Figure 7.8.

In the GH (Garrad Hassan) Bladed software, the aerodynamic characteristics of NACA 0015 and NACA 64–210 airfoil profiles are loaded as an initial dataset. The lift coefficient (C_L), drag coefficient (C_D), and pitching moment coefficient (C_M) are computed at different angles of attack (AoA), and the detailed experimental data set is available in Reference [16]. The computational turbine model prepared using the Bladed software is displayed in Figure 7.14. The variations of C_L, C_D and C_M are computed at the cut-in and cut-out wind speeds 4 m/sec and 23 m/sec respectively. The starting and ending limits of annual mean wind speed assigned for the simulation is about 4 m/sec to 11 m/sec respectively. Further, the Weibull shape parameter is maintained as 3 during the simulation. The variation of C_L and C_D along the length of the blade are shown in Figures 7.15 and 7.16 respectively. Similarly, the variation of pitching moment coefficient is presented in Figure 7.17.

The C_L and C_D distributions clearly indicate the potential of NACA airfoil configurations along the length of the blade in proportion to the pitch angles. Here, zero or negative pitch angle is maintained up to the hub wind speeds about 12 m/sec,

FIGURE 7.14   Computational turbine model prepared using the Bladed software.

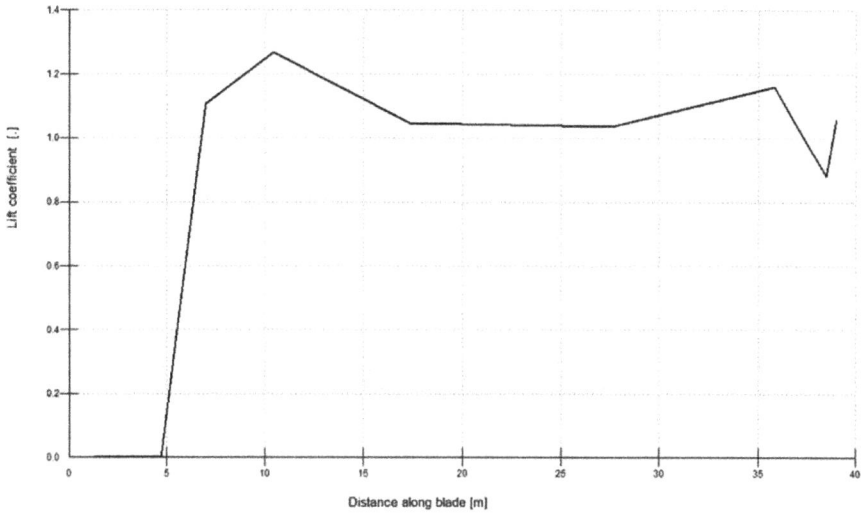

FIGURE 7.15 Lift coefficient along the length of the blade.

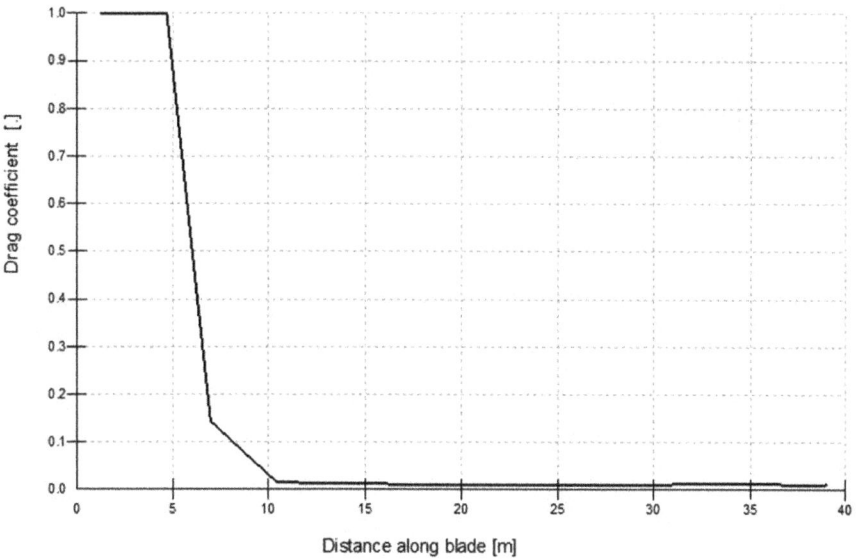

FIGURE 7.16 Drag coefficient along the length of the blade.

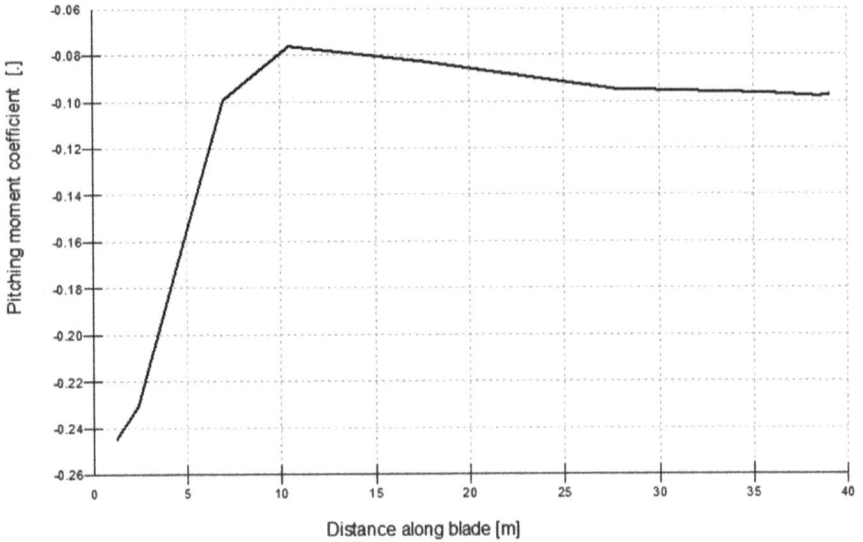

FIGURE 7.17   Pitching moment coefficient along the length of the blade.

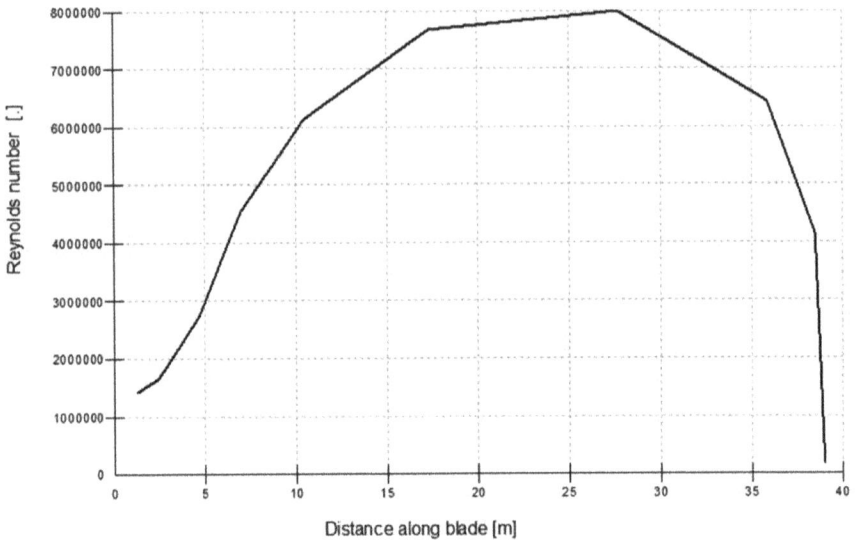

FIGURE 7.18   Reynolds number distribution along the length of the blade

and then the pitch angle will be increased in proportion to the wind speeds such that the drag is kept at a minimum [17]. The pitching moment also begins to decrease after the wind speed of 12 m/sec (rated speed), and thereby the aerodynamic torque is maintained as constant (Figure 7.19). The power coefficient (C_p)

also begins to decrease beyond the rated speed limit of the turbine, as shown in Figure 7.20. Hence, it is evident that the offshore turbines are able to deliver consistent power output with optimum tip speed ratio as compared to onshore turbines (Figure 7.21). The Reynolds number mapping along the length of the blade indicates

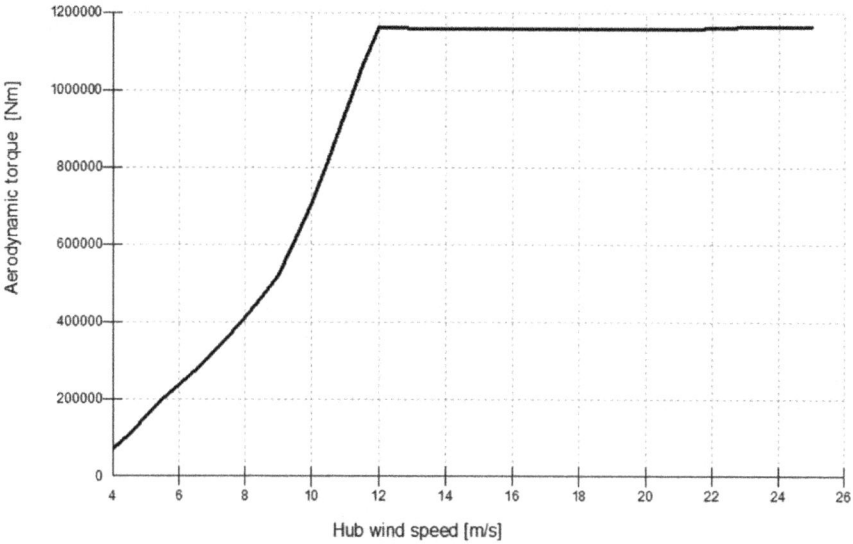

FIGURE 7.19   Aerodynamic torque distribution vs hub wind speed.

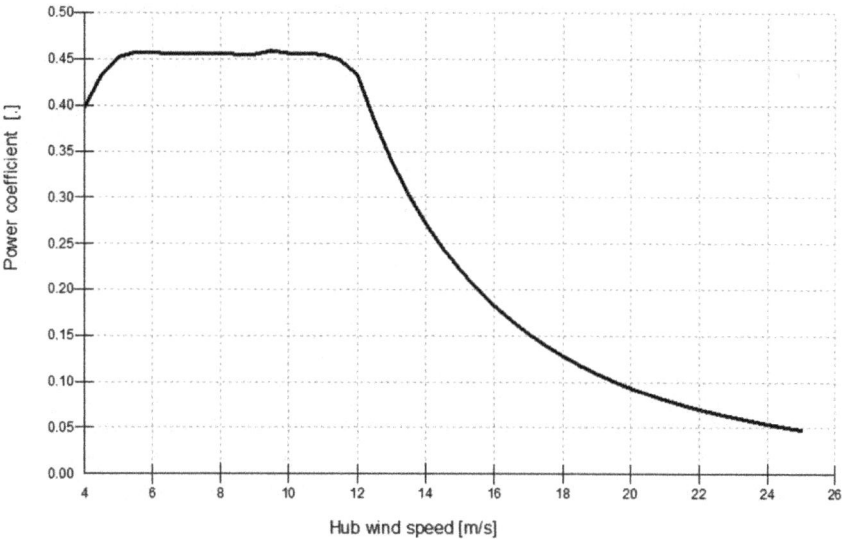

FIGURE 7.20   Power coefficient Vs hub wind speed.

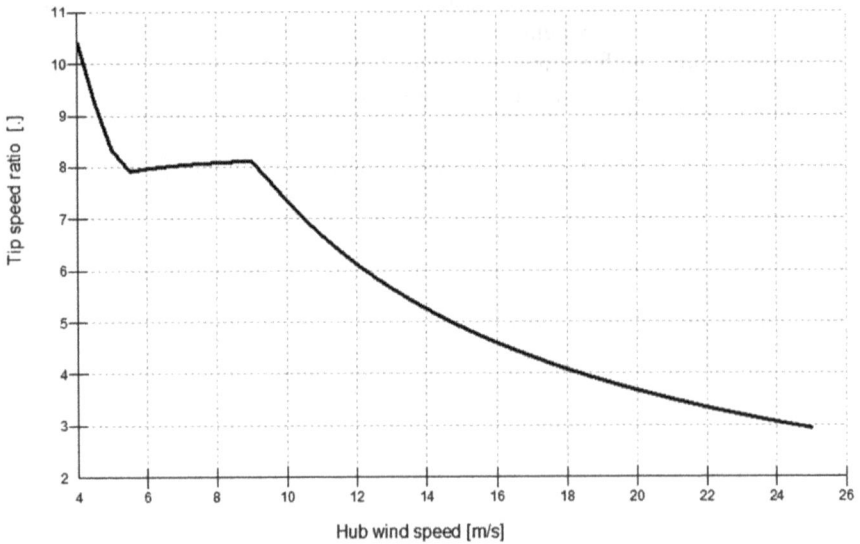

FIGURE 7.21   Tip speed ratio vs hub wind speed.

that the maximum inertial force is exerted across the major portion of the blade, as presented in Figure 7.17.

7.5   BLADE LOADING IN AN OFFSHORE ENVIRONMENT

Presently, only a few feasibility studies have been performed on the basis of comparative evaluation of onshore and offshore wind turbine modeling. The smart city consumer power demands can be satisfied efficiently by providing critical insights on offshore wind turbine performance with justified solutions [18]. From the aerodynamic performance viewpoint (section 7.4), it is observed that the power output of a 2 MW wind turbine in a shallow water environment is about 45% to 55% greater than onshore turbines of identical geometric features. Here, the blade material and dimensions are maintained as similar to onshore turbines. However, in the offshore environment, the structural integrity of the blade should be enhanced and condition monitoring systems with AI must be incorporated to minimize maintenance cycles [19]. Figure 7.22 shows the forces acting on the blade in x-direction at different wind speeds. The maximum force acting on the blade at the rated speed is about 93 kN and it is well below the ultimate load limit of the glass fibre reinforced epoxy materials. Further, the floating wind turbines may experience approximately 20% to 30% additional forces due to stronger winds present in waters deeper than 60 metres. In Japan and California, floating wind turbines simply remove the depth constraints, and it has been a great breakthrough in economic renewable energy sources.

Automated fault diagnosis and preventive maintenance significantly reduces the maintenance costs associated with wind generators. Wind energy is

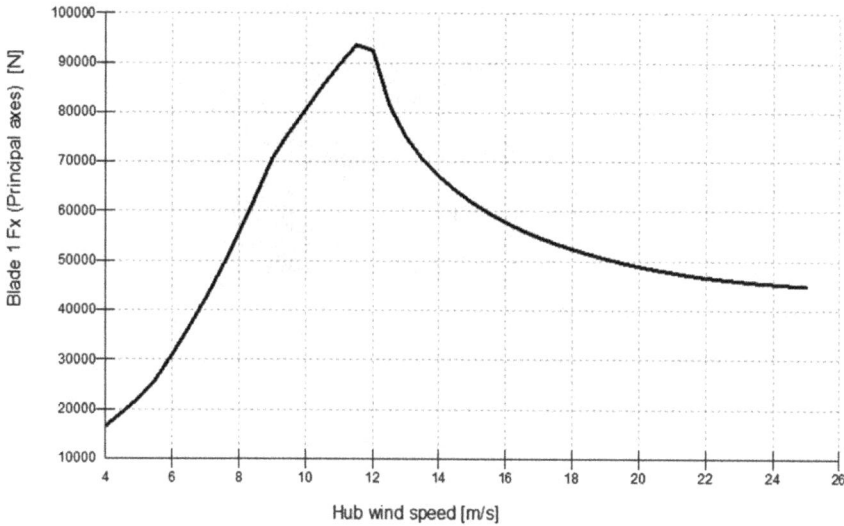

FIGURE 7.22   Force acting on the blade in *x*-direction at different wind speeds.

considered to be the primary RE source for green smart cities, and the application of AI in fault diagnosis becomes more imperative. Fault diagnosis and preventive maintenance techniques for wind turbine generators are in the premature
stage at present. The cost of wind energy production can be substantially reduced
if the structural failures are predicted in advance and turbine blade health is
monitored automatically.

7.6   AI-BASED CONDITION MONITORING SYSTEM

The IoT sensors collect data at periodic intervals and transmit the observed data
to cloud storage. A deep learning module is utilized to learn the observed data
and make periodic forecasts regarding the energy production and fault diagnosis. Currently, investigation is conducted using the data compiled by the SCADA
(Supervisory Control and Data Acquisition) system for fault prediction and energy
production in wind turbines [20] [21]. A schematic of computing technologies used
for conditional monitoring of wind turbines is presented in Figure 7.23. It comprises
IoT sensors deployed in the hub as well as the blades of the wind turbine. Zheng et
al. (2017) carried out a study on short-term wind energy prediction wherein wind
speed and generated power data acquired by the SCADA system of wind turbines
are used [22]. In their study, the authors propose a hybrid method based on a combination of different algorithms such as Hilbert-Huang Transform (HHT), genetic
algorithms (GA) and ANN [23]. The aforementioned hybrid method has shown
significant improvements in the accuracy of short-term wind energy predictions.
Moreover, deep learning methods can be effectively employed for determining the

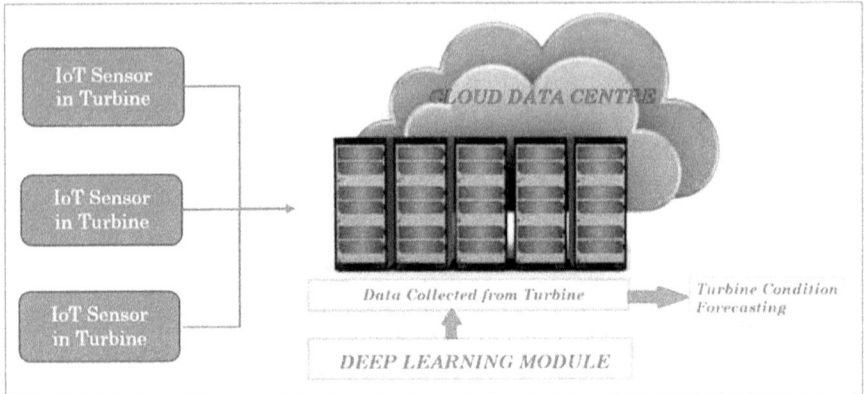

FIGURE 7.23 An IoT-deep learning framework model for energy prediction and condition monitoring of wind turbines in smart cities.

minor cracks that exist in the aged glass epoxy composite blades to avoid periodic manual inspection.

Presently, deep learning models are employed for forecasting wind energy production and fault diagnosis. The deep learning methods such as convolutional neural networks (CNN), multilayer perceptron and long short-term memory (LSTM) network would assist in identifying the faults through integrated non-destructive testing (NDT) methods in the near future. The CNN, multilayer perceptron and LSTM can be modified further for handling the multivariate data because wind energy prediction depends on the parameters such as atmospheric pressure, temperature, anemometer location, humidity, altitude, latitude, and solar radiation. In addition to this, sensors can also be employed to determine the regular corrosion issues occurring in the offshore environment with nacelle cabin pressurization.

7.7 CONCLUSIONS

In recent years, the demand for power has increased multifold due to rapid urban development and the establishment of smart cities. Renewable energy is a key source for the current generation relevant to world electricity, although its share is growing fast. In this chapter, an onshore wind turbine configuration is optimized for the enhanced power production at the offshore environment in a cost-effective approach. In a shallow water environment, a detailed field study has been conducted in Indian waters to customize the windmill configuration to generate maximum power output to fulfil the RE requirement of proposed green smart cities. Onshore regions experience very low to moderate wind speed ranges, from 2.5 m/s to 7 m/s, and power generation at the low wind speed regime is a tedious process. Hence, the wind farms should be developed in the offshore environment by bringing various stakeholders jointly from a multidisciplinary perspective. It has also been proved in different parts of the globe that offshore turbines yield consistent power output because of the presence of stronger winds.

Currently, AI techniques are widely used for the condition monitoring of wind turbines in the offshore environment. Previously, fault diagnosis has been done by machine learning (ML) techniques, and it has shown significant improvements in the automated data acquisition process. Using deep learning models, the parameters with respect to the deployed sensors, actuators and systems are observed, and the fault is diagnosed. It is a well-known fact that the fault occurs during the structural deflection of the blades, so the sensors are deployed in the blades to observe the deflections beyond the allowable threshold. Hence, the maintenance and monitoring costs would be reduced substantially to realize RE security for future green smart cities.

7.8   ACKNOWLEDGEMENTS

The author(s) express sincere thanks to Tamilnadu State Council for Science and Technology (TNSCST) for the Grant Ref.No. TNSCST/STP/AR/ET/2014–2015/1046: Dated: 11 April 2017, which was utilized for the field study and evaluation. Further, we are grateful to GH Bladed for the Educational licence shared and www.windy.com for the permission to use the statistical data.

REFERENCES

[1] M. Dolores Esteban, J. Javier Diez, Jose S. López, and Vicente Negro, "Review-why offshore wind energy?", *Renewable Energy* 36 (2011) 444–450.

[2] Prabir Kumar Dash, *Offshore Wind Energy in India*, Akshay Urja, April 2019, pp. 23–25.

[3] Md Abu S Shohag, Emily C. Hammel, David O. Olawale, and Okenwa Okoli, "Damage mitigation techniques in wind turbine blades: A review", *Wind Engineering*, 41, no. 3 (2017) 185–210.

[4] A.Z. Dhunny, M.R. Lollchund, and S.D.D.V. Rughooputh, "Evaluation of a wind farm project for a smart city in the South-East coastal zone of Mauritius", *Journal of Energy in Southern Africa*, 27, no. 1 (2016) 39–50.

[5] Otto Vik Mathisen, Maria E. Sorbye, Madhulika Rao, Gerrit Tamm, and Vladimir Stantchev, "Smart energy in smart cities: Insights from the smart meter rollout in the United Kingdom." In *Smart Cities: Issues and Challenges* (pp. 283–307). 2019. https://doi.org/10.1016/B978-0-12-816639-0.00016-8.

[6] Simon Watson, et al., "Future emerging technologies in the wind power sector: A European perspective", *Renewable and Sustainable Energy Reviews*, 113 (2019) 109270.

[7] Marlene Motyka, Scott Smith, Andrew Slaughter, and Carolyn Amon, "Renewables (em)power smart cities", *Deloitte Insights* (2019) 1–25.

[8] Angel G. Gonzalez-Rodriguez, "Review of offshore wind farm cost components", *Energy for Sustainable Development* 37 (2017) 10–19.

[9] ODE, Study of the costs of offshore wind generation. Technical Report URN number 07/779. Offshore Design Engineering (ODE) Limited, Renewables Advisory Board (RAB) & DTI. 2007.

[10] Adrian Stetco, Fateme Dinmohammadi, Xingyu Zhao, Valentin Robu, David Flynn, Mike Barnes, John Keane, and Goran Nenadic, "Machine learning methods for wind turbine condition monitoring: A review", *Renewable Energy*, 133 (2019) 620–635.

[11] H. Hersbach, et al., "The ERA5 global reanalysis", *Quarterly Journal of the Royal Meteorological Society* (2020). https://doi.org/10.1002/qj.3803.

[12] WINERCOST, TU1304, Memorandum of Understanding, COST Association, Brussels, November 22, 2013.

[13] Shuhua Wang, Xue Wang, Zhong Lin Wang, and Ya Yang, "Efficient scavenging of solar and wind energies in a smart city", *American Chemical Society, ACS Nano*, 10 (2016) 5696–5700. DOI: 10.1021/acsnano.6b02575.

[14] www.azocleantech.com/article.aspx?ArticleID=704, Accessed on 11 June 2020.

[15] Nenad Petrovic, and Dorde Kocic, "Data-driven framework for energy-efficient smart cities", *Serbian Journal of Electrical Engineering*, 17, no. 1 (2020) 41–63. DOI: 10.2298/SJEE2001041P.

[16] H.A. Ira Abbott, and Albert Edward von Doenhoff, *Theory of Wing Sections: Including a Summary of Airfoil Data*. New York: McGraw-Hill, 1949.

[17] Karim Oukassou, Sanaa El Mouhsine, Abdellah El Hajjaji, and Bousselham Kharbouch, "Comparison of the power, lift and drag coefficients of wind turbine blade from aerodynamics characteristics of Naca0012 and Naca2412", *Procedia Manufacturing*, 32 (2019) 983–990.

[18] Ruben Paul Borg, Neveen Hamza, Conor Norton, Christos Efstathiades, and Mantas Marciukaitis, "Urban wind energy: Social, environmental and planning considerations", *The International Conference on Wind Energy Harvesting*, 21–23 March 2018, Catanzaro Lido, Italy.

[19] K. Saravanan, E. Golden Julie, and Y. Herold Robinson, "Smart cities & IoT: Evolution of applications, architectures & technologies, present scenarios & future dream for the upcoming book series", *Intelligent Systems Reference Library*, 154 (2019), 135–151.

[20] Valentina E. Balas, et al. (Eds.), *Internet of Things and Big Data Analytics for Smart Generation*, 978-3-030-04202-8, 467407_1_En (7). www.springer.com/us/book/9783030042028.

[21] Z. Zhang, and K. Wang, "Wind turbine fault detection based on SCADA data analysis using ANN", *Advances in Manufacturing*, 2 (2014) 70–78. https://doi.org/10.1007/s40436-014-0061-6.

[22] D. Zheng, M. Shi, Y. Wang, et al., "Day-ahead wind power forecasting using a two-stage hybrid modeling approach based on SCADA and meteorological information, and evaluating the impact of input-data dependency on forecasting accuracy", *Energies*, 10, no. 12 (2017) 1–23.

[23] K. Saravanan, and P. Srinivasan, "Examining IoT's applications using cloud services." In P. Tomar & G. Kaur (Eds.), *Examining Cloud Computing Technologies Through the Internet of Things* (pp. 147–163). Hershey, PA: IGI Global, 2017. DOI: 10.4018/978-1-5225-3445-7.ch008.

8 COVID-19 Impact and Global Status of Renewable Energy
A Review

Parveen Kumar, Manish Kumar
and Ajay Kumar Bansal

CONTENT

8.1 INTRODUCTION

Continuous economic as well as exponential population growth is responsible for increased global demand for power, and a large portion of this demand is met by carbon-fossil-based energy sources, which have limited capabilities and have adverse effects on the environment. Concern about the environment has increased all across the world. Due to climate impact, cost and air quality issues, emphasis has been laid on the deployment of low carbon and more sustainable power resources. For control of electricity markets and prosperous growth in renewable and high-efficiency technologies, small, environmentally friendly dispersal of different forms called distributed generation (DG) have been created. The production units of DG have progressed in the direction of power sector development. Utility of distributed generation resources (wind energy, photovoltaic, fuel cells, biomass, small hydropower plants, tidal and geothermal etc.) in distribution grid is increasing due to its various techno-economical advantages. The regulatory power industry provides economic opportunities for investors and provides many potential benefits for utilities (peak shaving, loss reduction, asset use etc.), which encourage further incentives for the addition of DG [1, 2].

The emergence of wind, solar and other renewable technologies and their integration into the power system can be attributed to power sector reforms, policy support of the government, and direction-oriented guidelines toward the market in the last decade. It has created new business opportunities for independent power producers (IPPs), non-institutional private investors, for supplying electricity to the grid, which results in a huge flow of capital in the electricity sector.

The integration of renewable energy sources has contributed many technical and economic challenges to power producers in the smooth operation of the power system. Major challenges associated with integrating renewable energy (RE) sources in both transmission and distribution network systems include impact on power system operating costs and losses, power imbalances (scheduling and dispatch), transmission planning (congestion), nodal pricing with DG in distribution system etc. Thus, their integration into the power system is now an important issue to optimize resource usage and to increase the installation of renewable capacity in order to achieve the sustainability and security goals of the supply [3].

In this chapter, a detailed analysis of renewable energy potential across the worldwide has been provided. It also comprehensively elucidates why we are going closer to RE resources because of their economic, social and environmental impact and the demanding situations associated with RE systems. The impact of Covid-19 on the global renewable energy sector has been considered. Solar and wind energy are the main focus area of this chapter.

8.2 GLOBAL SCOPE AND STATUS IN RENEWABLE ENERGY

Energy is a fundamental requirement in the world of economic development and in every sector of the economy. In this way, it is essential that the nations have looked at new, clean and sustainable sources of energy around the world and implemented the new Renewable Energy Promotion and Energy Conservation Act. According to the annual energy report (MNRE, 2019), the capacity of renewable energy production has become 85908.37 MW.

The biggest part of current total energy capacity is from hydroelectricity (58%), then 23% wind energy and around 12% solar energy [4]. A major part of total energy capacity is wind power (37505 MW), then 33712 MW of solar power and more than 14533 MW of hydropower. This shows that solar photovoltaics (PV) has taken a lead role in energy.

Worldwide electricity generation in the year 2019 and the overall yearly growth rate of 2012 to 2018 have been shown in Figure 8.1 and Figure 8.2 respectively. Observe that overall renewable energy growth increased year by year and contribution of solar energy has increased more compared to the other renewable energy resources.

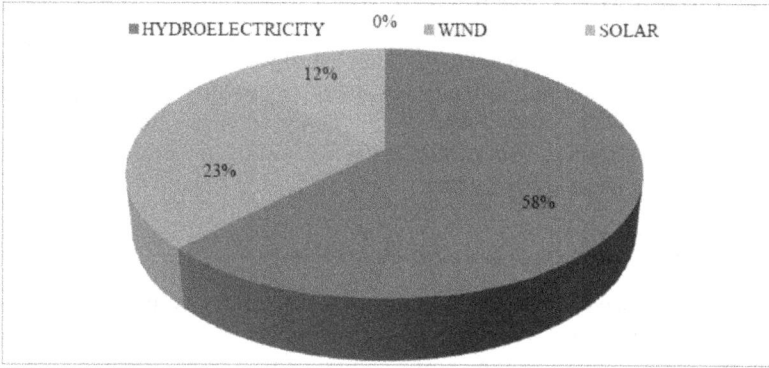

FIGURE 8.1   World electricity generation in year 2019 by MNRE source.

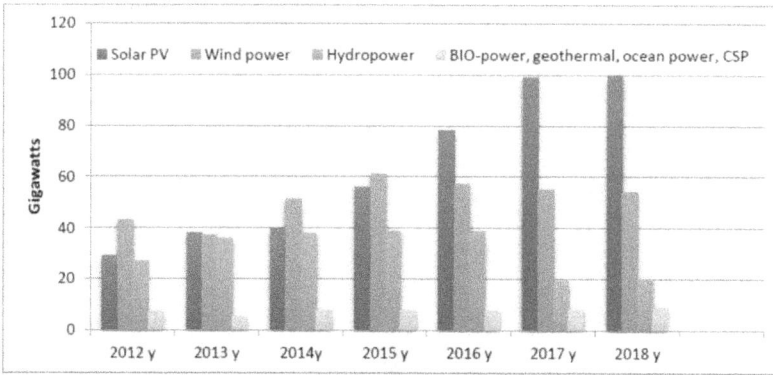

FIGURE 8.2   Renewable power capacity and yearly growth rates, 2012–2018.

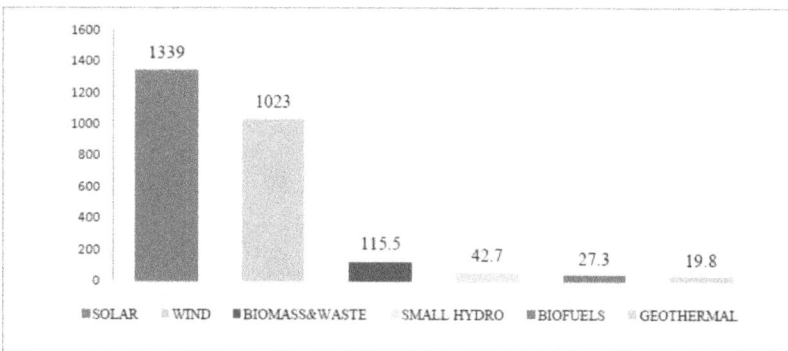

FIGURE 8.3   World investments in renewable by technology, 2010–2019.

Total investment in RE worldwide has been shown in Figure 8.3. Including early-stage and corporate-level funding as well as the financing of new capacity, this was $288.3 billion in 2018. This was 11% down from the 2017 record of $325 billion [5].

8.3 GLOBAL STATUS OF RENEWABLE ENERGY

Global status of RE has been shown in Figure 8.4. It represents the top 10 countries'
RE status and observes that Germany has the top status for RE, at approximately
12.74%, as compared to other countries [7]. Other countries also increased their
renewable energy resources. But similar to previous years, the growth of renewables
in the transport sector is very slightest.

8.3.1 GLOBAL STATUS OF SOLAR AND WIND POWER

Solar power increased rapidly in the last few years. Every country contributes to
solar power production. Total solar power production worldwide reached up to
580.1 GW, approximately. China plays a leading role in the solar power sector
as compared to other countries [7–8]. Other countries also increases their solar
power energy. India is seeing greater production in solar power in previous years
as compared to some other countries. The global status of the solar sector is
shown in Figure 8.5.

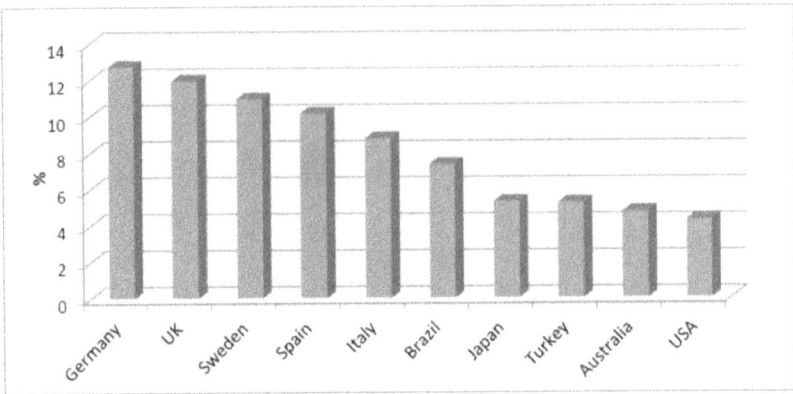

FIGURE 8.4   Country-wise status of renewable energy.

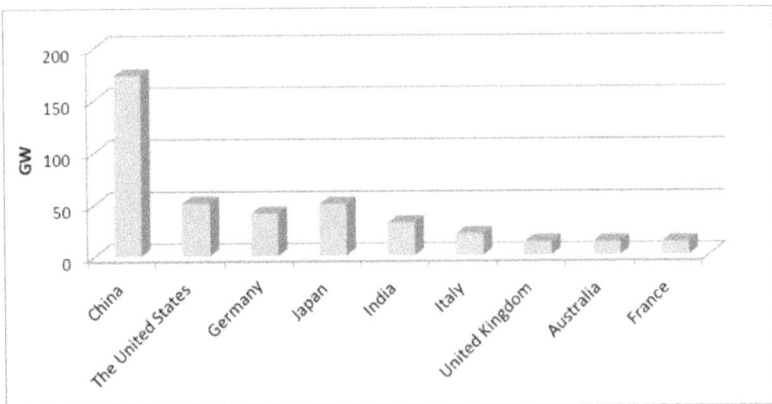

FIGURE 8.5   Global status of solar power.

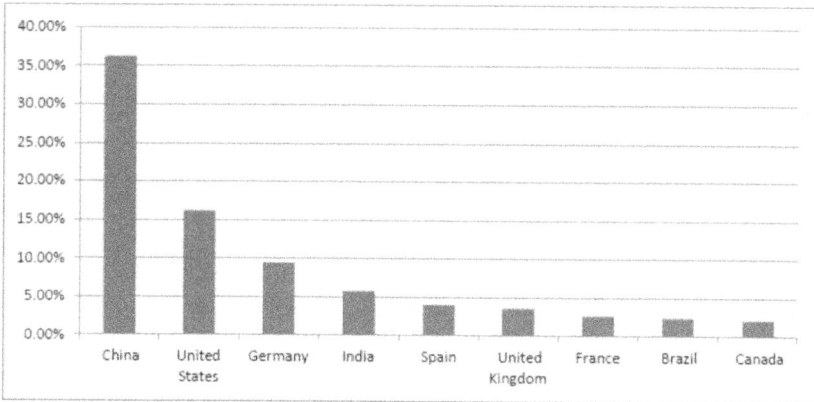

FIGURE 8.6   Global status of wind power.

Figure 8.6 represents wind power's contribution worldwide. It is observed that China has the greatest share in the field of wind power (36.3%) as compared to other countries. Wind power production worldwide reached up to 6508 GW [9].

8.4   COVID-19 IMPACT OF GLOBAL RENEWABLE ENERGY SOURCES

Renewable energy has been the strongest and easiest source of energy during the Covid-19 lockdown, as it was unaffected by global energy demand and the production of energy was unaffected and made proper use of renewable energy sources. Because of this, there was major growth in the use of renewable energy in Q1 2020, up 1.5% as related to Q1 2019. This will increase the global rate of using the renewable energy around the world. The data show there was an increase in the use of renewable energy by 1% in 2020 [10]. There is an expansion of renewable sources like solar, wind and hydro plants, predicted to increase the use of renewable energy generation by 5% by the end of 2020.

8.4.1   1ST QUARTER OF 2020

The use of renewable energy in Q1 2020 was 1.5% times more than the Q1 of 2019. This was done by the completion of solar PV projects and wind projects, respectively of 100 GW and 60 GW; these projects were completed in 2019, and with the help of these projects there was an increment of 3% in the generation of renewable energy. In the time of Covid-19 lockdown, there has been a major increase in the share of renewable energy sources, which was now around 28%; in the previous year it was around 26%. The increase in renewable energy demand came the decline of the share of coal and gas to 60% of electricity generation around the world.

8.4.2   2020 PROJECTION

According to the collective estimated data, renewable energy demand increased by 5% in the crisis of Covid-19. This actually leads an increase of global demand of renewable energy sources by 30% around the world. The hydro plants provide the largest output in generation, as hydroelectricity holds more than the 60% of renewable energy generation globally. It is dependent on rainfall, and there are no crucial conditions for generating power. Solar PV cells provide electricity with the sun as their main source. Solar PV is the fastest and widest-used renewable source of energy around the world [11]. This can be installed in large firms as well as small firms and is installed on their roofs. It also can be used by the individuals at their houses for generating a sufficient amount of energy. At the moment, the situation is not in our hands: due to the pandemic, so many projects were on hold or the rate of installation of solar PV has stopped or slowed down in the lockdown.

Wind power plants are growing in the field of power-generating sources steadily and smoothly. There are several wind projects launched around the world that generate a huge amount of power. The weather is expected to be windy in the first quarter of 2021 and will help the plants to generate the essential amount of energy and boost global energy demand. There are several effects on the recovery of renewable energy sources with a year or two to complete the recovery of renewable energy. The main cause that affects renewable energy sources is if weather turns out as predicted or not. This mainly affects the hydro plants, because if there is no rainfall from time to time, then there is a shortage of water. In wind plants, if there is no wind blowing, turbines could not possibly move and would fail. to generate electricity. In solar PV cells, the sun is the ultimate source [12]. This is why all the determinants are natural in renewable sources.

The annual growth for renewable electricity generation in the years of 2018–2020 is shown in Figure 8.7. It is observed that the impacts of the annual growth in 2020 is less as compared to the other years.

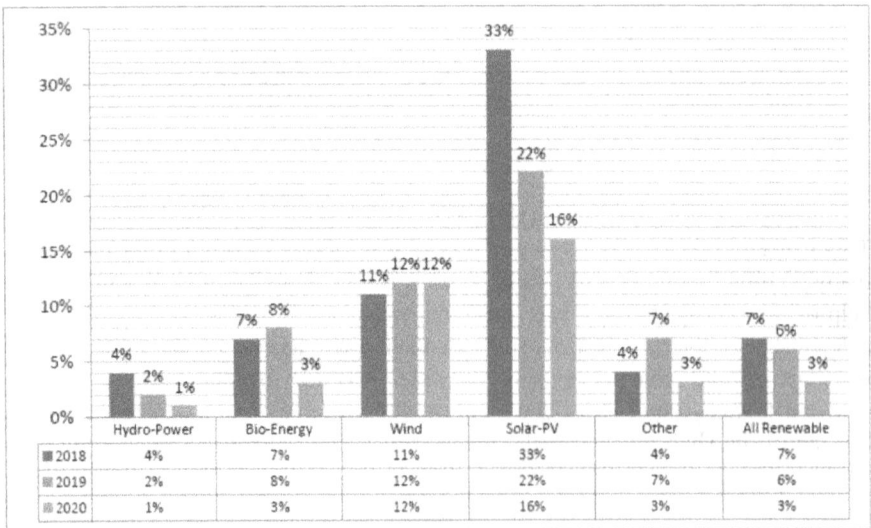

FIGURE 8.7 Annual growth for renewable electricity generation.

8.4.2.1 Impact of Electricity Demand

Global electricity demand dropped by 2.5% in the month of March 2020, due to the lockdown. There was a huge drop in several countries across the world; global energy demand dropped by 2.5% to 4.5% in Europe, Japan and Korea.

The sector most affected by the lockdown is the service sector, which includes retails shops, education institutes, hospitality, offices and tourism activities. All were restricted in the lockdown, and it caused a major economic crisis among the countries [13, 14]. Some of the industries and the factories have resumed work with precautionary measures to protect workers.

Daily reduction of electricity demand after the lockdown in several countries is shown in Figure 8.8. The global electricity demand is expected to fall by 5%. Every country has its own crisis due to the effect of Covid-19. China and India are not similar in terms of electricity demand; they both are dealing with the crisis on their own, and the electricity demand is reduced for both the countries.

8.4.2.2 Impact on Renewable Energy Projects

There was a major impact on renewable projects due to the Covid-19 outbreak. Countries such as China, Vietnam and Thailand are interdependent on the sources of renewable energy; more than 40% of their supply is reliant on the global sector in China. Many countries are dependent on raw material imports for wind turbine equipment around the world [15]. China and Europe hold more than 60% of it. The global wind industry imports the wind project equipment from different countries around the world.

Global reports say that the major concern about renewable sources revolved around the global demand of energy, because the renewable sources solar and wind faced issues of delaying their projects to be launched. On April 17, 2020, MNRE

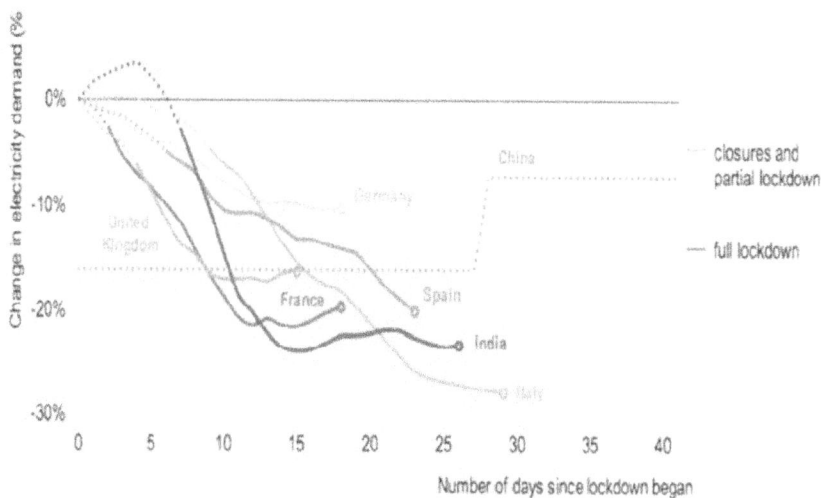

FIGURE 8.8 Reduction of electricity demand after lockdown.

declared that the projects that are ongoing for renewable energy sources are extended through the period of lockdown, and it buys them some more time for developing the project [16].

8.4.2.3 Impact on Global Solar and Wind Energy

Due to Covid-19, the market for renewable energy worldwide saw decreases in the rate of growth. Renewable energy plants were shut down because the gathering of people at worksites was prohibited by different nations to check the spread of the pandemic. Due to the lockdown, construction of ongoing sites got stopped, and laborers working in these sites were compelled to stay at their respective homes. Fall of production of solar power due to Covid-19 has been shown in Figure 8.9. It represents the fall in solar energy production in the top four countries due to Covid-19 and observes that the United States saw the greatest drop in solar energy, approximately 33% in comparison to the previous year's productions [17].

The global drop in the installation of wind power due to Covid-19 is 20%. The Fall of installation in wind power due to Covid-19 is shown in Figure 8.10. It shows that China saw the biggest reduction in installation, at50% as compared to the previous

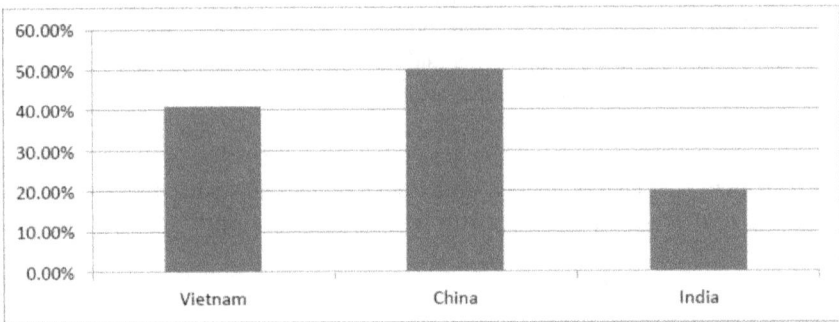

FIGURE 8.9   Fall in production of solar power due to Covid-19.

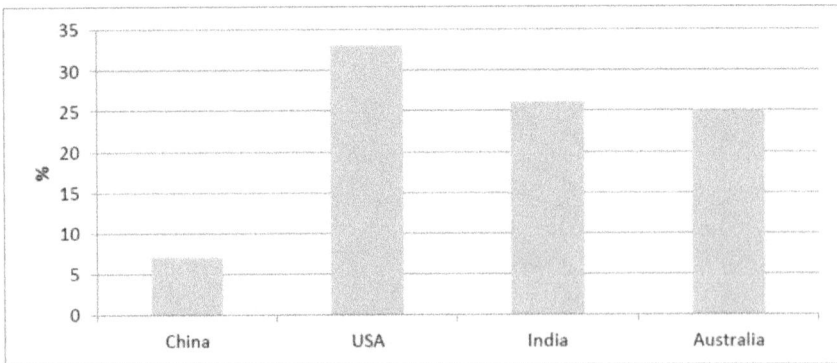

FIGURE 8.10   Fall in installation of wind power due to Covid-19.

year's installation. Other countries also fell in installation in wind power. In India, a total of 8 GW wind power installation was delayed due to Covid-19 [18].

8.5   CONCLUSION

In this chapter, the importance of renewable energy resources (RERs) for maintaining greenhouse gases in limiting CO_2 emissions, replacing fossil fuels and reducing the price of energy is described. First, the types and nature of different RERs were discussed, and then their presence in different countries was presented. Thereafter, the status of RERs across the world was described in detail. This chapter mainly focuses on the solar and wind sectors. It has been observed that every country plays a good role in promoting the renewable energy sector. China plays the biggest role in the production of wind and solar power globally. Many investors are interested in investing money in the solar energy sector. Also considered was the global impact of Covid-19 on solar and wind power. The United States rate of solar power installation dropped the most, and China faced the biggest fall in the wind power sector.

8.6   ACKNOWLEDGMENTS

I owe my special thanks to Dr. Manish Kumar for all his support and guidance, without which all this research was totally impossible. I am also thankful to my family and colleagues for their best wishes. I also owe special thanks to all my friends for their moral support.

REFERENCES

[1]   REthinking Energy 2017, Accelerating the global energy transformation. International Renewable Energy Agency (IREA 2017), Abu Dhabi. Available: www.irena.org.
[2]   International Energy Agency (IEA) 2019, Headline energy data, Paris. Available: https//.www.ira.org/newsroom/news/September/worldenergy-inverstment-2019-html.
[3]   International Energy Agency (IEA) 2019, World energy outlook 2019, IEA Paris.
[4]   REN21 (Renewable Energy Network for the 21st Century) 2019, Renewable global status report, Paris. Available: www.ren21.net/gsr.
[5]   Warren, Den 2019, May, Renewable Energy Country Attractiveness Index (RECAI), issue 47. Available: www.ey.com/Publication/vwLUAssets/EY-RECAI-47-May-2016/$FILE/EY-RECAI-47-May-2016.pdf.
[6]   Govt. of India Ministry of power center electricity authority New Delhi report 2020, January.   Available: www.cea.nic.in/reports/monthly/executivesummary/2020/exe_summary-01.pdf.
[7]   Renewable electricity capacity and generation statistics 2018, June. Retrieved 10 August 2020.
[8]   Solar-fuels & technologies, IEA. Retrieved 18 June 2020.
[9]   Countries-online access-the wind power-wind energy market intelligence, *The Wind Power*. Retrieved 13 January 2020.
[10] IEA based on U.S. EIA, POSOCO (India), RTE (France), TERNA (Italy), ELEXON (UK), China NBS, Red Electrica (Spain) and ENTSO-E.

[11] Global Energy Review 2020, The impacts of the Covid 19 crisis on global energy demand and CO2 emissions. Available: www.iea.org/corrigenda.

[12] OCCTO (Organization for Cross-Regional Co-ordination of Transmission Operators, Japan) 2020, Demand data provided by regional TSOs and sources provide therein. Available: www.occto.or.jp/index.html.

[13] NEA (National Energy Administration) 2020, National energy administration releases. Available: www.nea.gov.cn/2020-03/23/c_138908389.htm.

[14] RTE (Réseau de Transport d'Électricité) 2020, Electricity demand (website). Available: www.rte-france.com/en/eco2mix/eco2mix-consommation-en.

[15] Choudhary, S. 2020, GAIL expects gas demand to pick up soon, *The Economic Times*. Available: https://economictimes.indiatimes.com/industry/energy/oil-gas/gail-expects-gas-demand-to-pick-up-soon/articleshow/75145258.cms?from=mdr.

[16] EIA (Energy Information Administration) 2020, Natural gas weekly update, US Department of Energy. Available: www.eia.gov/naturalgas/weekly.

[17] Global Energy review 2020, The impacts of Covid-19 crisis on global energy demand and CO2 emissions. Available: www.iea.org/corrigenda.

[18] Covid-19 impact: Indian renewable sector. Available: www.saurenergy.com/research/care-ratings-report-on-covid-19-impact-indian-renewable-sector.

9 Green IoT for Sustainable Growth and Energy Management in Smart Cities

K. Suresh Kumar, T. Ananth Kumar,
S. Sundaresan and V. Kishore Kumar

CONTENT

9.1 INTRODUCTION

The industry of information and communication technology (ICT) briefly discusses the Internet of Things (IoT). It also gives the vision about how to connect virtually to any system and with every system. For a quality and comfortable life, ICT can be used. Resources that are used for sustainable development and also environmental management are facilitated by using the connectivity and technology related to sensors. When the IoT is implemented in a city, it provides various benefits such as conservation of water and other resources, improvement in efficiency, regular maintenance, implementation of a monitoring system through big data and conservation of energy, which results in the increase of quality of life [1]. For economic growth, the IoT plays a significant role by making possible the technical feasibility of smart cities. With this evolution of technology related to IoT, there is a dramatic increase in the economy, and because of this a smart city can be sustainable, intelligent, responsive and connected. The IoT contributes considerably to greener cities through energy management, water management, waste management, smart infrastructure, mobility management and supply chains. The primary and main concern of this environment based on IoT is security issues such as access control, storage of data, privacy and secured communication. The emerging issues that are related to green cities are IoT safety and liability against hacking and attacks [2]. For this, there exist various security methods designed to make the IoT system a secure one. The security issues, along with the environmental issues, are embedded in IoT development, which means it deserves considerable attention.

The major challenge in IoT-based green cities is classified based on the issues related to security, urban planning, privacy and cost with quality. All these issues are considered for environmental issues, also. This concept, which has been widely adopted without any management and assessment, provides a significant impact on the environment, which is often referred to as low GHG emission enabler IoT. This concept lays out the roles and issues of the IoT for the transformation of green cities [3]. This chapter deals with the roles of IoT in various sectors, the environmental footprints and energy consumption of IoT in green cities, the application of IoT and its benefits and the IoT's smart development contradictions. From the environmental perspective, the contribution of IoT devices in the cities is highly crucial. By the implementation of this IoT in smart cities, the backbone of the essential lifeline of the smart cities, that is, energy, will be highly benefited [4]. Like other devices, this IoT device also consumes power to operate properly on its given function. Based on the application used, the range of power consumed will be more than the normal range. This will lead to unavoidable power consumption, which provides energy wastage and generates an excess amount of heat dissipated from the device unnecessarily. However, in the IT sector, there is still a research gap found between the saving or offset value of energy to the footprint. Thus, verification is needed for the reduction of power consumption. To bring the smart concept into real time, continuous development and assessment are required. Energy efficient and environment friendly are the two essential characteristics that define the green IoT. Implementation of new techniques for energy efficiency is done on both hardware and software levels on the different platforms, because of which there exists a reduction in energy consumption, emissions of various gases and the greenhouse effect based on the different applications and services for which IoT is used [5].

The green IoT devices will get powered only when the devices need to perform any function, and during the idle state, the power will be reduced. This smart operation of the device will decrease the unwanted wastage of energy. To conserve energy, many different methods, such as data centres and conservation of energy using smart IoT, are implemented using green IoT. To save our environment and to make the system economically right to use, the excess heat generated and the wastage of energy need to be reduced to a considerably low range. This requirement leads to a new concept that is known as the green IoT. Within the context of the rapid urbanization and the challenges of developing the economy, management of resources and how to use energy efficiency and avoid pollution in the environment, the concept of smart cities was established. The smart city contains various components with sociological, political and ecological and many technological aspects. Smart cities, in general, include three domains as their essential domains. These are technology, people, and institutions to build the underlying infrastructure. The technology of information and communication is used for many applications for engaging citizens, urban system enhancement and service delivery in smart cities. In smart city development, some crucial terms act as catalysts for the designed system. They are a capital investment by humans, highlighting innovation and learning all available resources. This defines the second domain of people. The third domain is institutions, which is used to indicate the government, and its importance and the support provided for the development of smart cities.

9.1.1 GREEN IoT—AN OVERVIEW

In the fabric of the present world's information system, IoT plays a significant role. It is considered an environment for computing, an essential network for communication and a global, immersive, and invisible network based on the sensors, software, cameras and data for its development. Energy saving is the main idea adopted in using IoT to develop an environment with a full green campus. The elements involved in the development of the IoT have many advantages in architecture [6]. Architecture needs to be optimistic in order to develop a green environment with all the latest technologies for achieving efficiency and smartness. This green IoT involves many directions that are specified differently with its technical aspects.

In this smart world, it is necessary to have a system that mainly focuses on reducing the usage of energy in the IoT along with intelligent sustainability and also to decrease the amount of CO_2 released. The aspects that are considered in this green IoT include designing and leveraging aspects [7]. The elements used for designing the green IoT are referred to for the development of computing devices, energy efficiency, networking architecture and communication protocols. These IoT elements are used to reduce emissions of CO_2. There will be a reduction in the pollution level, and it also increases energy efficiency. This reduction will be enabled by the technology used for the green IoT. Figure 9.1 shows the architecture of Green IoT that is involved in various applications of sustainable smart city environments. The sensors integrated in the development of a smart city can communicate with other sensors and sense the data in the environment. The sensors connected will consume a considerable amount of power for performing the specified task. Here, the green IoT focuses on how to classify the position of relays and nodes used to fulfil energy savings and budget allocation. For the smart, sustainable future, green IoT plays an essential role in implementing IoT to minimize energy usage, reduce CO_2 emissions

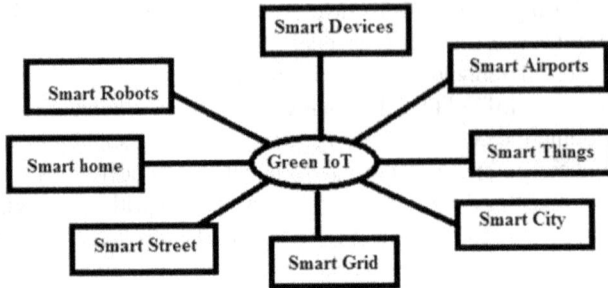

FIGURE 9.1 Architecture for green IoT in smart cities.

and reduce pollution. Thus, this environment will conserve the environment and reduce the consumption of power.

The green IoT mainly focuses on three concepts, namely designing, enabling and leveraging technologies. This technology is mainly used for devices, which are energy-efficient, interconnection, communication protocols and network architecture. This technology will be used for the reduction of carbon emission and for improving energy efficiency. This green ICT technology becomes an efficient method to reduce energy, reduce harmful emissions, reduce pollution and reduce the consumption of resources [8]. Thus, green IoT provides resource preservation, minimizing the technology and reducing the impact on the environment because the cost will be significantly reduced. Therefore, green IoT will concentrate on green architecture, green manufacturing, green disposal and green use. Green IoT designed for energy-efficient systems includes sound sensors, servers, computers and cooling equipment. Manufacturing computers and their modules together with non-impact or limited environmental impact subsystems will pave the way for green IoT. Obsolete computers, discarded devices and other design elements and equipment will be reused or recycled, minimizing and efficiently utilizing the computer and all other devices in the world.

9.1.2 IoT and Its Roles in Different Sectors

Green cities and smart cities are the concepts which are considered to be one of the key implementations in the IoT. For the quality life in the smart cities formed by using the IoT, ICT is used and plays a significant role. For obtaining quality in the smart cities, there are many modifications made [9]. They are using better resources and reducing emissions, lighting and heating buildings in a more efficient way, supplying water with a fully upgraded smart transport network and improving facilities for waste disposal. This also includes smart city administration, which is an interactive and responsive platform. It also acts as public spaces that are safer for the users and used to fulfil the needs of all the adult populations. There are two limitations when it is used for the different sectors of the smart cities: it limits intra-organizational management, flows of energy and materials in intersectoral areas and also the inter-organizational management within the sector and the governance of the entire management.

9.1.3 FINANCIAL AND CULTURAL CONCEPTS IN IoT

Handling, utilization and management have been influenced strongly by the concept of IoT since it is socially and culturally constructed. With a specific cultural, temporal and social context, the concept of waste is constructed, and this concept is not static since it is changing based on requirements. IoT acts as a powerful tool for the development of a green and smart world. Like other tools, it can be used more beneficially and produce a minimum amount of harm; its limit is of skilled operators [10]. The benefits are underlined as a strong need for education, which includes training, development in multidisciplinary areas and acceptance by the public. All these demands are understandable and familiar, which occurs based on agreements, including those between government, local, global, and regional politicians who can secure this with proper implementation and acceptance. There are a more significant number of tasks that can be covered by society with smart cities. There are many issues related to waste management in the present economy. The main questions to resolve are whether it is necessary to close or minimize the currently active infrastructure, how to perform recovery of waste and treatment of the same and how emissions mitigation, sequestration and capture are done. The idea of 3R is to reduce, recycle and reuse. But the 6R can be used for the moderation and minimization of the waste. It is given as rethink, reduce, reuse, repair, recycle, and rebuy. In well-functioning cities, the issues related to waste management have been traditional and are still considered cornerstones. By the process of detection and optimization, IoT can extent the contribution towards waste management. The smart sensors used in waste management will facilitate waste minimization, distribution, collection and selection of treatment, storage and logistics.

The process involved in this waste management is predictable simply because of its nature, such as waste amount detection, the characteristics and supply of waste, and the safety of the sensors involved in this process. Sensors used for smart cities are rapidly evolving, and there are significant construction, manufacturing and even transportation components. For energy utilization and to avoid unnecessary energy usage in smart cities, IoT is used to adjust the conditioning of air, adjust the lighting and develop ventilation based on the temperature and occupants. This process also includes the monitoring of pollution, water flow, waste amount and leakage for corrective action to be taken if necessary. It also focuses on decreasing the cost of operation and increasing efficiency by minimizing the issues related to the environment. Reduced traffic and improved transport network and parking will benefit smart mobility more. Another keystone in this is the person's mobility and the raw material used as the source and carriers for the energy and products involved in this smart city.

9.1.4 THE CONTRADICTIONS IN IoT AND SMART DEVELOPMENTS

The basic knowledge obtained from this may serve as to access the importance of IoT impartially. With the help of this technology, the user can implement any accessibilities by preserving much energy and thus enhance the functionalities. Some inconsistency is inevitable in the assessment because it includes production, energy consumption, creation of a frictionless market and life management. Figure 9.2 indicates the various strategic functions that are highlighted in sustainable green city environments in contradiction with the electronic commerce applications. The energy

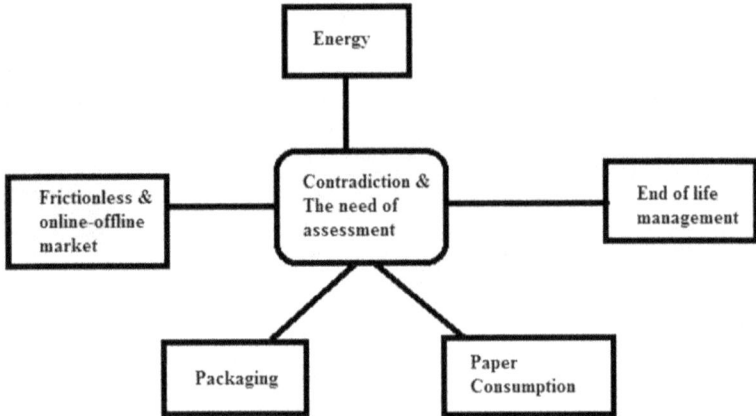

FIGURE 9.2   Contradiction of IoT development.

consumption and waste generated through this sector expansion will not be consistent when the IoT sectors are analyzed [11]. So this needs to be addressed by the IoT sector at the onset, and then when the result is analyzed. The demand for electricity consumption increases, as there is the integration of IoT based on daily practice. However, the energy saving is based on the consumption of energy by all the sectors present in the IoT. The IoT implementation for the reduction of energy consumption seems to be a given. It provides a market without friction, which will be used to increase the demand, rate of purchasing and packaging. Some new highlights are used to stimulate new energy consumption in the rebound effect of e-commerce. The door-to-door delivery of the service of environmental sustainability is not carried out throughout the design. Hence, the awareness level of sustainable environmental practice in this profession is very low. To ensure the functions' sustainability, assessment and a strategic approach are used.

9.2   ROLE OF IOT IN SUSTAINABILITY

In the field of sustainable growth, IoT is defined as and believed to be an effective method in ICTs involved in smart city development. Smart cities have IoT applications in order to be deeply involved in the context, for the development of the IoT technology serves as the model for developing all those applications. The designers can easily correlate smart cities and sustainability. Theoretically, "smart cities with sustainability" defines smart cities as those that extend the sustainability required in order to achieve the maximum benefits for users who rely on technology— but with minimal costs and reduced impact on side-effects that are considered the sustainability target. The smart cities are divided into multiple divisions through the "green city layer". This layer is also used for the representation of the potential outcome. So the concept of smart cities applies to environmental improvement. Through the implementation of the smart city's technology, there can be a reduction of greenhouse gas emissions up to 40%. However, some information indicates that the link between smart cities and sustainability is weak, so this smart city technology can be effectively used for the purpose of marketing instead of using it for infrastructural needs [12].

The development of these smart cities within the urban context with the base for IoT application can be used for connectivity with sustainability. First, it is necessary to know about IoT's initial position throughout the process involved in sustainable development. After this, the evaluation of the performance of the IoT in the sustainable urban environment can be made. This IoT application is performed within a sustainable environment based on the data collected by the connected sensor. After the data has been collected by the sensor network, it can be used to evaluate the conditions of the environment. It can also be used for tracking any specific devices and optimizing environmental measurements. The primary stage of solving all these problems is collecting the needed data involved with that specific IoT application [13]. This IoT technology is considered a critical step used for the collection of information by the sensor, which is connected to the sensor networks and IoT technologies. These data are crucial to know the performance of the environment and to know about resource consumption. Thus, this IoT technology is known as the primary technology used in the sustainable development process of the smart cities.

9.2.1 WASTE MANAGEMENT FOR URBAN SUSTAINABILITY AND ITS IOT APPLICATIONS

The traditional use of IoT technology in urban development is for waste management. An intelligent waste container is deployed to track waste loads using this technology. It is also used for the optimization processing of garbage and to potentially decrease pollutions. In this system, the data based on sensors in the garbage bins are sent to the control center, which then determines the collection time and routes the trucks for the collection of garbage. It is also used to decrease the wastage of food items just by tracing the food garbage weight [14]. Figure 9.3. shows the centralized

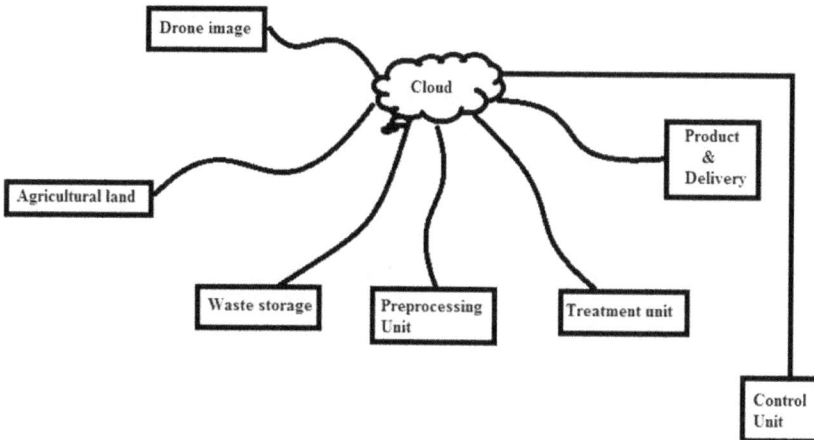

FIGURE 9.3 Proposed cloud-based smart waste management system.

management of connected buildings and agriculture sectors which involves the identification and control of waste from various sources. The primary use of this smart
waste management system is to improve the time required in terms of the data tracking coverage and in terms of density and to identify the level of complexity of the
system, followed by routing the collection trucks in the correct route and reducing
the installation costs [15]. The proposed system's scope is identified. It can also be
used to capture the data, which can be different and used both to detect levels in
garbage receptacles filling and in the identification of toxic substances. Hence, this
waste management system can be combined with another system for functions such
as detection of a toxic substance. Thus, this system will make a massive contribution
to urban sustainability.

9.3   SMART IOT SYSTEMS

9.3.1   SMART STREET LIGHTS

The next issue related to the sustainable environment is the usage of smart street
lights. The function of the smart street light is based on varying the intensity of
brightness. Many places where these smart street lights are deployed make use of
wireless nodes to monitor and check the lighting levels for the optimization of energy
consumed. In addition to this, the sensors implemented can collect environmental
data such as emissions of carbon dioxide, levels of humidity and noise occurrence
[16]. Thus, this adaptive system will be an essential step in forwarding sustainability,
since this system is built on infrastructure where there is no excessive modification
needed on the currently available urban environments.

The main challenge in this smart street light is fault tolerance. It generally differs
from the waste management system in that the waste management system is used
only to capture the data, but these smart street lights are essential in this infrastructure during the night time. If there exists any failure in this light control system,
it will lead to many safety-related issues. Having an alternative system which can
be used as a solution for the failure faced in this smart street light system is very
advisable.

9.3.2   SMART HOMES AND INTELLIGENT RESIDENTIAL DISTRICTS

The next urban management technology is the smart home, which is used as a part of
a smart residential area by IoT for sustainability in the residential region. This smart
home control involves controlling temperature, lighting and humidity to reduce
energy consumption [17]. For planning in urban areas, the process of data tracking is
potentially connected with the smart grid, which can be used to allocate electricity
resources efficiently in nearby cities. Many issues other than energy usage are related
to the smart home [18]. They are issues like security and medical care, etc. For all
these issues, the only solution is the smart home, which is considered to be part of
sustainability. Based on this, it can be inferred that the smart home plays an essential
role in IoT technology. This smart home definition will extend to both cities and

suburban areas. As the method of increasing the homes' energy efficiency, the local scales make use of integrated IoT technology, including household energy operations. When viewed on a large scale, the smart home implemented for an individual house can be used to create a network that intelligently tracks the performance of the designers, who work to analyze the energy used at a local scale.

9.3.3 Methodology and Implementation for Measuring the Accuracy Overdrain Measurements in Greenhouse Horticulture

IoT for greenhouse horticulture was purely developed using a device with Internet access to measure the variables and store the measured variables automatically from previously selected data. The environmental variable and the local variables are tracked frequently by the researchers. Those variables are light intensity, moisture, temperature, pH value, conductivity and overdrain quantity. The final price of the system is identified as the critical aspect of this system. So this system is mainly designed for the economically limited users, and it also restricts the use of costly data ranging in the order of several thousand dollars. Based on these requirements, a system based on the commercial off-the-shelf is selected. It is also used to promote the tools used that are available at the cloud server at a little cost [19].

For the accurate measurement of the overdrain value, the liquid flow meter is used. From this measurement, the user is able to know about adequate irrigation. Daily, in the early morning, this measure is carried out. In the areas of tropical countries, this liquid flow meter is used to provide data that are very close to the overdrain data, even if it is measured under environmental conditions which are protected, such as greenhouses, and environmental conditions which suffer quick variation. To calculate the frequency, it is necessary to assume some conditions and allow the flow rate to be 7.5, from which the liquid flow meter will be able to generate a pulse rate of 450 pulses per litre. These measurements are done manually. Every day, parameters like humidity and temperature are observed by the LCD hygrometer and thermometer. This data acquired will be used to compare to the data which is automatically captured from the cucumber plantation. Figure 9.4. depicts the integral associations of four different parameters such as security, productivity, comfort and convenience, which forms the building services.

The assistant handling the system can collect the information related to overdrain collection, conductivity and pH of the system. This data is also necessary for the researchers for the determination of irrigation periodicity. This system automatically collects the data of the variables, and it will get uploaded to the cloud server, which can be accessed easily by the researchers. Diagram researchers can access the application by using the Internet connection through the web browsers. The researchers can access this platform with a user name and password. There are a higher number of data samples available for research. The training process involved in this is based on the Kalman filter. Through the filters implemented using the data acquisition system, the Kalman gain of the system is measured. This data obtained will have good coverage of data over 300.71 with a ±1% error.

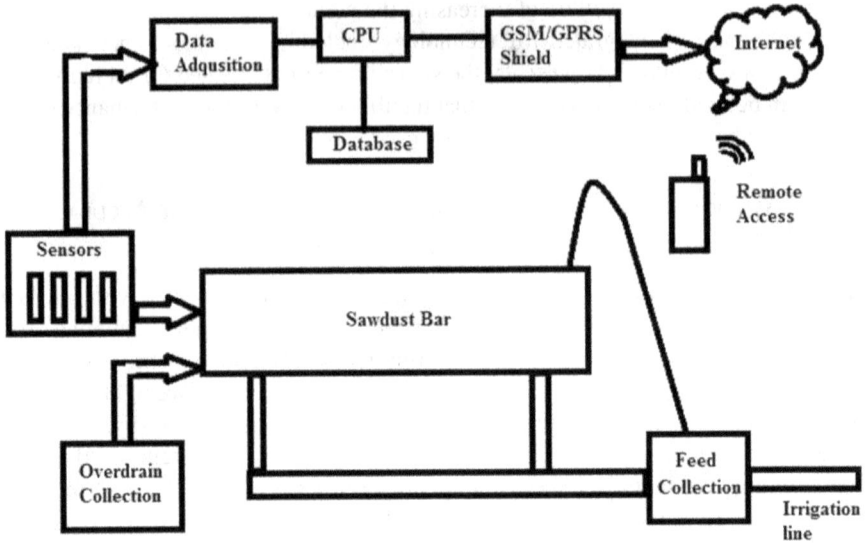

FIGURE 9.4   Schematic representation for green horticulture.

TABLE 9.1
Comparison between the Measured and Estimated Values for Each Sample Taken in Kalman Filter

Sample numbers	Measured value	Estimated value
0	3.2	3.01
2	3.15	3.05
4	2.9	2.98
6	3.05	3.02
8	3	3.06
10	2.92	3.04
12	2.95	3.07

9.4   IOT FOR GREEN AGRICULTURE

A smart agriculture system is developed for a plantation. The rate of germination is the most significant in this process. In this example, the plants are grown higher than 8 cm. Cameras will capture images of the system. These photos are then transmitted to the preprocessing unit for quality improvement of the image and to remove the noise. The mean filter is then used as the filtering technique through which the pixel points in the image will be detected. The outcome would then identify whether the pixels belong to the plantation or not. Based on the number of cameras used, more images can be captured and stored in memory. These acquired data will act as the resource for the next round. Thus, the plant can be monitored easily and can send a warning message based on the observation [20].

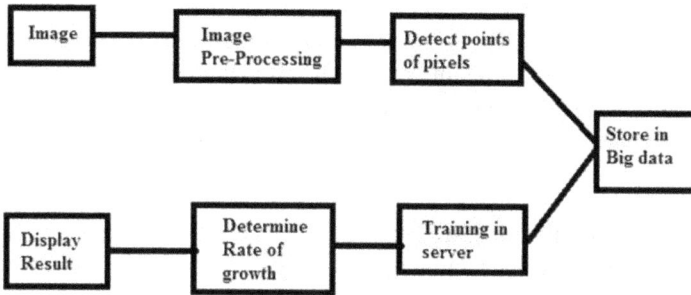

FIGURE 9.5   Green farming: an image processing-based approach.

The personal computer is the main module that controls all the peripheral devices, motor, sensors and relay block. Using the MQTT communication protocol, the server program operates with the Linux operating system. To recognize the cultivating condition, the CPU collects data sensed by the sensors. Later, based on the data collected, the personal computer controls the pump to provide a better environment and give a notification to the farmers. Figure 9.5 demonstrates the process of green farming that gets an effective control over various scenarios based on image acquisition and enhancement methods.

9.4.1   SMART IoT FARMING

In the implementation of smart agriculture, digital information handling and the technologies related to communication are much needed to improve the monitoring sector to develop the various processes and operations of agriculture. The sensors used in this smart system might help in catching information such as soil moisture, temperature rate and the use of fertilizers. Thus, it supports farmers in providing sufficient real-time periodical access to data. The analysis is performed on the availability of crops. The livestock gives the farmers real-time access to the information that helps in the analysis of the logistics via the cellular wireless network. The performance of the operation is boosted by the implementation of the smart farm to analyze and keep acting on the data collected so that there might be some increase in productivity, which helps streamline operations [21]. We can see the various advantages predicted by the implementation of the IoT based technology in the numerous farms. The smart farms built by using advanced technology can access using 5G speed capability and the bandwidth. Various applications generated by the IoT implementation led us to reach out achievement in the field of agriculture. Countless applications are using IoT technology. Farming and agriculture can be improved by using IoT. This is done by analyzing the case details of the blueberry's irrigation, as one example of crop growth. The applications used for smart agriculture might be multifunctional, and hence can handle a number of aggravations. This can be analyzed by using case studies conducted from the various parts of the world where the IoT technology-based smart farming was is already in use.

Crop diseases are considered to be the major threats which are being faced consistently. An analysis was done by conducting the case study in the Punjab farmers regarding an outbreak of disease, and information was gathered from remote sensors which are IoT enabled. Conditions like the rainfall, humidity and temperature condition are monitored periodically. That might help the farmers to identify the crops that are disease sensitive, and the responses related to that are predicted instantly. Considering the Indian population, 1.2 billion, people are doing an outstanding job in the area of agriculture. This is a huge number when compared to other countries. These huge numbers of farmers are doing a tremendous job in doing farming with reduced risk of disease infection. The IoT based smart system offers an improved version of food security and the most significant improvement in the agriculture-based economy.

9.4.2 APPLICATIONS OF IoT TECHNOLOGY IN SMART FARMING

The IoT applies to machines with unique characters and capacities concerning remote detecting, tracking and incidentally recording those information records. IoT is at a cross-stage where PCs are getting smarter, and they are figuring out a more intelligent and coordinated effort, which is considered to be more informative. The IoT frameworks can likewise impart information to different gadgets and programming, straightforwardly, or in a roundabout way, progressively. The number of strategies utilized in smart cultivating is mind-boggling, emerging from the complexities of ranchers' exercises. Sensor-based water system frameworks give ranchers another suitable option. IoT innovation will limit costs and improve sensor-based water system frameworks by an assortment of sensor information arrangements. IoT is an overall system of standard correspondence conventions. This consolidates numerous information stockpiling procedures, from physical volumes to IoT applications [22]. The IoT receives a wide scope of supporting innovations, for example, remote sensor systems, cloud organizing, extensive information, coordinated frameworks, security conventions and interfaces, correspondence conventions and web administrations. The key highlights and favourable circumstances are the essential considerations in each of the IoT configurations in shrewd agribusiness. In smart cultivating, numerous IoT ventures show the enormous capacity and significance of IoT-based advancements and applications. In reference [20], the creators recommended a decision-support system (DSS) concentrated on remote sensor incorporation and wireless sensor and actuator network innovation. The suggested measures were planned for limiting water issues and expanding yields during different climate conditions. In [23], the scientists proposed a framework, AgriTech, to computerize numerous horticultural administrations (water, fertilizers, bug sprays and physical work) in agribusiness utilizing IoT. The farmer may follow harvests and farmland remotely utilizing a handheld terminal framework. In [24], a smart nursery model was recommended that helped farmers naturally complete farmstead work. Utilizing dribble fertigation procedures, the right measure of nitrogen, phosphorus, potassium and different minerals was included in utilizing soil well-being card data. In [25], creators proposed a portable PC program to follow distinctive soil attributes distantly. They used the invasion of soil protection schemes that protect soils from soil moistness to measure soil sogginess content. Numerous IoT improvement frameworks have been effectively applied to handle the water system and control of water quality [26], constructing and presenting a programmed water system framework. The machine utilized a cell phone to catch and process soil analysis close to the yield's root area and optically gauge water content. In [27], creators actualized an independent trickle water system organized, fueled and constrained by an ARM9 processor. The machine informed shoppers of any unfavourable conditions, for example, loss of dampness, temperature increment, and CO_2 fixation. A constant input control module was designed to follow and deal with all dribble water system frameworks. In [28], a remotely managed water system framework was implemented to limit optional water system volume and timing choices. In [29], the creators contemplated an IoT-based front wetting framework (IoT-WFD). The structure was divided into two sub-frameworks, a sensor hub and a cloud application. In [30], analysts manufactured a mechanized system utilizing IoT advances to track and gather continuous yield development information. The focal unit gathered information to build up the yield development model, anticipated harvest necessities in water for different development cycles and applied water system choice.

Open Platform Communications (OPC) was explored in brilliant cultivating for horticultural apparatus telemetry. In [31], the appropriateness of Open Platform Communications Unified Architecture (OPC UA), the most recent OPC programming neology for horticultural apparatus telemetry applications, was examined. The creators introduced both the worker side structure responsible for the consolidated reaper and the controller application program. Nevertheless, the time taken by both the worker and the consumer in same region is 250 ms. In smart development, IoT advancement frameworks were additionally utilized in various rural yield perspectives. In [32], authors built up a smart IoT gadget to control wheat infections, vermin and weeds. The program could analyze and conjecture about wheat diseases, bugs and weeds, yet could also caution farmers. In [33], an IoT framework was created to reduce bug sprays and fungicides. The program gave information on illness and nuisances utilizing pre-phrasing models dependent on relationship data, so farmers could oversee them without any problem. In [34], the creators applied dampness sensors to quantify dampness in a Lingzhi mushroom field. Normal humidity hit 90–95%. A CCTV (closed-circuit television) framework controlled the working condition of sprinklers and hose siphons, while a microcontroller was utilized to control on and off tasks. IoT applications were additionally utilized with domesticated animals. Using IoT and wireless body area networks (WBANs), the well-being status of dairy cows was examined. Off-body remote channel (long range) was characterized at 868 MHz.

The results prove that the selected model for handling disaster may precisely hold the massive scope blurring. In [35], researchers introduced a field snail recognition WSN. The recommended system could either cause a snail following caution or incorporate snail recognition prediction models via ecological factors, such as temperature or humidity. Numerous IoT programming applications in farming are currently financially practical. Such frameworks incorporate information assortment, information handling and examination. Table 9.1 sums up the smart cultivating area's most basic IoT arrangements. Such IoT frameworks have a comparable objective: to mechanize information and use information from all channels, utilizing a standard application programming interface (API). These frameworks frequently consolidate sifting and conglomeration highlights. The benefits of these IoT applications in smart horticulture remain unclear [36]. There are a few obstructions that, despite everything, preclude across-the-board utilization of IoT for the smart water system, for example, modern IoT gadget frameworks, propelled sensor incorporation, etc. [37]. In any case, new IoT frameworks can be utilized to capture enormous volumes of information. This information would then be able to be broken down to process and evaluate customized farm proposals.

9.4.3 APPLICATIONS OF UAV TECHNOLOGY IN SMART FARMING

Using unmanned aerial vehicle (UAV) technology, we can collect some base knowledge to help farmers determine whether there is currently a big research trend. Agricultural UAVs have been used for various agricultural issues. In [38], the writers introduced a novel method for taking UAV photographs of agricultural crops. The model proposed by them shows a technology where they established a procedure to coordinate on-the-field tridimensional point clouds, reconstructing crop 3D models to monitor plant-level growth parameters. A similar approach was used in [39] to measure field maize and sorghum plants' crop height. In one example, they used a novel approach for monitoring the height of the crop of sorghum plants by using UAV and 3D model reconstruction [40, 41]. The authors reported an average individual sorghum height root mean square error (RMSE) with manual field data of 0.33 m.

9.5 CURRENT TRENDS AND FUTURE CHALLENGES

The IoT and the smart city environment are mutually correlating with each other. The advancements in the smart city are enhanced with the help of the IoT. In providing a green environment, the IoT is paving a wonderful way to develop and advance some unique tools and techniques for the development of the smart green city environment.

9.5.1 IN AGRICULTURE AND HORTICULTURE

The transformation in the agriculture sector might recognize the new situation that held the application of smart cities in its current models. Horticulture has created from heritage choice emotionally supportive networks with predefined time booking usefulness to another period of cultivating frameworks consolidating different advanced advances, for example, IoT, UAV, human-made consciousness, AI, and so on. A significant number of these frameworks are in theoretical (non-business) mode and, for the most part, tackle a particular development (or assortment) process. None of these structures joins in development strategies, or even the total yield (from planting to the gathering). A considerable lot of the significant empowering advancements, true to form, have a larger number of advantages in various development farms than others, yet they are generally challenging to consolidate for end-clients (farmers). Actualizing IoT advancements in different cultivating exercises has upgraded complete harvest measurements regarding item creation, quality and amount, and expanded benefits. IoT has started to reveal its likely incentive to end-clients as it can bolster and encourage the dynamic.

9.5.2 UAV FRAMEWORKS

On the other hand, IoT must overcome a few obstructions to multifaceted mechanical nature, definition, ease of use, conveyance, productivity and PCs' capacity. Inevitably, the critical issue for mishandling the enormous scope IoT innovations in agribusiness is handling better rural strategies that limit explicit focuses for farmers. UAV frameworks are utilized for control and observation in agribusiness. Even though UAV has numerous entanglements, principally about force self-governance and correspondence productivity, that despite everything should be examined, the advantages of utilizing this innovation have shown up ahead of schedule. In any case, academics, specialized labourers, and additionally, farmers understand the numerous benefits of applying UAV innovation to various agrarian economy angles. Second, UAV plays (and will maintain) an essential job in weed distinguishing proof and control.

9.5.3 INNOVATIONS IN AGRICULTURE

Along with automated vehicle airborne capacity, this agrarian issue by definition enables end-clients to oversee weeds in agribusiness productively. UAVs' utilization of AI methods to create multifaceted picture information has additionally upgraded UAV's creation capacity. Second, the capacity of users to gather distinctive plant files utilizing UAV innovation and multi-phantom imaging uncovered the advantage of such advancements in development rehearses. Field-level phenotyping is another essential part of the agribusiness industry to which UAV innovation proficiently contributes. Utilizing UAV frameworks and field-level phenotyping would permit producers to appraise full plant development and foresee the last yield more precisely. UAV frameworks exhibited this present innovation's mechanical edge in the field of utilizing static cameras. Using UAV advancements in the field, a few complex farming issues are settled (early). Agrarian assembling, and in the end, the farming economy, is unquestionably a dominant worldwide market with high potential. For example, in the rising turns of events, IoT and UAV will assume a significant job later on. Regardless, other horticulture/development issues should be handled, for example, weed mediation, field-level phenotyping and multi- and hyperspectral disease-control imaging, water system power, manures, supplements, plant-based development and yields, 3D plant displaying and observing, crop quality and yield improvement, among other different rural concerns. Key supporting advancements, for example, IoT and UAV, will contribute effectively and productively to smart cultivating rehearses in this troublesome condition by meeting the accompanying imperatives: framework ease of use and versatility, ease of use, quick conveyance and upgraded benefits. These improvements will bit by bit transform conventional agribusiness into a new farming biological system by tending to human needs in urban and rustic situations.

9.6 CONCLUSION

Thus, this green IoT technology is a tremendous benefit that has been developed with many advantages all over the world. This growing technology has high energy efficiency with the intention of mitigating e-waste and hazardous emissions. This chapter identifies the technology used in green IoT, which helps keep the environment smarter and greener. Based on this technology, the things which are found around us will become more energized for performing autonomous tasks. It also gives a new type of communication medium between human beings and things, among the users with maximum utilization of bandwidth, with reduced hazardous emissions and reduced power consumption. Many new revolutions can be made in day to day life just because of this green Internet of things, which completes the vision of green ambient intelligence. In the upcoming year, there may be a development of massive sensors that can communicate with each other easily through their intelligence and provide green support to all its users for handling their tasks. Smart farming also has many agronomic aspects to tackle. This involves UAVs for precise irrigation with minimizing salts, prosperous and fair fertilization, and use of pesticides by significantly reducing polluting aquifers, adequate weed control, increasing crop yields, and efficient management of crop diseases in the field. Thus, recent developments found in the area of Green IoT have been identified and discussed here which can be very much useful for the researchers to come up with new ideas in this area.

REFERENCES

[1]	Klemeš, Jiří Jaromír, and Yee Van Fan. "Internet of Things for green cities transformation: Benefits and challenges." In 2019 4th International Conference on Smart and Sustainable Technologies (SpliTech), Split, Croatia, pp. 1–6. IEEE, 2019.
[2]	Ahmad, Rafeeq, Mohd Asim, Safwan Zubair Khan, and Bharat Singh. "Green IoT—issues and challenges." In Proceedings of 2nd International Conference on Advanced Computing and Software Engineering (ICACSE), Kamla Nehru Institute of Technology, Sultanpur, India, 2019.
[3]	Gadre, Monika, and Chinmay Gadre. "Green Internet of Things (IoT): Go green with IoT." International Journal of Engineering Research & Technology (IJERT) ICIOT 4, no. 29 (2016).
[4]	Liu, Limin. "IoT and a sustainable city." Energy Procedia 153 (2018): 342–346.
[5]	Zhang, Shufan. "The application of the Internet of Things to enhance urban sustainability." Agora, 1 no. 1, (2017): 102–111.
[6]	Alsamhi, S., O. Ma, M. S. Ansari, and Q. Meng. "Greening Internet of Things for smart everythings with a green-environment life: A survey and future prospects. arXiv 2018." arXiv preprint arXiv:1805.00844.
[7]	Carrasquilla-Batista, Arys, Alfonso Chacón-Rodríguez, and Milton Solórzano-Quintana. "Using IoT resources to enhance the accuracy of overdrain measurements in greenhouse horticulture." In 2016 IEEE 36th Central American and Panama Convention (CONCAPAN XXXVI), San José, Costa Rica, pp. 1–5. IEEE, 2016.
[8]	Saravanan, Krishnan, E. Golden Julie, and Y. Harold Robinson. "Smart cities & IoT: evolution of applications, architectures & technologies, present scenarios & future dream." In Internet of Things and Big Data Analytics for Smart Generation, pp. 135–151. Springer, Cham, 2019.
[9]	Ben-Daya, Mohamed, Elkafi Hassini, and Zied Bahroun. "Internet of Things and supply chain management: A literature review." International Journal of Production Research 57, nos. 15–16 (2019): 4719–4742.
[10]	Rong, Ke, Guangyu Hu, Yong Lin, Yongjiang Shi, and Liang Guo. "Understanding business ecosystem using a 6C framework in Internet-of-Things-based sectors." International Journal of Production Economics 159 (2015): 41–55.

[11] Boyes, Hugh, Bil Hallaq, Joe Cunningham, and Tim Watson. "The Industrial Internet of Things (IIoT): An analysis framework." *Computers in Industry* 101 (2018): 1–12.

[12] Bibri, Simon Elias. "The IoT for smart sustainable cities of the future: An analytical framework for sensor-based big data applications for environmental sustainability." *Sustainable Cities and Society* 38 (2018): 230–253.

[13] Kao, Yu-Sheng, Kazumitsu Nawata, and Chi-Yo Huang. "Systemic functions evaluation based technological innovation system for the sustainability of IoT in the manufacturing industry." *Sustainability* 11, no. 8 (2019): 2342.

[14] Morrissey, Anne J., and John Browne. "Waste management models and their application to sustainable waste management." *Waste Management* 24, no. 3 (2004): 297–308.

[15] Bibri, Simon Elias. "The IoT for smart sustainable cities of the future: An analytical framework for sensor-based big data applications for environmental sustainability." *Sustainable Cities and Society* 38 (2018): 230–253.

[16] Yoshiura, Noriaki, Yusaku Fujii, and Naoya Ohta. "Smart street light system looking like usual street lights based on sensor networks." In 2013 13th International Symposium on Communications and Information Technologies (ISCIT), Ho Chi Minh City, Vietnam, pp. 633–637. IEEE, 2013.

[17] Srivatsa, Deepak K., B. Preethi, R. Parinitha, G. Sumana, and A. Kumar. "Smart street lights." In 2013 Texas Instruments India Educators' Conference, pp. 103–106. IEEE Computer Society, Washington, DC, 2013.

[18] Lynggaard, Per, and Knud Erik Skouby. "Complex IoT systems as enablers for smart homes in a smart city vision." *Sensors* 16, no. 11 (2016): 1840.

[19] Dion, Pierre-Paul, Thomas Jeanne, Mireille Thériault, Richard Hogue, Steeve Pepin, and Martine Dorais. "Nitrogen release from five organic fertilizers commonly used in greenhouse organic horticulture with contrasting effects on bacterial communities." *Canadian Journal of Soil Science* 100, no. 2 (2020): 120–135.

[20] Ferrag, Mohamed Amine, Lei Shu, Xing Yang, Abdelouahid Derhab, and Leandros Maglaras. "Security and privacy for green IoT-based agriculture: Review, blockchain solutions, and challenges." *IEEE Access* 8 (2020): 32031–32053.

[21] Glaroudis, Dimitrios, Athanasios Iossifides, and Periklis Chatzimisios. "Survey, comparison and research challenges of IoT application protocols for smart farming." *Computer Networks* 168 (2020): 107037.

[22] Huang, Kai, Lei Shu, Kailiang Li, Fan Yang, Guangjie Han, Xiaochan Wang, and Simon Pearson. "Photovoltaic agricultural Internet of Things towards realizing the next generation of smart farming." *IEEE Access* 8 (2020): 76300–76312.

[23] Lowry, Gregory V., Astrid Avellan, and Leanne M. Gilbertson. "Opportunities and challenges for nanotechnology in the agri-tech revolution." *Nature Nanotechnology* 14, no. 6 (2019): 517–522.

[24] Freeborn, John R., David L. Holshouser, Marcus M. Alley, Norris L. Powell, and David M. Orcutt. "Soybean yield response to reproductive stage soil-applied nitrogen and foliar-applied boron." *Agronomy Journal* 93, no. 6 (2001): 1200–1209.

[25] Abu-Hamdeh, Nidal H. "The effect of tillage treatments on soil water holding capacity and on soil physical properties." In Conserving soil and water for society: sharing solutions. ISCO 13th international soil conservation organization conference, Brisbane Australia, paper, no. 669, pp. 1–6. 2004.

[26] Sobsey, Mark D., Christine E. Stauber, Lisa M. Casanova, Joseph M. Brown, and Mark A. Elliott. "Point of use household drinking water filtration: A practical, effective solution for providing sustained access to safe drinking water in the developing world." *Environmental Science & Technology* 42, no. 12 (2008): 4261–4267.

[27] Khan, Gulnaj, Kanchan Dhakate, Shivani Kambe, Shraddha Meshram, and Akhilesh Lunge. "A review on Arduino based smart irrigation system." *International Journal of*

Scientific Research in Science and Technology (IJSRST) 4, no. 2 (2018). Print ISSN: 2395-6011 | Online ISSN: 2395-602X

[28]   Hering, Janet G., T. David Waite, Richard G. Luthy, Jorg E. Drewes, and David L. Sedlak. "A changing framework for urban water systems." *Environmental Science & Technology* 47, no. 19 (2013): 10721–10726.

[29]   Mateen, Ahmed, Qingsheng Zhu, and Salman Afsar. "IoT based real time agriculture farming." *International Journal of Advanced Smart Convergence* 8, no. 4 (2019): 16–25.

[30]   Duvick, Donald N. "The contribution of breeding to yield advances in maize (Zea mays L.)." *Advances in Agronomy* 86 (2005): 83–145.

[31]   Hansch, Gerhard, Peter Schneider, Kai Fischer, and Konstantin Böttinger. "A unified architecture for industrial IoT security requirements in open platform communications." In 2019 24th IEEE International Conference on Emerging Technologies and Factory Automation (ETFA), Zaragoza, Spain, pp. 325–332. IEEE, 2019.

[32]   Pierce, George E. "Pseudomonas aeruginosa, Candida albicans, and device-related nosocomial infections: implications, trends, and potential approaches for control." *Journal of Industrial Microbiology and Biotechnology* 32, no. 7 (2005): 309–318.

[33]   Sammons, Philip J., Tomonari Furukawa, and Andrew Bulgin. "Autonomous pesticide spraying robot for use in a greenhouse." *Australian Conference on Robotics and Automation*, 1, no. 9. (2005).

[34]   Ali, Ahmad, Yu Ming, Sagnik Chakraborty, and Saima Iram. "A comprehensive survey on real-time applications of WSN." *Future Internet* 9, no. 4 (2017): 77.

[35]   Larios, Diego F., Julio Barbancho, José L. Sevillano, Gustavo Rodríguez, Francisco J. Molina, Virginia G. Gasull, Javier M. Mora-Merchan, and Carlos León. "Five years of designing wireless sensor networks in the doñana biological reserve (Spain): An applications approach." *Sensors* 13, no. 9 (2013): 12044–12069.

[36]   Fensel, Dieter, and Christoph Bussler. "The web service modeling framework WSMF." *Electronic Commerce Research and Applications* 1, no. 2 (2002): 113–137.

[37]   More, Ratnamala Prakash, and Anil S. Hiwale. "IoT based wireless sensornode for mine safety application." *Communications* 5, no. 9 (2016).

[38]   Boursianis, Achilles D., Maria S. Papadopoulou, Panagiotis Diamantoulakis, Aglaia Liopa-Tsakalidi, Pantelis Barouchas, George Salahas, George Karagiannidis, Shaohua Wan, and Sotirios K. Goudos. "Internet of Things (IoT) and agricultural Unmanned Aerial Vehicles (UAVs) in smart farming: A comprehensive review." *Internet of Things* (2020): 100187.

[39]   Malambo, Lonesome, Sorin C. Popescu, Seth C. Murray, E. Putman, N. Ace Pugh, David W. Horne, G. Richardson et al. "Multitemporal field-based plant height estimation using 3D point clouds generated from small unmanned aerial systems high-resolution imagery." *International Journal of Applied Earth Observation and Geoinformation* 64 (2018): 31–42.

[40]   Hu, Pengcheng, Scott C. Chapman, Xuemin Wang, Andries Potgieter, Tao Duan, David Jordan, Yan Guo, and Bangyou Zheng. "Estimation of plant height using a high throughput phenotyping platform based on unmanned aerial vehicle and self-calibration: example for sorghum breeding." *European Journal of Agronomy* 95 (2018): 24–32.

[41]   Saravanan, K., and P. Srinivasan. "Examining IoT's applications using cloud services." In *Examining cloud computing technologies through the Internet of Things*, pp. 147–163. IGI Global, Delhi, 2018.

10 For a Better Tomorrow
Smart and Sustainable Cities Using Artificial Intelligence

Aditya Punia and Navdeep Mor

CONTENT

10.1 INTRODUCTION

Smart and sustainable cities are the two terms being used increasingly in the past two decades. 'The cities of the future' is the term used to refer to such developments. These are technologically enhanced and sustainable planned utopias springing up over all parts of the world. The goal for such a green smart city should be that it creates delight when entered, serenity and health when it is occupied and regret on departing. As compared to traditional cities, these incorporate advanced and latest technologies, for example, artificial intelligence (AI) techniques such as machine learning (ML), deep learning (DL), pattern recognition (PR), and Internet of Things (IoT) techniques such as machine to machine (M2M) communication. These techniques are implemented in a smart way to ease and minimize the daily problems of traditional cities such as traffic congestion, air quality monitoring, public transport availability, and providing last-mile connectivity.

With a growing population, the increase in the number of vehicles has been exponential, which leads to frequent traffic congestion during peak hours in almost every city in the world. This results in more travel time and fatigue for commuters during what could have been a productive time, along with the high and inefficient burning of fuel, causing more pollution. Green smart cities equipped with smart transportation systems minimize this problem by using AI for dynamic traffic signal management. Nowadays, this technique is further assisted with artificial neural networks (ANN) and pattern recognition (PR) for better results. Another major aspect of green smart city is water management, which is indirectly related to energy consumption because it needs to be pumped, treated, and stored before being pumped again for supplying to residents. So using AI and IoT for efficient water management, monitoring pipeline leakages at an early stage and spikes in usage to predict outages, can help in the conservation of this scarce resource.

The initial cost of such green smart cities is higher, but it will surely be paid back after some time in the tangible form of savings and non-tangible forms like the peace of mind and well-being of its residents right from day one. This chapter explores the use of AI and IoT in the field of green smart cities to achieve the desired sustainable goals, which primarily focus on conservation of energy in all aspects. After going through this chapter, one will be able to answer the following questions:

- What is the need for green, smart and sustainable cities?
- What is the role of the increasing urban population in climate change and the need to control it?
- In which domains of a smart city can AI and IoT be exploited to achieve the desired sustainable results?

10.2   NEED FOR GREEN, SMART, AND SUSTAINABLE CITIES

The need for green, smart and sustainable cities is apparent from the Agenda 2030 for Sustainable Development, adopted in 2015 [1]. The complete timeline for the formulation of the agenda is presented in Figure 10.1. Some of the major goals to be achieved in this agenda include ending poverty and deprivation, improving education and health, and spurring economic growth with its mitigation strategies. All these goals were formulated keeping in mind the preservation of our forests and oceans and tackling the adverse effects of climate change. Economic development policies and international commitments to climate change mitigation are centered on the twin targets of spurring rapid market growth with minimal environmental impact. Out of

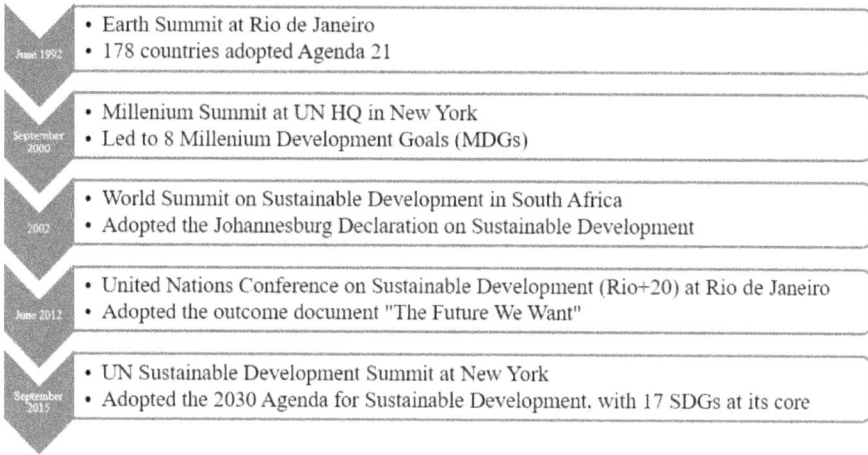

- Earth Summit at Rio de Janeiro
- 178 countries adopted Agenda 21

June 1992

- Millenium Summit at UN HQ in New York
- Led to 8 Millenium Development Goals (MDGs)

September 2000

- World Summit on Sustainable Development in South Africa
- Adopted the Johannesburg Declaration on Sustainable Development

2002

- United Nations Conference on Sustainable Development (Rio+20) at Rio de Janeiro
- Adopted the outcome document "The Future We Want"

June 2012

- UN Sustainable Development Summit at New York
- Adopted the 2030 Agenda for Sustainable Development, with 17 SDGs at its core

September 2015

FIGURE 10.1 A timeline to the formulation of SDGs.

Agenda 2030's 17 Sustainable Development Goals (SDGs), seven are relevant to this chapter:

SDG 3—"Good Health and Well-being: Ensure healthy lives and promote well-being for all at all ages."

SDG 6—"Clean Water and Sanitation: Ensure availability and sustainable management of water and sanitation for all."

SDG 7—"Affordable and Clean Energy: Ensure access to affordable, reliable, sustainable, and modern energy for all."

SDG 9—"Industry, Innovation, and Infrastructure: Build resilient infrastructure, promote inclusive and sustainable industrialization, and foster innovation."

SDG 11—"Sustainable Cities and Communities: Make cities and human settlements inclusive, safe, resilient, and sustainable."

SDG 12—"Responsible Consumption and Production: Ensure sustainable consumption and production patterns."

SDG 13—"Climate Action: Take urgent action to combat climate change and its impacts."

However, no green smart city can meet the sustainability criteria unless the buildings of the smart city are also smart, energy-efficient, and sustainable, because the building sector is not only one of the largest consumers of energy, it is among the largest producers of greenhouse gases (GHGs). Smart buildings deploy smart and advanced applications of energy efficiency and renewable energy to operate, thus lowering the operation and maintenance costs in addition to the energy savings and drastic reductions in emission of GHGs. This ultra-high-energy performance is achieved by the incorporation of smart technology by embedding it into the building so that everything is connected to a central hub and individually to each other, thus enabling intelligent control according

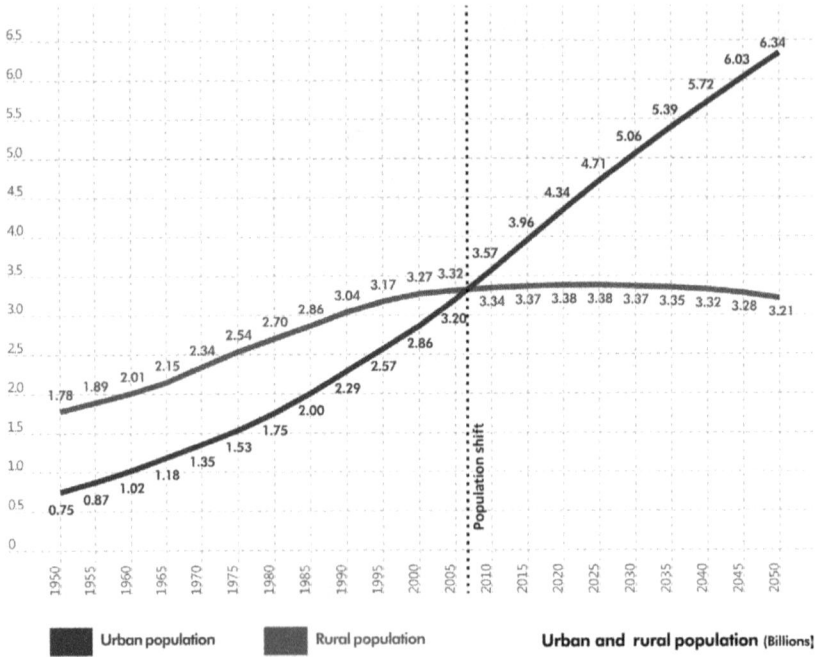

FIGURE 10.2 Urban and rural population worldwide.

Source: United Nations, 2017. World Population Prospects. https://esa.un.org/unpd/wpp/publications/Files/WPP2017_KeyFindings.pdf.)

to the real-time requirements of the building. This reduces the errors, the amount of human effort, and the need for manual monitoring and control of energy consumption. In smart cities, a platform of information and communication technologies (ICT) solutions is deployed to achieve urban sustainability targets.

In 2007, the worldwide urban population exceeded the rural population for the first time in history [2], as can be seen from Figure 10.2. This demographic change, which happened very rapidly, changed the picture that rural populations characterize the history of civilization. All this started in the 1950s, with people migrating from rural to urban areas. The obvious reason for this was the availability of professional, social, economic, and personal growth opportunities the urban environments were offering. In 2014, 54% of the total world population was urban occupants [2]. This global trend is continuing and is even expected to continue. The world population and the number of people living in urban areas are expected to rise. Such an increase in population comes with substantial challenges and difficulties in urban environments, which will soon become incapable of meeting basic needs in a sustainable manner. This demographic change in the scenario has resulted in growing inefficiency levels and calls for an overhaul of urban environments by sustainable planning [3].

New approaches to urban planning by embedding advanced technologies in the infrastructure of the city are required to achieve the objectives of urban sustainability.

The invention of wireless technologies such as Wi-Fi, Bluetooth, mobile data, and radio frequency identification (RFID) has allowed uncountable electronic devices to be connected to the Internet.

> By the dawn of the twenty-first century, though, inexpensive, ubiquitous wireless connections were linking whole new classes of things into networks. Very tiny things, very numerous things, very isolated things, highly mobile things, things deeply embedded in other things and things that were jammed into tight and inaccessible places [4].

In the coming few decades, the Internet will be populated with a large amount of real-time data from network sensors embedded in the built environment [5]. In terms of technological revolution, IoT is amongst the most beneficial innovations since the invention of the Internet; a large number of domains have already benefitted by using IoT.

10.3 APPLICATIONS OF AI AND IOT IN GREEN AND SMART SUSTAINABLE CITIES

AI has a vast application area, ranging from research on a microscopic level about microbes, bacteria, and viruses to the macroscopic level about space technology, defense systems, and complex areas such as weather forecasting. With the increasing use of embedded devices driven by AI and IoT, it is set to become the ultimate driver of sustainability. Allowing us to offload repetitive tasks, AI gets the job done quickly and efficiently than humans [6].

In AI, many methods such as machine learning (ML), deep learning (DL), and artificial neural networks (ANN) are used. DL allows the input data to go to the learning algorithm without extracting the featured data [7], while ANN works according to the weights assigned to its elements, which are adjusted automatically and continuously during training according to a prespecified learning rule until the desired task can be performed successfully [8]. ANN is well suited to perform pattern recognition (PR) to identify, and classify objects and signals in vision, speech, and control systems. IoT is called the nervous system of a smart city [9] because it is used for the collection of immense data required for the proper functioning of the AI systems in smart cities. In addition to providing real-time data, IoT can also provide usage trends over months or years. Different protocols like MQTT (Message Queuing Telemetry Transport) components or RESTful (representational state transfer) interfaces are used by IoT to establish communication.

Databases and file systems such as XML (Extensible Markup Language) files in HDFS (Hadoop distributed file system), RDBMS (relational database management system), and NoSQL (Not only SQL), according to the type of IoT data, such as structured, semi-structured, and unstructured data, can be used for storing the big data [10]. The most-used ML models are Gaussian process regression (GPR), which is a kernel model using a non-parametric probabilistic approach and support vector machines (SVM), which is computationally demanding classification algorithm [11].

AI can be used in green smart cities in the domains of smart transportation, smart water management, smart energy management, smart pollution control, smart buildings, and smart public services, etc. Some of the domains of smart cities are shown in Figure 10.3, while the application areas and the technology they use are listed in Table 10.1. These domains and their respective application areas that can benefit from AI and IoT in smart and sustainable cities are discussed in the following

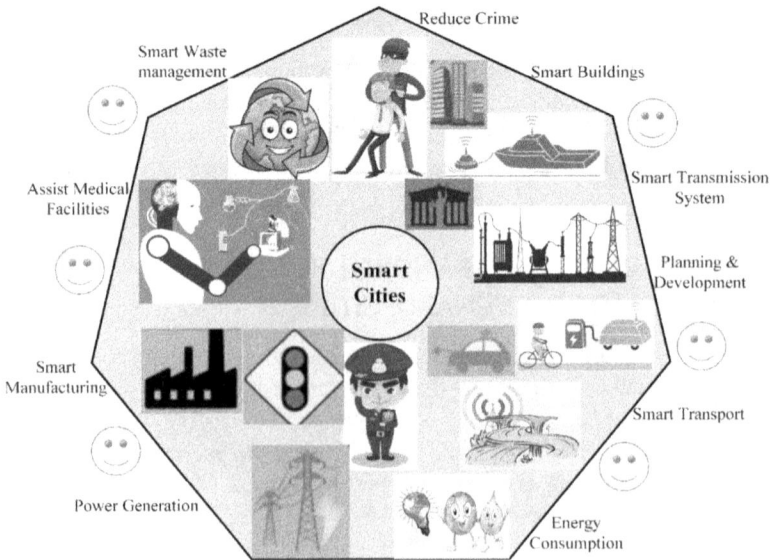

FIGURE 10.3   Smart cities: an idea.

TABLE 10.1
Suitable Technology for Smart City Applications

Domain of Smart Cities	Applications	Suitable Technology	
		AI	IOT
Smart transportation	Traffic flow management	x	
	Driver assistance in fog	x	x
	Parking management	x	
Smart water management	Minimizing losses in supply		x
Smart energy management	Renewable energy integration	x	x
	Plug load management		x
Smart pollution control	Air quality monitoring	x	x
	Measuring vehicular emissions		x
Smart buildings	Hvac operation	x	
	Artificial lighting management		x
Smart public services	Citizen safety	x	x
	Lower emergency response time	x	x

TABLE 10.2
AI/ IoT Methods Used for the Analysis of Typical Problems in a Smart City

Name of Authors	Problem Analyzed	Domain of Problem	AI/IoT Method Used
Nisha Pahal, Deepti Goel, and Santanu Chaudhury [12]	Intelligent traffic monitoring	Smart transportation	Dynamic Bayesian Networks (DBN)
Meryeme Boumahdi and Chaker El Amrani [26]	Outdoor air purification	Smart pollution control	Levenberg-Marquardt ANN algorithm
Mohamed El Khaili, Abdelkarim Alloubane, Loubna Terrada, and Azeddine Khiat [44]	Traffic flow management	Smart transportation	Model Predictive Control (MPC)
Naoufal Ainane, Mohamed Ouzzif, and Khalid Bouragba [45]	Data security	Smart public service	Zigbee and Z-Wave
Imane Sahmi, Tomader Mazri, and Nabil Hmina [46]	Threats in iot	Smart Public Service	MQ Telemetry Transport Application Protocol
Aroua Amari, Laila Moussaid, and Saida Tallal [15]	Parking management	Smart transportation	Smartphones as Sensors
Jihane Kartite and Mohamed Cherkaoui [18]	Renewable energy electrification	Smart energy	Backtracking Search Optimization Algorithm (BSA)
Zakaria Boucetta, Abdelaziz El Fazziki, and Mohamed El adnani [39]	Citizen safety	Smart public service	Mobile Crowdsensing
Alae Labrini, Nabila Rabbah, Hicham Belhaddoui, and Mounir Rifi [47]	Smart home control	Smart buildings	State Machine Strategy (SMS) & Equivalent Consumption Minimization Strategy (ECMS)

sections. In the literature survey, it was observed that an immense variety of AI and IoT techniques have been used in applications that can be beneficial for smart cities. Some methods used by authors in their analysis are included in Table 10.2.

10.4 SMART TRANSPORTATION

With the rapid growth of the urban population, as already discussed earlier in this chapter, the vehicles in the urban areas are also increasing exponentially, and as a result, traffic congestion can be seen often on the roads. To tackle this complex problem of managing vehicular traffic on the roads and in helping authorities in proper planning of future road expansion plans, a smart traffic management system using AI and IoT can be deployed in a sustainable smart city. The various applications of AI and IoT in transportation management are discussed in this section.

10.4.1 TRAFFIC FLOW MANAGEMENT

The traffic management system is one of the major fields of a smart city. For optimizing traffic flow, appropriate algorithms are used to manage different traffic conditions. An ontology framework for traffic flow management uses a network of IoT

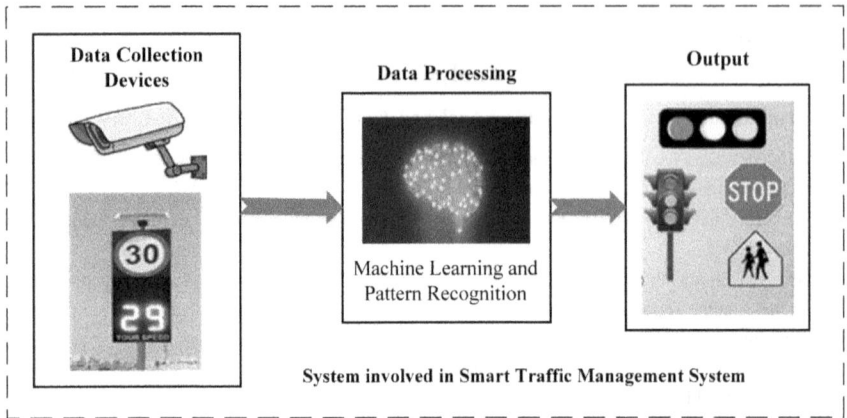

FIGURE 10.4 System model of traffic management system.

embedded sensors and cameras, to produce audio, text, and video outputs to detect traffic density, predict jam conditions, instances of a vehicle collision, and speeding in real-time. Multimedia Web Ontology Language (MOWL), which exploits dynamic Bayesian networks (DBN) for real-time modeling, can be used for this purpose [12]. The continuous monitoring of collected data can provide potential threat alerts to nearby drivers, pedestrians, security agencies, and police on display boards or as text messages and e-mails.

A traffic management system using AI and IoT, a system model of which is shown in Figure 10.4, works by being able to change the signal timing dynamically according to the vehicular density on a particular road. This is done to reduce the average waiting time at the signal, which can be achieved by using SUMO (Simulation of Urban Mobility). It has been observed that dynamic signals reduce the waiting time at signals when compared to fixed time signals [13].

10.4.2 Driver Assistance in Fog

To enhance road safety and help drivers, driver assistance systems (ADAS) embedded with sensors are used. These systems are dependent on image data processing captured by the sensors, so in adverse weather conditions like heavy rain or fog, the functioning of this system is affected because the image is not sharp. To eliminate fog in the captured data, convolutional neuron networks (CNN) and fast R-CNN architecture is used so that nearby traffic lights, vehicles, and pedestrians can be detected by restoring image sharpness. CNN allows the removal of haziness in images for better object detection in varying environmental conditions [14].

10.4.3 Parking Management

With increasing vehicles, it is as important to have proper parking management as part of traffic management. Smart parking management lets people locate and pay for a parking space by using advanced technologies. Priority is given to people with vulnerable health conditions, to provide the closest parking space. For this purpose, IoT sensors and smartphones are used to provide real-time information about parking availability, cost, walking distance, waiting time, and driving time. AI algorithms are used to assess the health condition of drivers before allocating parking space while ensuring efficient management of parking facilities [15].

10.5 SMART WATER MANAGEMENT

Water is a scarce but vital resource for us; even today, a large portion of rural and tribal people do not have access to potable water or have to travel several kilometers daily to access water for their daily needs. Every effort possible should be done for conserving this vital resource. This section discusses the areas of water management that can benefit from using IoT and AI.

10.5.1 MINIMIZING LOSSES IN WATER SUPPLY

Deploying a water management system that reduces human intervention and is operable in both rural and urban conditions, keeping in view the sustainability factor, can help overcome the community hardship where water is yet not accessible. Recent work on smart water management systems include investigations into a sustainable water distribution strategy with smart water grid [16], explaining possible methods to incorporate AI and IoT solutions to enhance efficiency of the system and to bridge the gap between different types of water resources. A representation of a smart water management system is shown in Figure 10.5. The water from rainwater harvesting systems, treated gray water, and different water sources are used as integrated water sources for the system for an optimized and sustainable environment [17].

Smart meters based on IoT for water supplies have significant advantages because they provide information about leaks and other kinds of pipeline failures well in advance, thus preventing not only water wastage but also potential structural damage. Moreover, smart meters show spikes in usage, alerting about required future upgrades and potential outages. For landscape irrigation and farming, ground

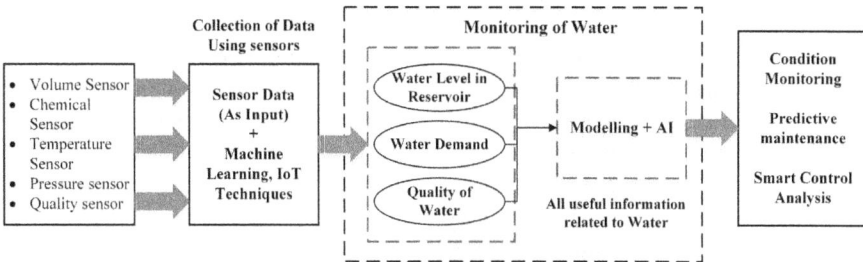

FIGURE 10.5 Smart water management system.

humidity sensors and environmental data, such as the rain forecast, are used to know when and where to irrigate with the right amount of water. Also, sensor-based and low-flow taps with aerators have proven to save a substantial amount of water for similar usage compared with traditional taps; moreover, sensor-based taps provide a no-touch operation, thus maintaining the highest possible standards of hygiene.

10.6   SMART ENERGY MANAGEMENT

Energy is the thing that ultimately drives everything around us but also significantly contribute to the emissions. So there is a need to manage energy consumption and production and also to increase use of renewable energy sources. This section discusses the areas of energy management that can benefit from using IoT and AI.

10.6.1   RENEWABLE ENERGY INTEGRATION

Renewable energy, despite its random fluctuations, can be used to power remote areas with decentralized production and meet their energy demand. This also reduces production costs, along with a reduction in the emission of greenhouse gases. This concept of integrating renewable energy with the traditional grid is called a smart grid that follows a hybrid renewable energy system (HRES) approach. This approach uses a hybrid solar and wind energy generation system with battery storage. Two optimization algorithms, backtracking search optimization (BSA), and improved backtracking search optimization (IBSA) can be used to predict the power generation required from different renewable energy sources to meet the energy demands of a location. The simulation using these algorithms yields the capacity of solar PV, wind-powered generation turbines, and battery storage for the desired hybrid system [18].

10.6.2   PLUG LOAD MANAGEMENT

Plug loads comprise a substantial amount of energy consumption, and according to 2010 annual energy forecasts by information administration, energy consumed by plug loads can see a rise of 36% to 65% by 2030 [19]. The guide to good practices in operating ICT devices [20] and actionable feedback from occupants [21] have shown the immense energy-saving potential of over 130,000 kWh and 311 kWh annually respectively. The plug loads can be divided into two types of loads: first, loads that can be shifted, such as usage of printers, external drives, document scanners, etc., which can be used in off-peak hours; second, all other plug loads, such as usage of personal computers, laptops, mobile chargers, that is, non-shiftable loads. Non-shiftable loads are less flexible, but their energy consumption can be regulated by effective control. Plug load management using a graded solution to weigh the financial feasibility and operation of equipment can be done [22]. Time series subsequence mining can be used to gather information about appliance operational states to be able to effectively characterize plug load appliances [23]. The overall savings with respect to the cost and energy of plug loads can be as high as 21%. AI can improve this by learning consumer behavior in terms of plug load operations [24].

10.7 SMART POLLUTION CONTROL

Pollution is an unwanted by-product of industrialization; the growing number of industries and vehicles have contributed a lot to polluting the environment we live in with harmful and toxic chemicals. In order to be able to limit the concentrations of these pollutants to a safe level, we must be able to measure their concentration on a large scale. This section discusses the areas of pollution control that can benefit from using IoT and AI.

10.7.1 AIR QUALITY MONITORING

For something as important as air, monitoring is absolutely required, because according to WHO, air pollution and poor air quality cause 5.5 million unnecessary deaths in a year. By using the power of AI and IoT, real-time air quality data such as the presence and concentration of NO_x (oxides of nitrogen), CO_2 (carbon dioxide) and black carbon can be recorded. This is achieved by using the blanket of IoT sensor devices to gather information about changes in the air quality over time, areas with high concentrations of pollutants, and insights to their possible source. Black carbon is the most significant contributor to air pollution in urban areas. It is known to cause respiratory and cardio-vascular problems as well as possible congenital disabilities. This air quality monitoring is necessary to pursue better public health. This can also help to identify the possible point sources of pollution, and necessary actions can then be taken to curb the menace at specific point sources. Air quality monitoring (IoT network sensors) and prediction of air quality (AI and ML) are the building blocks of an air quality system [25].

To design a photocatalysis-based air purifier for outdoor streets at low cost, several techniques to optimize the purification scenario can be simulated in COMSOL and Matlab using AI. The best purification model is selected by comparing the results of neural networks and polynomial regressions. Using an early stopping procedure for the Levenberg Marquardt-based ANN is a good approach for designing outdoor air purification systems [26].

10.7.2 MEASURING VEHICULAR EMISSIONS

There are negative impacts of transportation on the environment, which are mostly caused by vehicular emissions resulting from traffic congestion. So the relation between emissions and congestion needs to be studied. This can be done by modeling a road with the Lighthill Whitham Richard (LWR) model for traffic distribution using a Runge-Kutta scheme [27]. It is known that heavy congestion leads to CO_2 emissions. These emissions are also proportional to vehicle speed and fuel consumption. Europe targeted reducing the average CO_2 emissions of all cars sold in 2020 to 95 grams and 75 grams in 2025 from the present value of 140 grams per kilometer [28].

10.8 SMART BUILDINGS

The buildings that use advanced technologies to best utilize the environmental and climatic conditions for their benefit are referred to as smart buildings. These buildings use a network of IoT sensors to collect real-time data to be processed by an

AI-based building management system to maximize energy efficiency while maintaining the comfort of the occupants. This section discusses the areas of smart buildings that can benefit from using IoT and AI.

10.8.1 HVAC OPERATION

HVAC systems are a part of our daily life in modern-day society. Just 1% improvement in the energy efficiency of HVAC systems adds up to millions of dollars in annual savings at the international level [29]. Incorporation of AI techniques such as fuzzy logic and ANN in the design, operation, and maintenance of HVAC systems depicts the trend towards effective, efficient, intelligent, and smart HVAC methodologies. ANN can be used to evaluate the plant dynamics to be able to estimate future plant outputs, plant sensitivity, and other information for neural control adaptation [30]. The nonlinear nature and functional mapping properties of ANN are the reason for their use in control and identification, for which two techniques known as forwarding modeling and inverse modeling are used. The procedure used for training an ANN to represent the forward dynamics of a plant is referred to as the forward system identification approach [31]. This depicts a supervised learning approach, for which the system provides the outputs directly in the output coordinate system of the learner [32]. The entire process takes place in the controller of the HVAC system, parameters of which are then adjusted in order to minimize the cost index. Following the criteria of minimum energy consumption and providing maximum thermal comfort to the occupants, the HVAC system makes the best use of AI technologies. It switches to a low power mode when buildings are not occupied and learns to work on a schedule or learns about the occupancy levels of a building and then operates accordingly.

10.8.2 ARTIFICIAL LIGHTING MANAGEMENT

IoT sensors integrated with the lighting system can provide immense energy savings by turning lights on or off according to the presence of people in the building while enabling them to override the sensors for emergency purposes. The proposed architecture of the IoT sensor network of a smart building is shown in Figure 10.6. To fulfill the demand arising from user preferences about air quality, ambient room lighting, and occupancy and from the point of view of efficient consumption of energy [33], classical Bayesian networks [34] and dynamic Bayesian networks, which are also called Markov chains, can be used to implement this model of lighting control. This allows probabilistic reasoning for dynamic scenarios, which depend not only on the real-time present state but also on past states.

The AI building management system uses sensor information to learn and adapt to the requirements of the user; during the interaction of the user with manual override systems, such as by switching a light on, the user informs the system that present ambient lighting is inadequate. This is called an implicit feedback collection system, more details of which are contained in [35]. This system is implemented using nodes equipped with sensors providing real-time data about ambient light exposure and level of noise deployed at various strategic locations. The energy consumption is

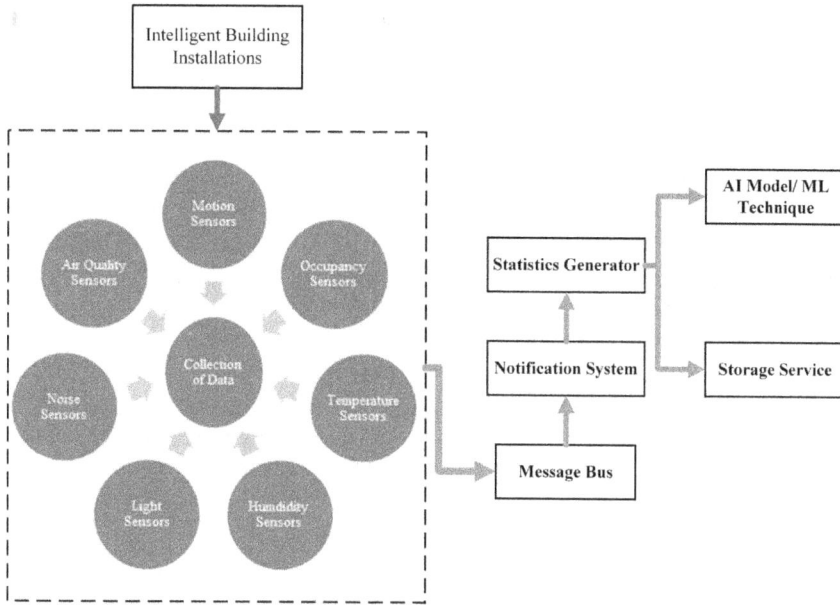

FIGURE 10.6 IoT network for a smart building.

monitored to gather real-time information about active power, voltage, and current. By using this information, the AI system is able to plan its actions to satisfy predefined energy consumption patterns.

Window shades can also contribute to energy efficiency by operating in sync with the indoor lighting requirements and the outdoor lighting conditions, keeping in view the effect of opening or closing of blinds depending on the indoor and outdoor temperature of the building and thus asking for the required change in the HVAC system to keep the building at a maximum thermal comfort level and minimum energy consumption level. Exterior solar panels can be embedded in motorized window shades so that they generate the energy required to open and close them.

10.9   SMART PUBLIC SERVICES

Smart cities succeed in their objective if the citizens living in it feel secure and safe. It will be very much appreciated if the public service delivery is quick and oriented toward well-being of citizens. This section discusses the areas of public services that can benefit from using IoT and AI.

10.9.1   Citizen Safety

Smart monitoring and surveillance systems can help reduce crimes or anticipate them before they happen. Since manual monitoring of video surveillance of entire cites is almost impossible and wastes time and effort, AI and ML come into the picture. This smart monitoring system consists of CCTV cameras providing a video feed to trained ANNs that have capabilities of detecting suspicious activities and alert police or emergency services without manual intervention [36]. A huge network of IoT-enabled interconnected cameras and sensors connected to the cloud is deployed for data collection. This system can be used to monitor vehicle movement, traffic status, hit-and-run accidents, actions of the public, etc., in real-time. This helps in detecting traffic violations such as speeding, which may result in accidents, and can detect people carrying weapons or showing suspicious behavior [37]. This also helps in traffic flow management, as already discussed.

Crowdsensing, a new approach toward citizen safety and public service delivery based on mobile devices, has emerged in recent years. Using crowdsensing, we can detect road quality deterioration, location of potholes, etc., by analyzing GPS data and driver behavior [38]. This data can be sent to the responsible authorities for road maintenance or to traffic departments to penalize drivers with reckless driving behavior. This is a mobile application-based architecture that uses the HyperText Transfer Protocol (HTTP) and JSON for data exchange [39].

10.9.2   REDUCING RESPONSE TIME OF EMERGENCY VEHICLES

In the year 2005, the response time for a life-threatening emergency call was found to be nearly 18 minutes [40]. This is almost twice the value suggested by the national performance standards of the UK government, which specifically indicates that 3 out of 4 life-threatening emergency calls should be attended within 8 minutes [41,42]. This can be improved by using emergency response vehicles equipped with GPS and RFID and connected to the smart traffic management system. For this purpose, real-time routing is done by the system using GPS equipment embedded in the emergency response vehicles. Similar equipment is used to control traffic lights and to collect real-time data about current traffic density and flow volume at intersections and paths the emergency vehicle will be following [43]. The actual routing algorithm decides routes with the help of ANN by estimating the expected arrival time of all the possible routes the emergency vehicles can follow. Real-time data is used to train the ANN, which results in the routing algorithm working even faster. This enhances the safety of emergency response vehicles by reducing emergency response time [44].

10.10   CONCLUSION

Ease of living, comfort, and sustainability are things that have a delicate relationship; if we focus on only one, the others will be lost. Hence, these must be worked on together. This can be achieved only when the stakeholders from multiple domains work on the complex requirements of the design and operation of AI-based smart, green, and intelligent cities right from the planning stage for new cities and work out a plan to embed this system in existing cities. In this chapter, we discussed how AI and IoT can be integrated into various applications of a smart city. We also elicited and discussed some important data about the growing urban population and the effect it has on our environment along with the global commitments to prevent climate change and promote sustainability.

In summary, enabling smart cities with the help of AI and IoT to achieve sustainability targets requires complex systems. These systems are embedded with advanced features capable to interact with many elements. Developing standards for this kind of large-scale system will require a systematic approach. Countries, both developing and developed, are showing increasing interests in smart cities that are also green, energy-efficient, and sustainable in the fight against the current unsustainable urban development paths. This may seem like a troublesome and challenging task, but with the help of technologies advancing every day, it is getting easier than ever, and there is light at the end of the tunnel.

REFERENCES

[1] Sustainable Development Goals, https://sustainabledevelopment.un.org/, last accessed 2020/05/12.
[2] United Nations (2017). World population prospects: The 2017 revision, key findings and advance tables. ESA/P/WP/248. United Nations. https://esa.un.org/unpd/wpp/publications/Files/WPP2017_KeyFindings.pdf, last accessed 2020/05/10.
[3] Untangling Smart Cities, https://doi.org/10.1016/B978-0-12-815477-9.00002-5.
[4] Ericsson (2015). Ericsson mobility report: On the pulse of the networked society. November 2015. Ericsson. www.ericsson.com/assets/local/mobility-report/documents/2015/ericssonmobility-report-nov-2015.pdf, last accessed 2020/04/13.
[5] Mitchell, W. J. (2003). *Me++: The cyborg self and the networked city.* The MIT Press, Cambridge, MA.
[6] Machine Learning, https://in.mathworks.com/solutions/machine-learning.html, last accessed 2020/05/15.
[7] Boton-Fernandez, V., Lozano-Tello, A., Perez Romero, M., & Cadaval, E. R. (2013). Consumer electronics intelligent decision support system for the efficient energy management in household environments. 2013 8th Iberian Conference on Information Systems and Technologies (CISTI), pp. 1–6, 19–22 June.
[8] Neural Network, https://in.mathworks.com/discovery/neural-network.html, last accessed 2020/05/07.
[9] Zou, Y. (2016). The internet of things: Nervous system of the smart city., In S. McClellan et al. (eds.) *Smart cities* (pp. 75–96). Springer International Publishing AG, Cham.
[10] Cai, H., Xu, B., Jiang, L., & Vasilakos, A. V. (2016). IoT-based big data storage systems in cloud computing: Perspectives and challenges. *IEEE Internet of Things Journal*, 1.
[11] Wu, X., Kumar, V., Ross, Q. J., Ghosh, J., Yang, Q., & Motoda, H. (2008). Top 10 algorithms in data mining. *Knowledge and Information Systems*, 14(1), 1–37.
[12] Pahal, N., Goel, D., & Chaudhury, S. (2019). Environment monitoring system for smart cities using ontology. Innovations in Smart Cities Applications. SCA 2018, LNITI, pp. 44–56.
[13] Rida, N., & Hasbi, A. (2019). Dynamic traffic lights control for isolated intersection based wireless sensor network. Innovations in Smart Cities Applications. SCA 2018, LNITI, pp. 1036–1044.
[14] Samir, A., Mohamed, B. A., & Abdelhakim, B. A. (2019). Driver assistance in fog environment based on Convolutional Neural Networks (CNN). Innovations in Smart Cities Applications. SCA 2018, LNITI, pp. 1028–1035.

[15] Amari, A., Moussaid, M., & Tallal, S. (2019). A new service in smart parking management. Innovations in Smart Cities Applications. SCA 2018, LNITI, pp. 1159–1165.

[16] Byeon, S., Choi, G., Maeng, S., & Gourbesville, P. (2015). Sustainable water distribution strategy with smart water grid. *Sustainability*, 7(4), 4240–4259.

[17] Narendran, S., Pradeep, P., & Ramesh, M. V. (2017). An Internet of Things (IoT) based sustainable water management. 2017 IEEE Global Humanitarian Technology Conference (GHTC), San Jose, CA, USA.

[18] Kartite, J., & Cherkaoui, M. (2019). Towards 100% renewable production: Dakhla smart city electrification. Innovations in Smart Cities Applications. SCA 2018, LNITI, pp. 1146–1156.

[19] Boton-Fernandez, V., Lozano-Tello, A., Perez Romero, M., & Cadaval, E. R. (2013). Consumer electronics intelligent decision support system for the efficient energy management in household environments. 8th Iberian Conference on Information Systems and Technologies (CISTI), Lisboa, Portugal, 2013, pp. 1–6.

[20] Kamilaris, A., Ngan, D. T. H., Pantazaras, A., Kalluri, B., Kondepudi, S., & Wai, T. K. (2014). Good practices in the use of ICT equipment for electricity savings at a university campus. 5th International Green Computing Conference (IGCC), Dallas, TX, USA, 2014, pp. 1–11.

[21] Kamilaris, A., Neovino, J., Kondepudi, S., & Kalluri, B. (2015). A case study on the individual energy use of personal computers in an office setting and assessment of various feedback types towards energy savings. *Energy and Buildings*, 104, 73–86.

[22] Vuppalaa, S. K., & Kumar, K. (2013). HPLEMS: Hybrid Plug Load Energy Management Solution. *Energy Procedia*, 42, 133–142.

[23] Kalluri, B., Kamilaris, A., Kondepudi, S., Kua, H. W., & Tham, K. W. (2016). Applicability of using time series subsequences to study office plug load appliances. *Energy and Buildings*, 127, 399–410.

[24] Sharma, P., Reddy, S., Shrivastava, S., & Kumar, R. (2015). Strategic plug load management system for smart buildings with rooftop photovoltaic system. 2015 Annual IEEE India Conference (INDICON), New Delhi, India, 2015, pp. 1–6.

[25] Schürholz, D., Kubler, S., & Zaslavsky, A. (2020). Artificial intelligence-enabled context aware air quality prediction for smart cities. *Journal of Cleaner Production,* 271.

[26] Boumahdi, M., & Amrani, C. E. (2019). Outdoor air purification based on photocatalysis and artificial intelligence techniques. Innovations in Smart Cities Applications. SCA 2018, LNITI, pp. 94–103.

[27] Chergui, S., & Agoujil, S. (2019). Finite differences-Runge Kutta schemes for vehicle occupancy-aggregate emission rate relationship. Innovations in Smart Cities Applications. SCA 2018, LNITI, pp. 1045–1053.

[28] Berdigh, A., Oufaska, K., & Yassini, K. E. (2019). Connected car & CO2 emission overview: Solutions, challenges and opportunities. Innovations in Smart Cities Applications. SCA 2018, LNITI, pp. 1000–1013.

[29] Shoureshi, R. (1993). Intelligent control systems: Are they for real? *ASME. Journal of Dynamic Systems, Measurement, and Control,* 115(2B), 392–401.

[30] Teeter, J., & Mo-Yuen Chow. (1998). Application of functional link neural network to HVAC thermal dynamic system identification. *IEEE Transactions on Industrial Electronics*, 45(1), 170–176.

[31] Hunt, K. J., Sbarbaro, D., Zbikowski, R., & Gawthrop, P. J. (1992). Neural networks for control systems-a survey. *Automatica*, 28, 1083–1112.

[32] Jordan, M. I., & Rumelhart, D. E. (1991). *Forward models: Supervised learning with a distal teacher.* Center for Cognitive Science, Mass. Inst. Technol., Cambridge, MA, Occasional Paper 40.

[33] De Paola, A., Gaglio, S., Re, G. L., & Ortolani, M. (2009). An ambient intelligence architecture for extracting knowledge from distributed sensors. Proceedings of the 2nd International Conference on Interaction Sciences Information Technology, Culture and Human—ICIS '09, Seoul, Korea, 24–26 November 2009.

[34] Cowell, R. (1999). *Probabilistic networks and expert systems*. Springer-Verlag, New York.

[35] De Paola, A., Farruggia, A., Gaglio, S., Lo Re, G., & Ortolani, M. (2009). Exploiting the human factor in a WSN-based system for ambient intelligence. 2009 International Conference on Complex, Intelligent and Software Intensive Systems, Fukuoka, Japan, 2009, pp. 748–753.

[36] Mor, N., Sood, H., & Goyal, T. (2020). A statistical model to prioritize selected Northern-Indian States/UT of India based on accident data. *Journal of Discrete Mathematical Sciences and Cryptography*, 23(1), 305–312.

[37] Mor, N., Sood, H., & Goyal, T. (2020). Application of machine learning technique for prediction of road accidents in Haryana-A novel approach. *Journal of Intelligent & Fuzzy Systems*, (Preprint), 38(5), 6627–6636.

[38] Kumar, A., & Mor, N. (2020). Prediction of accuracy of high-strength concrete using data mining technique: A review. In *Proceedings of International Conference on IoT Inclusive Life (ICIIL 2019)*, NITTTR Chandigarh, India (pp. 259–267). Springer, Singapore.

[39] Boucetta, Z., Fazziki, A. E., & Adnani, M. E. (2019). Crowdsensing based citizen's safety service. Innovations in Smart Cities Applications. SCA 2018, LNITI, pp. 1014–1027.

[40] eHealth IMPACT. (2006). City of Bucharest Ambulance Service, Romania—DISPEC tele triage and dispatch system. empirica Communication and Technology Research, Bonn.

[41] Department of Health, UK. (2008) Improving Ambulance Response Times: High Impact Changes and Response Times Algorithms, report no. 8048.

[42] London Ambulance Service. (2008). www.londonambulance.nhs.uk.

[43] Nandal, M., Mor, N., & Sood, H. (2020). An overview of use of artificial neural network in sustainable transport system. In *Computational methods and data engineering* (pp. 83–91). Springer, Singapore.

[44] Mor, N., Sood, H., & Goyal, T. (2020). A statistical model for prediction of road accidents in the State of Punjab. *Journal of Interdisciplinary Mathematics*, 23(1), 229–236.

[45] Ainane, N., Ouzzif, M., & Bouragba, K. (2019). Data security of smart cities. Innovations in Smart Cities Applications. SCA 2018, LNITI, pp. 702–712.

[46] Sahmi, I., Mazri, T., & Hmina, N. (2019). Security study of different threats in the Internet of Things. Innovations in Smart Cities Applications. SCA 2018, LNITI, pp. 785–791.

[47] Labrini, A., Rabbah, N., Belhaddoui, H., & Rifi, M. (2019). An optimized control method of an energy source renewable with integrated storage source for smart home. Innovations in Smart Cities Applications. SCA 2018, LNITI, pp. 1079–1092.

11 Machine Learning and AI Techniques in Green Cities

S. Dilip Kumar, R. Rajesh, S. Narendiran and M. Balamurugan

CONTENT

11.1 INTRODUCTION

The global urban population is projected to cross 66% or 70% of total population by 2050, according to statistical estimates. This tremendous surge in urbanization would have dramatic impacts on the climate, management, and protection of cities. Many countries have introduced the idea of smart sustainable green cities to efficiently manage the resources and maximize energy usage to handle the meteoric rise in urbanization. A city where both nature and living things are in equilibrium is called a green zone. The idea of the green city offers new solutions to issues posed by the distributed model of urban growth: cities are greener, livable, and less distributed. Green cities refer to three pillars of the theory of sustainability and other related issues like wellness, prosperity, resilience, and greenery. To meet a smart city's requirements, it is highly important to make efficient use of information and communication technologies with some computational techniques to efficiently manage data processing and data communications and effectively execute complex strategies to ensure a smart city's smooth and safe operation. The Internet of Things (IoT) is the most critical and significant component of smart city applications that generate huge amounts of data. In the presence of such large and complex data quantities, it is difficult to determine precisely which actions are most efficient and successful. Big data can be best analyzed using advanced techniques such as artificial intelligence (AI) and machine learning (ML) to arrive at an optimal choice. The preceding techniques consider a long-term target and may result in the best or near-optimal control decisions possible.

The accuracy and precision of these techniques can be further improved by increasing the amount of training data to improve their learning capabilities and hence the automated decision efficiencies. Algorithms for AI and ML have become increasingly an integral part of many industries. They are now finding their way to green technology projects with the goal of automating and progressing local operations and activities in general. A city that is recognized as a green building usually means that it uses some kind of IoT and ML machines to collect data from different points. A green city has different use cases for AI-driven and IoT-enabled technology, from preserving a healthy environment to advancing public transport and safety. A city should prepare for better smart traffic solutions by using AI and machine learning algorithms together with IoT to ensure that inhabitants are as healthy and move as effectively as possible from one point to another. Machine learning gathers data from various points and transmits it all for further application to a central server, and it has to be used to make a city smarter until more data is obtained.

These are the few approaches pertaining to AI/ML techniques used in green cities:

S.No	Approach	Summary
1.	Hybrid fuzzy with deep reinforcement learning	To decrease the computational difficulty of the testing process in green city projects.
2.	Deep reinforcement learning	Effective consumption of the energy storage applications with varying tariffs structures in a sustainable city.
3.	Deep learning neural network	To support the service benefactors in acquiring energy resources from different customers to balance the energy variation and improve smart grid reliability.
4.	Hybrid genetic algorithm and simulated annealing	Support optimal communication and 5G networks.
5.	Q learning with echo state network	Power consumption control in green cities.
6.	Deep learning neural network	Improvement in network coverage in sustainable cities.
7.	Neural network	Drug discovery application.
8.	Representational state transfer with artificial neural network	Remote patients care using tele-monitoring and ontology regulations.
9.	Deep learning neural network	Traffic flow prediction in green cities and traffic clearance.
10.	Random forecast machine learning technique	Improved security in IoT devices and optimized data storage.

In this chapter, utilization of AI and ML in ICT, health care, air quality monitoring, and intelligent transport systems are discussed.

11.2 ICT MODULES

In recent years, it has been very difficult to maintain the balance between the power supply and demand and also the quality of power due to the introduction of distributed and renewable energy sources in the traditional power grid. The drawback of the conventional power grid is its inability to handle bidirectional power flow, and also renewable energy sources such as solar and wind will provide reverse power flow from distributed power generation. In electricity networks, maintenance of power flow is difficult due to the intermittent nature of renewable energy sources [1]. To overcome these issues, the conventional power grid can be transformed into an intelligent grid by incorporating information and communication technologies along with AI and ML techniques.

The smart grid is a digital technology that allows two-way communication between the customers and the utility. The smart grid provides benefits to operators and consumers in the generation, transmission, and distribution sectors [2]. The smart grid provides an opportunity for the energy industry to move into a new era of availability, efficiency, and reliability of electrical power. Some of the benefits of the smart grid are that the security of the grid is improved, transmission of electricity is more efficient, and the peak demand is reduced, which will further reduce the electricity rate to some extent. In addition, even after frequent power failures, the quick restoration of electricity is assured, and there will also be considerable increase in the integration of renewable energy systems on a large scale and a reduction in cost for management and operation for utilities, which ultimately reduces the power cost for consumers. The integration of the power generation system along with the renewable energy system is better compared to conventional grids. Machine learning and artificial intelligence techniques play a vital role in demand management, network stability, network reliability, empowering consumers, and integration of renewable energy resources of the smart grid.

Artificial intelligence and machine learning techniques will enable the smart grid to make decisions based on customer demand, power outages, and any sudden rise or drop in renewable energy output. The most important application of machine learning in the power grid is to forecast energy demand and power generation of intermittent renewable energy sources. In this chapter, the discussion is about how the machine learning algorithms like artificial neural networks and block chain technology will help in demand-side management and network reliability and security, respectively [3].

In demand side management, the objective is to efficiently operate the system by controlling the load on the consumer side. The feature of a smart grid is to provide more emphasis on the management of energy on the consumer side. The major challenge is to implement the electric system with the transformation of large data in the grid. This drawback has been overcome by introducing the artificial intelligent technique called artificial neural network (ANN). The purpose of ANN is to classify the pattern of load curve and also select the most suitable data for consumers to reduce the peak demand and electricity bill [4]. Information and signal processing systems are associated with ANNs that consist of a greater number of processors and are called cells or neurons; they are used in digital circuits or programs to emulate

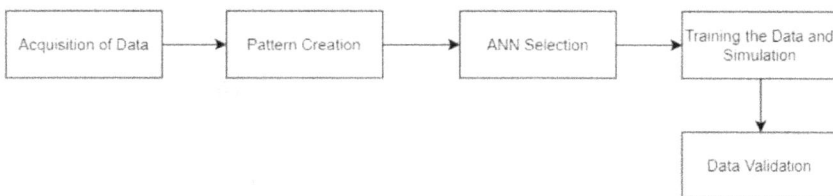

FIGURE 11.1   Classification Stages of Demand Side Management.

the biological nervous system. One more component is called synapses, which have been interconnected through a large number of neurons through direct connections. ANNs are mostly used in systems where there is no accurate measurement or no mathematical system is involved. The features of ANN are stability, fault tolerance, and high robustness, and also, it will handle a huge volume of data associated with data classification [5]. ANN further is recommended for interpolating, analyzing, mathematical modeling, and sorting the data.

At first, the data can be acquired from the energy distribution company and the measurements are recorded every four hours in a day for price calculation. The data obtained from the weekdays will differ from the data obtained during the weekend because of more energy consumption at the weekend. From the processed data, the median curve has been obtained for each consumer.

The data has been obtained from the previous step, and an array of data for simulation software has been created. The simulations are carried out to create different curves by using the k-means method. Based on the features, four different load curves have been created: the type 0 curve shows the energy consumption of the day, and it is constant except for a few hours in a day; the type 1 curve represents peak consumption two or more times in a day; the type 2 curve presents the sharp peak consumption of the day; and the type 3 curve shows the consumption during the night. Based on the type of load curves, the policies of companies have been generated, and the pattern of load curve is given in Figure 11.2. From the observations, it has been proved that the ANN technique is most suitable for demand side management problems in the smart grid.

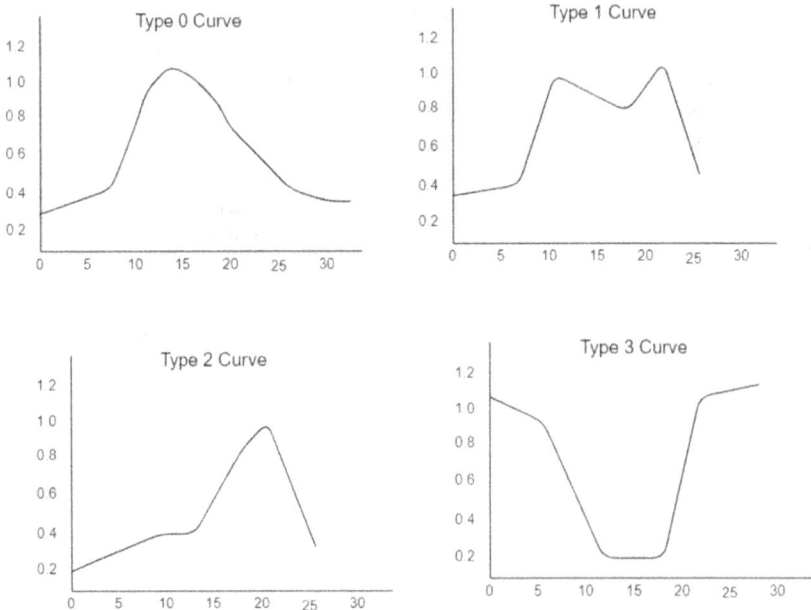

FIGURE 11.2   Pattern of load curve.

In the smart grid, since all the information has been provided online, there is a security threat in handling the data. Therefore, reliability and security are a big challenge in a smart grid. To overcome this problem, blockchain technology is used in a smart grid to avoid the cyber threat as well as electricity theft. The data has been encrypted in blockchain technology with a hash, and also it has been interlinked with previous node data that has been stored in a block, and it will act in the form of a chain. Because of these features, it is very difficult to vary the data in the blockchain. All the transactions that have been done by using this technology is transparent, and it is the most secure technology [6].

11.3   HEALTH CARE

The average age of humans increases day by day. This leads to an increase in the demand of providing and improving the services of health care. The information and communication technology advances in the recent past lead to the development of smart cities and their components [7]. One of the major components in smart city development is smart health/intelligent health care. It is used in improving its field by providing services such as early diagnosis, patient monitoring, etc. Recently, there are many AI techniques that can provide these health care services to society [8]. Figure 11.3 depicts the components of intelligent health care system.

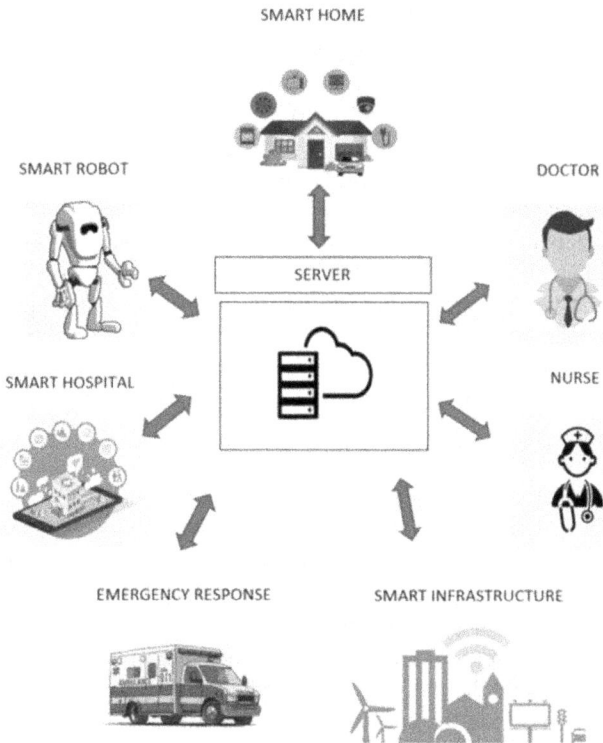

FIGURE 11.3   Components of an intelligent health care system.

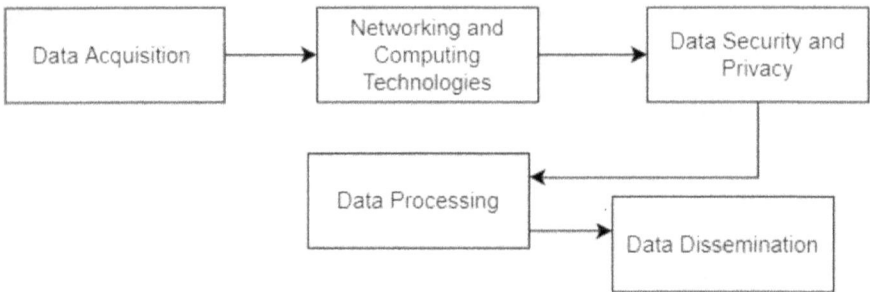

FIGURE 11.4   Pipeline of intelligent health care system.

This intelligent health care system forms the pipeline for data acquisition, net-working, computing techniques, data security and privacy, data processing, and data dissemination. The goal for health care is to become more personal, predic-tive, and participatory, and AI will make major contributions in this field [10], so AI will become a powerful tool for smart health applications. The applications of AI in the field of biomedical research are used for diagnosis, assistance, informa-tion processing, and research. The major area of health care using AI techniques is taking care of elderly and disabled people to improve the quality of their life. The major applications of AI in health care also include health care delivery, changing health care needs, performing existing tasks, process requirements, natural lan-guage processing, image analysis, and predictive analysis. Recently, based on the research available, there are three phases of scaling available for the solutions in health care, namely, addressing low-end priorities, home-based care, and clinical practice from trials in clinics. Therefore, AI is the top of the mind decision-maker in health care.

The particular health care application is used for signal and image processing, prediction of human organ functional changes such as control of the urinary bladder, strokes, etc. The promising areas of AI include self-care, wellness and prevention, triage and diagnosis, diagnostics, clinical decisions, care delivery, and chronic care management [11].

If a human is suffering from neurological disease or spinal cord injury, there will be a failure in the storage and urination function of the bladder. Today, prac-tical restoration of such abnormalities can be done by neuro simulation. In such cases, a new AI method [12] has been developed for sensing both the pressure and fullness of the bladder using neural activities from the regular neural roots of the bladder. The hybrid AI and logistic regression [13] is used to evaluate a glaucoma diagnosis. It uses the data collection process from the hospital and preprocesses it by equalizing the size of the image. It uses the feature extraction by scale-invariant and local binary, so during the end process, it achieves 60.84% accuracy; by using the local binary pattern technique, it achieves 62.67% accuracy; whole image one branch CNN achieves 71.36%; extracted image one branch CNN pattern achieves

74.95%; 2 branch CNN achieves 74.89%; and by using five layers it achieves 81.69% accuracy [14].

11.4 INTENSIFICATION OF AIR QUALITY MONITORING IN GREEN CITIES

The quality of living for people living in a country depends on the health of the environment, which comprises land, water, and air. Air pollution often creates irreversible damage to the human respiratory system and health. Potential sources of air pollutants are from power plants, industry, and automobiles for transportation, which are a part of economic growth in cities. Essential monitoring with an air quality monitoring network (AQMN) is usually a first step in tackling pollution through air [15].

The primary purpose of an AQMN is to differentiate the areas where pollutant levels violate an ambient air quality standard. Thus, the air quality index can be found. Green cities provide environmental benefits and socioeconomic impetus and can support human health. Green infrastructure deployed properly will reduce pollution and can reap benefits [36]. Air pollution is a mixture of nano- to micro-sized particles and can be gaseous pollutants. Particulate matter of various sizes such as PM10, PM2.5, PM1, and ultrafine particles [UFP] should be monitored, and their source should be controlled [16]

Knowing the severe impacts of particulate matter, which hampers human health, spurred policymakers to frame severe restrictions in emission standards and thus paved the way for green infrastructure [17]. World Health Organisation guidelines for quality living are not met by many countries due to the pollutants in the air, especially particulate matters. Green infrastructure plays a crucial role in tackling pollution [37]. The main objective of green infrastructure is not only removing pollutants but also controlling their distribution. Urban vegetation helps in removing a small percentage of emissions by deposition. Due to the presence of the wind, if there is no green infrastructure, pollutants will travel to a greater extent and create havoc, but with urban vegetation, emissions are diluted by the way of dispersion. Green infrastructure can be created by identifying types of street canyons. In street canyons where there is less traffic, air quality at street level will be higher than the surrounding buildings. Trees that are abundant in the area can provide a green corridor. In street canyons with moderate or heavy traffic, air quality at street level is worse than the above surrounding buildings [18]. The addition of green open space to one side is always beneficial. Apart from this, AI-based system for air quality prediction paves the way for proactive behavior in monitoring and control of air quality in a city [19]. By acquiring the information from sources, a prediction algorithm can be formulated; thus, the user's health profile can be monitored and the quality index can be traced.

The proposed context-aware AQ prediction system consists of three blocks as depicted in Figure 11.5, namely context modeling, situation reasoning, and prediction model. Context modeling: [n context modeling, a mathematical model will be formulated that should be a robust and simple tool to realize real-time data. All the variable parameters are to be taken into account in the model. These data can be

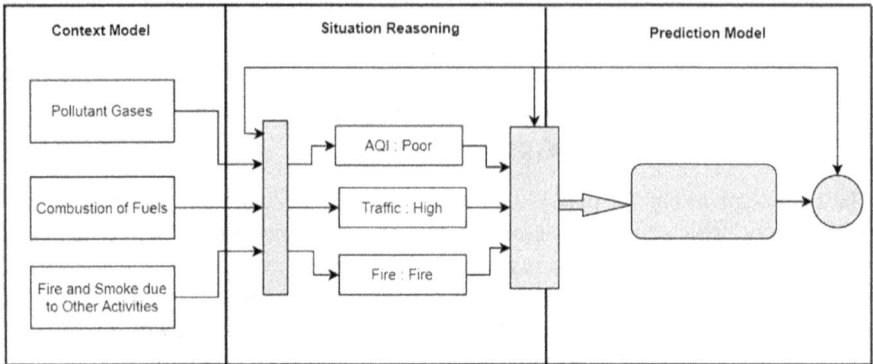

Context Model	Situation Reasoning	Prediction Model
Pollutant Gases	AQI : Poor	
Combustion of Fuels	Traffic : High	
Fire and Smoke due to Other Activities	Fire : Fire	

FIGURE 11.5   Building block of the context-aware AQ prediction system.

obtained from sensors and allied sources that are normally termed as context attributes [20]. Attributes are AQ attributes will involve gases, particulate matters (PM), meteorological attributes, extended external attributes, and user attributes such as geolocation, timestamps, and pollutant sensitivity. Situation reasoning: following the context modeling approach, situation spaces must be clearly defined. Situations can be designated "n" numbers, for example, based on AQI, based on traffic density, or fire incidents and other activities. So in the situation reasoning stage, based on the prevailing conditions, a necessary control or alert is to be made. This is done in the next section. Prediction: having a prediction model that consists of AQ and extended context attributes to predict the situation requires a suitable algorithm [21]. For prediction, ANN and DNN techniques can be adopted; based on the data set, the prediction model can be formulated by deep neural network technique. Providing accurate and customized air quality predictions is a highly critical task. Having a DNN prediction model paves the way by providing the solution to the monitoring stations, and AQI can be maintained at the limits by enforcing control actions from the source of pollutants.

11.5   INTELLIGENT TRANSPORTATION SYSTEM

In the fastest-growing countries, the generation of greenhouse gases is the biggest problem. The Indian transport sector is the third most greenhouse gas emitting sector, particularly from road transportation. According to a WHO survey of 1,650 world cities, the air quality in Delhi (the capital of India) is the worst of any major city in the world. India has the world's highest death rate from chronic respiratory diseases and asthma, according to the WHO, and it is assessed to kill about 2 million people every year. In the current world situation, the energy consumption rate and cost of energy have increased exponentially due to the increase in the population rate. In 2012, the supply of 13.371 million tonnes of Greenhouse gases is more than 200% of the supply from 1973, the year of the first oil crises [22]. To move towards sustainable development and a green city, energy consumption and greenhouse gas generation control are one of the major objectives. By employing a multi-passenger sharing system, an on-demand transportation system using some recent technologies like IoT, AI, ML, and DS will be able to achieve a smart sustainable green environment.

The tag "intelligent transport system (ITS)" is used to term the combination of control and ICT with transport infrastructure. ITS incorporates all methods of transport, traffic flow, the possible interaction of all constitutions of the transport system, safety, emission control, and energy consumption to produce valuable and optimal information [23]. The ITS idea was proposed by scholars in the United States in the 20th century [24]. After the introduction of AI and DS into the ITS, it has now attracted an enormous deal of attention from academia and industry because such systems not only progress city traffic condition but also improve safety, efficient travel, passenger inconvenience due to waiting, energy consumption, prediction of travel patterns and reduction of greenhouse gas generation. In the real-time traffic problem analysis or by using a traditional traffic system, it is very difficult to fill the uncertainties and gaps within the data sets. Therefore, the transport system uses AI to analyze the uncertainties between the cause and effect of diverse real-time scenarios by merging the accessible data with assumptions and probabilities.

The combination of AI with the traffic system is very effective in the green cities and developing countries because the different traffic patterns, energy consumptions, and levels of generation of greenhouse gas are completely different from the normal cases, and they cannot be met by the traditional traffic system.

Traffic flow prediction with AI in modern transportation is important for optimal path planning, and it helps travelers make a better choice of route; reduces energy consumption; predicts where and when congestion will occur, which is a great help for precautionary action; and avoids risk, generation of greenhouse gases, and travel delays. Usually, prediction about future conditions is based on prior knowledge and with a set of data, but in the smart sustainable cities, data collection, transmission, storage, and mining are the huge task. Small errors in the data collection will reflect on the predicted information.

The recent development in the hybrid big data and artificial intelligence [25] brings extraordinary chances for attaining estimates with the highest precision. The shallow structured model has been used in many of the situations with small sample data [26], but it has limitations to meet the huge data and efficiently handling historical data. But the deep learning-based neural network can handle complex applications and large data sets such as images, languages, video, and audio [27]. Recently, a significant amount of proof reveals that the deep learning-based algorithms are a capable tool for traffic flow prediction and optimal path planning compared to fuzzy logic and genetic algorithm because it is a difficult task in space and time [28–29]. The overview of the traffic prediction system with a neural network is given in Figure 11.6. In the AI-based optimal flow prediction, the input is taken from many resources such as historical data, social media, and real-time feeds (road segments, speed, traffic flow) to predict how traffic conditions evolve in the next minute, the next few hours, and the future. These inputs feed into the artificial deep learning-based traffic prediction algorithm to predict and identify the traffic flow patterns, optimal path, and blockage of roads. Also, it can analyze what-if scenarios through simulation to understand the effect of random instances in the probable traffic situations, and so more informed conclusions will be taken in case of real occurrence. These decisions may include adjusting the signal timing and message signs as well as closing some roads. These predictive analyses and optimal decisions save energy consumption in urban development and green cities. Figure 11.6 depicts the schematic flow diagram of traffic flow prediction.

Today, most of the travelers depend on the private vehicle, which is highly reliable, liberating, and more convenient compared to the public transport system. But an increase in the usage of the private vehicle develops many challenges to the government for sustainable development during population growth such as the production of CO_2, the need for the extension of the road to accommodate the increased number of vehicles,

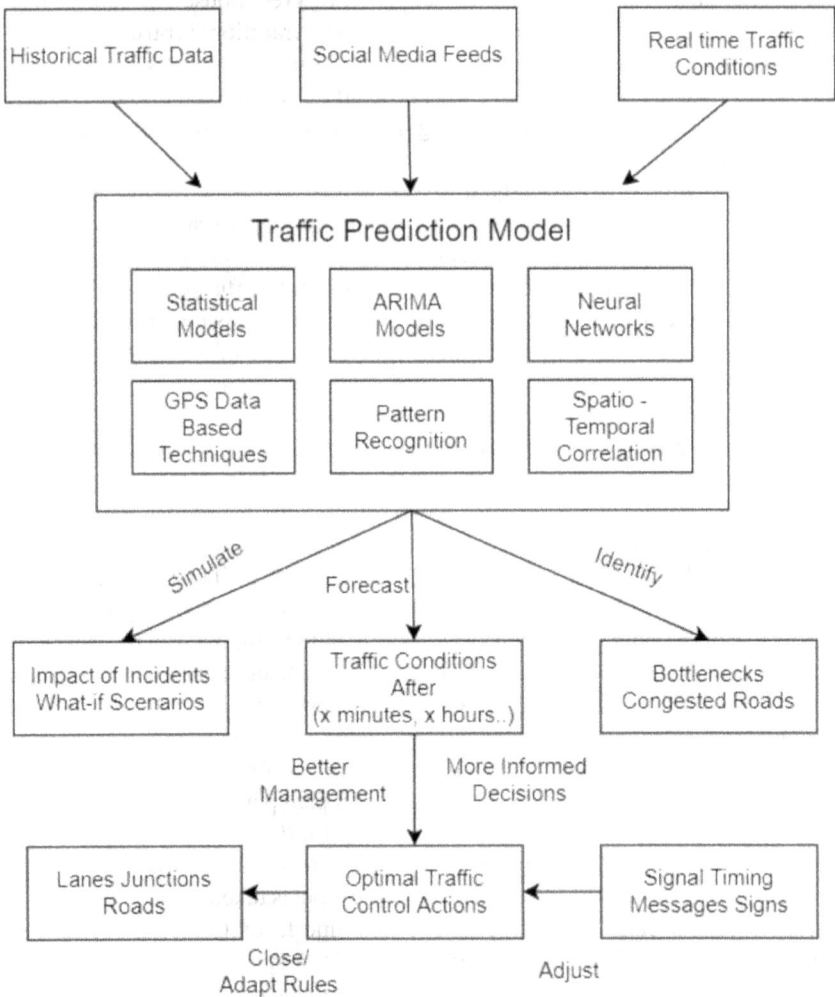

FIGURE 11.6   Schematic flow diagram of traffic flow prediction.

energy consumption, traffic congestion, etc. This increase in traffic population destroys the livelihood of urban cities and creates health issues. In Melbourne, the metropolitan traffic bottleneck was calculated to cost around $3 billion in the year 2015, which was expected to reach its double in 2020 [30]. Currently, an AI and IoT enabled on-demand multi-passenger systems [31] has been proposed to replace the traditional public transport system to meet current population growth transportation needs.

The traditional passenger transport system runs on predefined schedules, routes, and timing, but an on-demand approach is more convenient, like private vehicles, because it does not have prescheduled times; instead it works through apps [32]. Depending on the passenger's needs and availability, these vehicles work at a lower cost and provide better quality than privately owned vehicles. This AI inbuilt multi-passenger system even provides better-optimized routes, trying to meet all the passengers' needs and ensure vehicle availability with historical data and predicted results [33]. This multi-passenger system also developed in the taxi and auto-rickshaw to share the vehicle with the other passengers on the same route (carpooling) [34], which reduces the cost of traveling and is particularly healthier in the green cities to achieve energy consumption goals. The AI-based traffic system also ensures safety issues of passengers [35–36].

11.6   SUMMARY

The need for green energy use plays a crucial role in the coming days as we began developing smart cities. The role of machine learning and artificial intelligence is enormous and has been addressed in detail, including the successful use of ICT, health care, air quality management, and transport systems. We conclude that these strategies would be useful in the realization of green cities in the days to come.

REFERENCES

[1]   Azad, Salahuddin, Sabrina, Fariza and Wasimi, Saleh, 2019. Transformation of smart grid using machine learning. IEEE 29th Australasian Universities Power Engineering Conference (AUPEC), Nadi, Fiji.
[2]   Fang, X., Misra, S., Xue, G. and Yang, D., 2011. Smart grid: The new and improved power grid: A survey. *IEEE Communications Surveys & Tutorials*, 14(4), pp. 944–980.
[3]   Machine learning: The power and promise of computers that learn by example, 2017. *The Royal Society*.
[4]   Sun, M., Konstantelos, I. and Strbac, G., 2008. A deep learning-based feature extraction framework for system security assessment. *IEEE Transactions on Smart Grid*. doi:10.1109/TSG.2018.2873001
[5]   Macedo, M.N.Q., Galo, J.J.M., de Almeida, L.A.L. and de Lima, A.C., 2019. Demand side management using artificial neural networks in a smart grid environment. *Renewable and Sustainable Energy Reviews*, 41, pp. 128–133.
[6]   Xu, Yueqiang, Ahokangas, Petri, Louis, Jean-Nicolas and Pongracz, Eva, 2019. Electricity market empowered by artificial intelligence: A platform approach. *Energies*, 12, pp. 1–21.
[7]   Rayan, Zeina, Alfonse, Marco and Salem, Abdel-Badeeh M., 2019. Machine learning approaches in smart health. *Procedia Computer Science*, 154, pp. 361–368.

[8] Rong, Guoguang, Mendez, Arnaldo, Assi, Elie Bou, Zhao, Bo and Sawan, Mohamad, 2020. Artificial intelligence in healthcare: Review and prediction case studies. *Engineering*, 6(3), pp. 291–301.

[9] Jiang, F., Jiang, Y., Zhi, H., et al., 2017. Artificial intelligence in healthcare: Past, present and future. *Stroke Vasc Neurol*, 2(4), pp. 230–243.

[10] Knickerbocker, J.U., et al., 2018. Heterogeneous integration technology demonstrations for future healthcare, IoT, and AI computing solutions. 2018 IEEE 68th Electronic Components and Technology Conference (ECTC), San Diego, CA, pp. 1519–1528.

[11] Bennett, K.P., 2019. Artificial intelligence for public health. 2019 IEEE International Conference on Bioinformatics and Biomedicine (BIBM), San Diego, CA, USA, pp. 1–2.

[12] Xu, S., Hu, C. and Min, D., 2019. Preparing for the Ai Era under the digital health framework. 2019 ITU Kaleidoscope: ICT for Health: Networks, Standards and Innovation (ITU K), Atlanta, GA, USA, pp. 1–10.

[13] Kumar, J. Naveen Ananda and Suresh, S., 2019. A proposal of smart hospital management using hybrid cloud, IoT, ML, and AI. 2019 International Conference on Communication and Electronics Systems (ICCES), Coimbatore, India, pp. 1082–1085.

[14] Qadri, A. Nauman, Zikria, Y.B., Vasilakos, A.V. and Kim, S.W., 2020. The future of healthcare internet of things: A survey of emerging technologies. *IEEE Communications Surveys & Tutorials*, 22(2), pp. 1121–1167, Secondquarter.

[15] Shareef, Mohammed Mujtaba, Husain, Tahir and Alharbi, Badr, 2016. Optimal air quality monitoring network for green cities. *International Journal of Environment and Sustainability [IJES]*, 5(2), pp. 72–88, ISSN 1927–9566.

[16] Kumar, Prashant, Druckman, Angela, Gallagher, John, Gatersleben, Birgitta, Allison, Sarah, Eisenman, Theodore S., Hoang, Uy, Hama, Sarkawt, Tiwari, Arvind, Sharma, Ashish, Abhijith, K.V., Adlakha, Deepti, McNabola, Aonghus, Astell-Burt, Thomas, Feng, Xiaoq, Skeldon, Anne C., de Lusignan, Simon and Morawska, Lidia, 2019. *Environment International*, 133(Part A).

[17] Artmann, Martina, Kohler, Manon, Meinel, Gotthard, Gan, Jing and Ioja, Ioan-Cristian, 2019. How smart growth and green infrastructure can mutually support each other: A conceptual framework for compact and green cities. *Ecological Indicators*, 96(Part 2).

[18] Yang, Jun, Yu, Qian and Gong, Peng, 2008. Quantifying air pollution removal by green roofs in Chicago. *Atmospheric Environment*, 42(31), pp. 7266–7273.

[19] Hewitt, C. Nick, Ashworth, Kirsti and MacKenzie, A. Rob, 2020. *Using green infrastructure to improve urban air quality (GI4AQ)*. Springer, New York, pp. 62–73.

[20] Bottalico, Francesca, Chirici, Gherardo, Giannetti, Francesca, De Marco, Alessandra, Nocentini, Susanna, Paoletti, Elena, Salbitano, Fabio, Sanesi, Giovanni, Serenelli, Chiara and Travaglini, Davide, 2015. Air pollution removal by green infrastructures and urban forests in the city of Florence-Florence "sustainability of well-being international forum". 2015: Food for Sustainability and Not Just Food, Florence SWIF.

[21] Schurholz, Daniel, Kubler, Sylvain and Zaslavsky, Arkady, 2020. Artificial intelligence-enabled context-aware air quality prediction for smart cities. *Journal of Cleaner Production,* 271.

[22] IEA—International Energy Agency, 2015. www.iea.org/publications/freepublications/publication/KeyWorld2014.pdf. Accessed 17 August 2015.

[23] Barbaresso, J., Cordahi, G., Garcia, D., Hill, C., Jendzejec, A., Wright, K. and Hamilton, B.A., 2014. USDOT's Intelligent Transportation Systems (ITS) ITS strategic plan, 2015–2019 (No. FHWA-JPO-14-145). United States. Department of Transportation, Intelligent Transportation Systems Joint Program Office.

[24] Da Rasa, A.V., 2013. *Fundamentals of renewable energy processes*. Third edition. Academic Press, Oxford.

[25] Ran, B., Jin, P.J., Boyce, D., Qiu, T.Z. and Cheng, Y., 2012. Perspectives on future transportation research: Impact of intelligent transportation system technologies on next-generation transportation modeling. *Journal of Intelligent Transportation Systems*, 16(4), pp. 226–242.

[26] Krizhevsky, A., Sutskever, I. and Hinton, G.E., 2012. Imagenet classification with deep convolutional neural networks. In *Advances in neural information processing systems*. NIPS'12: Proceedings of the 25th International Conference on Neural Information Processing Systems – Volume 1, December 2012, pp. 1097–1105.

[27] Ran, B., Jin, P.J., Boyce, D., Qiu, T.Z. and Cheng, Y., 2012. Perspectives on future transportation research: Impact of intelligent transportation system technologies on next-generation transportation modeling. *Journal of Intelligent Transportation Systems*, 16(4), pp. 226–242.

[28] Dairi, A., Harrou, F., Sun, Y. and Senouci, M., 2018. Obstacle detection for intelligent transportation systems using deep stacked autoencoder and $ k $-nearest neighbor scheme. *IEEE Sensors Journal*, 18(12), pp. 5122–5132.

[29] Taylor, M.A., Zito, R., Smith, N. and D'este, G.M., 2005. Modelling the impacts of transport policies to reduce greenhouse gas emissions from urban freight transport in Sydney. *Journal of the Eastern Asia Society for Transportation Studies*, 6, pp. 3135–3150.

[30] Liyanage, S., Dia, H., Abduljabbar, R. and Bagloee, S.A., 2019. Flexible mobility on-demand: An environmental scan. *Sustainability*, 11(5), p. 1262.

[31] Kaufman, R., 2016. Chasing the next uber: Next city: Which city will be the first to crack on-demand mobility?

[32] Lucic, P. and Teodorovic, D., 2002, November. Transportation modeling: An artificial life approach. 14th IEEE International Conference on Tools with Artificial Intelligence, 2002. (ICTAI 2002). Proceedings, pp. 216–223, IEEE.

[33] Atasoy, B., Ikeda, T., Song, X. and Ben-Akiva, M.E., 2015. The concept and impact analysis of a flexible mobility on demand system. *Transportation Research Part C: Emerging Technologies*, 56, pp. 373–392.

[34] Faghri, A. and Hua, J., 1992. Evaluation of artificial neural network applications in transportation engineering. *Transportation Research Record*, 1358, p. 71.

[35] Saravanan, K., Golden Julie, E. and Herold Robinson, Y., 2019 Smart cities & IoT: Evolution of applications, architectures & technologies, present scenarios & future dream, for the upcoming book series. *Intelligent Systems Reference Library*, 154, pp. 131–151.

[36] Saravanan, K. and Srinivasan, P., 2018. Examining IoT's applications using cloud services. In P. Tomar and G. Kaur (Eds.), *Examining cloud computing technologies through the internet of things* (pp. 147–163).

12 Machine Learning and Artificial Intelligence Techniques in Smart Health Care Systems

K. Padmavathi

CONTENT

12.1 INTRODUCTION

Health care is important in human life, that impacting aspects of quality of life like physical, social, and mental health status. Health care is used to enhance quality of human life by enhancing health. Health care is of great significance for public health. Healthcare can be defined as the diagnosis, treatment, prevention, and management of disease and the preservation of the physical and mental health of humans. Health care can use various technologies that focus and help to improve the disease diagnosis and increase the survival rate of a human's life. Computer science and their related technologies play an important role in the health care system and also increase the influence of its practical applications in the medical field. In health care, dangerous diseases can be diagnosed in their initial stage by using technology. The development of computer science and associated technologies has helped to increase the survival rate of humans. Hospitals, clinics and health care medical centers are using these technology developments to save lives. The invention of new technologies has helped to diagnosis the various diseases and mental disorders of humans in early stage. Doctors can use them to give the best treatment to the patients at the right time using the technology. Technological development of health care systems has helped to improve patients' health, providing positive effects to society. Computer science and its modern technologies are used to give treatment to patients effectively. Common uses of these technologies in health care are:

- Patient diagnosis: Computers are used to manage and maintain patient records, which can help to make an accurate diagnosis.
- Medication and treatment: Technologies are used to maintain patient records properly, which is used to prevent and reduce human error. These can be used to retrieve the patient information at anytime.
- Surgical procedures: During complex operations, doctors can use computer assistance. Computers can create the simulation model of the surgical procedure that is used by surgeons during surgery.
- Information sharing: Technologies can help to share information between doctors and patients. Patients can discuss their issues and get advice via computers.

In recent decades, the interrelationship between health care and technology has been developed, and modern health care uses the technology advancements such as artificial intelligence, biomedical, biotechnology, etc., in various fields of medicine. In health care, modern technologies are used to take decisions and have increased the importance of patient care. Health care systems use different technologies and methods such as descriptive, perspective analysis to measure the disease level. These are also used to store and dispose of patient information. The modern technologies can be used to measure the quality of health care and provide reliable information to health care services [5].

Modern computer science technologies like artificial intelligence and machine learning provide advancement opportunities that are used to help identify dangerous diseases of human at an early stage. A health care system generates, maintains, and stores a large amount of data for analyzing and identifying diseases. Recent computer science technologies like artificial intelligence and machine learning are used to extract insights from data that can help to analyze and identify the level or depth of diseases. These technologies also play a role in health promotion, interventions, and recommendations. The evolution of these technologies can increase public health's capacity and improve the health of the population. The interdependent relationship between these technologies and health care can be used to promote the development of reliable health care and to increase the accuracy of the health care system [6].

12.2 EVOLUTION OF TECHNOLOGIES IN SMART HEALTH CARE

The evolution of technology in health care systems improves the quality of the results. The digitization of health care uses various new technologies to assist the doctors, patients, etc. Modern technology plays an important role in improving and saving human life all over the world. Heath care is a broad field that uses modern technology in sustaining health. The significant contributions of technology in various fields such as bioinformatics and biotechnology improve the health of people all over the world. Technology developments have a positive impact on health care. In health care systems, the innovations of new ideas using modern technology improve disease diagnosis, surgical methods, and patient health care and also give better results.

Modern technology has introduced new tools for private practices of doctors and hospitals to connect doctors and patients through telecommunications. This can be used to save time and money for patients. Patients can get health information from any specialist or any geographic location. Doctors and physicians can access any type of information related to their studies and patients' information by using these tools. The new scanning devices and better monitoring systems minimize the recovery time of patients and improve the quality of life. The development of modern technology helps scientists and doctors to examine cellular-level diseases and produce vaccines against them. These vaccines are used to prevent the spread of diseases and save human lives all around the world. Due to the development of modern technology, the concept of smart health care is gradually growing. Modern technological innovations such as artificial intelligence, machine learning, and big data analytics have turned the traditional health care systems into smart health care systems. Smart health care enhances the diagnosis process, improves the treatment level, and improves quality of life.

Generally, smart health care includes adopting new technologies for disease diagnosis and also exchanging the information among doctors, giving better management of medical data. The smart health care system uses smart health technologies for storing patients' health information, computing this information, and delivering personalized advice to patients. Smart health care uses the new technological developments such as artificial intelligence, machine learning, big data, and cloud computing. These developments are used to transform traditional health care into more efficient and convenient health care. The evolution of smart health care creates multilevel changes in the medical field. These changes are embodied in the following aspects:

- Patient care.
- Common information construction.
- General medical management.
- Preventive health care.

These changes focus on improving the efficiency of health care and enhancing health care services.

Smart health care is a health service system that uses modern technologies to access information, manage health information, and respond to the medical information. Smart health care can

- Improve communication between all people.
- Help to make decisions.
- Facilitate the collection of resources.
- Increase the level of services.

From a patient's perspective, modern technologies monitor patients' health at all times, provide medical assistance, and implement remote services to patients at all levels. From a doctor's perspective, modern technologies are used to make intelligence decisions and improve diagnosis. The current trends in health care systems propose new directions to explore various paradigms.

12.3   IMPORTANCE OF AI AND ML IN SMART HEALTHCARE

Artificial intelligence offers traditional and modern techniques to make decisions in clinics. Artificial intelligence uses learning algorithms that provide more precise and accurate results interacting with a training data set. Learning algorithms allow gaining unprecedented insights into disease diagnostics, patient care, and outcomes. Artificial intelligence is a collection technology that has important relevance in the health care field. Some particular artificial intelligence technologies have high importance in health care systems. Techniques in artificial intelligence and machine learning are applied in health care to transform many aspects of patient care and administrative processes. Machine learning is a collection of statistical techniques that contains fitting models and training models. In healthcare, machine learning uses traditional and complex models to predict disease levels and treatment protocols. The traditional models use training data sets for predicting diseases and treatment protocols. The complex models use various neural networks for determining diseases and treatment levels. The most complex models use deep learning for prediction.

AI plays an important role in healthcare with machine learning, which has the primary capability to develop precision medicine, medical image analysis, bioinformatics, etc. [2]. AI and ML act as an assistant to find the right treatments at the right time for capturing patient data, cancer identification, and drugs levels and expediting the clinical trials. The main goal of these applications (a combination of AI and ML) is to analyze the relationships between disease prevention and patient outcomes or to analyze the relationships between treatments and patient outcomes. These are applied in various areas such as disease diagnosis, treatment and drug development, and patient monitoring and care. Disease diagnosis is a complex task in healthcare. The technologies AI and ML are used to enhance the functionality of diagnosis to improve patients' health and increase the service level.

12.3.1   Uses of AI and ML in Health Care

AI and ML techniques are used in the following areas to find and analyze aspects of human health:

- Diagnosing cancer using blinded validation study.
- Detecting mental health conditions using neuroscience.
- Diagnosing heart disease using cardiac images and providing remedy through automated ventricle segmentations.
- Restoring the movement of patients via spinal motor neurons.
- Helping to manage information about the disease and patient.
- Helping and encouraging a healthy human lifestyle.

- Helping to medical practitioner collaboration.
- Aiding drug research and discovery.
- Implementing automated administrative tasks.

AI and ML can help to diagnose, plan treatment, analyze clinical reports, and identify genetic information. Generally, AI and ML applications and their advanced computing ability are used in healthcare for assisting clinicians to make the right decisions at the right time.

12.4   AI AND ML APPLICATIONS IN HEALTHCARE

AI and ML give great hopes and expectations in healthcare. Techniques in AI and ML help to improve efficiency, reliability, and accuracy of decision making. Algorithms in AI and ML are used to:

- **Access training data in different environments**.
  - Techniques in AI and ML allow accessing the data using training data sets and algorithms. In biomedical image analysis, cancer imaging is an important source for analyzing the depth of the disease.
- **Provide and demonstrate the clinical data with training data set.**
  - During the development process of algorithms, researchers need to focus on the problems of patients. The demonstration of algorithm with a training data set provides a more effective and efficient way to develop the algorithms.
- **Integrate and deploy the data with a training data set**.
  - The deployment and integration of these technologies provide the capability to handle third-party applications effectively and easily.

Machine learning techniques can be used as the platform for analyzing data, ordering a test, suggesting a disease screening, and integrating data. In medical field, machine learning techniques are helped to solve critical problems, because of these techniques can be used to detect and discover patterns, identify suspicious spots on the skin, lesions and tumors. AI and ML tools are helped to transform the quality of human lives around the world. This section describes the importance of AI and ML techniques in one of the important application area of healthcare.

12.4.1   IMAGING ANALYTICS

AI and ML techniques help to discover hidden insights and extract meaningful information from inaccessible and unstructured data sets. Medical imaging data is an important source in a healthcare system. Medical imaging uses various techniques and methods for diagnostics, treatment, and decision making. Medical imaging refers to the different imaging techniques such as MRI (magnetic resonance imaging), X-ray radiography, ultrasound, and CT (computed tomography) for diagnostics. The modern technologies (AI and ML) are used to improve the computational capabilities of traditional medical imaging techniques with better accuracy.

AI and ML improve the accuracy of medical imaging techniques with the following aspects:

- Automation: improves the automation level of radiology.
- Productivity: increases the computational capabilities.
- Processes: AI and ML tools can use big data and statistical analysis techniques that give more efficient results.
- Diagnosis: provides accurate disease diagnosis results in cancer diagnosis.
- Computation: ability to compute quantitative and quantitative data.

Techniques of ML and AI have played important roles in healthcare like medical imaging, image interpretation, image fusion, image registration, image segmentation, image-guided therapy, image retrieval, and analysis. ML techniques are used to extract information from the images and represent information effectively and efficiently. ML and AI techniques help doctors to diagnose and predict accurate risk of diseases and prevent and treat them at the right time. These techniques enhance the abilities of doctors and researchers to understand how to analyze the genetic variations that will lead to disease. One of the most important implementation of AI and ML is cancer detection using medical image analysis. Cancer is a heterogeneous disease that involves abnormal or uncontrolled cell growth or division of cells, with the potential to spread to other parts of the human body. Biopsy helps to determine cancer stage and cancer type. It also helps to determine the treatment protocols based on the type and stage of the cancer. The applications of machine learning techniques are to identify the stage and treatment protocol of cancer. ML tools detect features of cancer using complex data sets. A combination of techniques, including artificial neural networks (ANNs), Bayesian networks (BNs), support vector machines (SVMs), recurrent neural network (RNNs), and decision trees (DTs) have been widely used in cancer diagnosis for developing predictive models, which are helpful in effective and accurate decision making. Various stages are involved to analyze and diagnose the depth of the cancer. Each stage uses different techniques to disease diagnosis. They are:

- Preprocessing.
- Segmentation.
- Feature extraction.
- Classification.

12.4.1.1 Preprocessing

The captured medical images have some common characteristics such as noise, poor contrast, and unclear boundaries. These unwanted characteristics affect the content of the medical images. These unwanted characteristics are rectified by preprocessing techniques. The preprocessing techniques increase and improve quality of the medical images, which are used for segmentation and classification. The preprocessing techniques, such as noise removal, increase the brightness, and special mark removal help to enhance the image quality.

Figure 12.1 depicts the various pre-processing stages are used in Imaging Analytics.

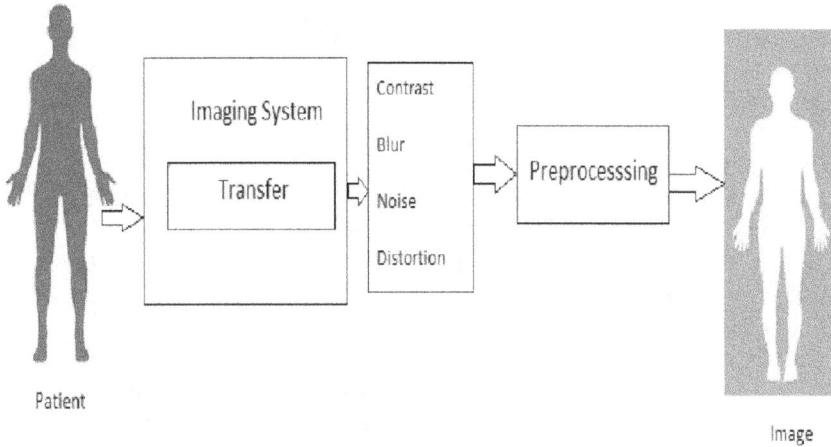

FIGURE 12.1 Preprocessing.

The main aim of the preprocessing is to improve the quality or reduce the surplus parts in the background of the cancer. A common preprocessing technique is filtering. The goal of the filter is used to improve the image quality for human viewers [8]. Filtering techniques are classified into four categories based on their characteristics and usage:

- Conventional techniques.
- Region-based techniques.
- Feature-based techniques.
- Fuzzy enhancement techniques.

These techniques are used to improve the quality of cancer medical imaging to achieve high diagnostic performance. The conventional technique tends to blur the images and noise removal. This technique has the ability to preserve features in the cancer image while reducing the noise. The region-based technique is mostly used to enhance image contrast of specific regions without any addition of artifacts. This technique is well suited for identifying tissue density. The feature-based technique keeps only the higher frequency threshold and suppresses low frequencies during the image enhancement process. At the end of the enhancement, the image contains only highest frequencies of lesion regions of images. The fuzzy techniques are used to enhance the contrast and suppress the noise using the entropy principle [4].

12.4.1.2 Segmentation

Segmentation is used to extract the specific portion that is used to detect the abnormal tissues on the cancer. The segmented region is important for feature extraction and classification. Segmentation helps to separate the cancer region from the image and also preserve the margin characteristics of the cancer image. The segmentation technique is categorized as threshold-based segmentation, region-based segmentation, and feature-based segmentation.

12.4.1.3 Feature Extraction

Feature extraction helps to identify key features of the image based on requirements. Feature extraction is helpful to reduce the number of unnecessary resources without losing important or relevant information. It can also reduce the amount of redundant data in the image. In cancer diagnosis, feature extraction helps to discriminate between benign and malignant cancer [3]. Machine learning has a number of statistical and optimization techniques and also detects and recognizes patterns from different data sets. The pattern recognition results assist medical practitioners in cancer disease diagnosis. Principal component analysis (PCA) is a ML-based feature extraction technique that is used for extracting features from cancer image.

The PCA algorithm reduces dimensionality using second order statistical information on the cancer image [7]. PCA is a linear dimensionality reduction method that produces linear combinations of original features and reduces the feature space by capturing linear dependencies among various features. PCA calculates principal components (PCs) that are linear combinations of original attributes. The PCs are orthogonal to each other and capture the maximum amount of variance in the data. Several variants of PCA use a correlation matrix for calculating eigen values. Compared to other feature extraction techniques, PCA has low computational cost and low noise sensitivity. PCA depends upon the scalability of data and relies on the assumption that features covering maximum variance.

12.4.1.3.1 Steps Involved in Principal Component Analysis

- Calculate mean vector.
- Assemble data samples in a mean adjusted matrix.
- Calculate the covariance matrix.
- Calculate the Eigen vectors and Eigen values.
- Calculate the basis vectors.
- Represent a linear combination of basis vectors.

Let us consider that each image is x pixels by y pixels. The image vector of size is $x \times y$ without loss of information and the image has point in $x \times y$ dimensional space. In a large image, the PCA technique reduces the image dimensionality significantly. PCA reduces the image dimensions mathematically. Then the high variance dimensions are selected by reducing the set of dimensions.

12.4.1.3.1.1 Calculate Mean Vector

Compute the mean vector for n variables using the formula:

$$S = \left(S_1 + S_2 + S_3 + \ldots + S_N \right) / N \tag{1}$$

12.4.1.3.1.2 Assemble Data Samples in a Mean Adjusted Matrix

The mean adjusted vector of the image vector is calculated using the following formula:

$$S_i = \left(S_i - \bar{S} \right) \tag{2}$$

Assemble the mean adjusted matrix of mean adjusted vector is:

$$S_{mean} = \left[\overline{S}_1 - \overline{S}_n \right] \tag{3}$$

12.4.1.3.1.3 Calculate the Covariance Matrix

The covariance matrix is used to represent the covariance measurement of sample in the data set. This is computed between two random variables and the variables relationship with one another. If the two variables increase together and decrease together, the covariance is positive. If the variables are inversely proportional, the covariance is negative. The covariance matrix displays the relationships between each pair of samples in the dataset.

$$C = \begin{bmatrix} COV_{0,0} & COV_{0,1} & \ldots & COV_{0,n} \\ COV_{n,0} & \ldots & \ldots & COV_{n,n} \end{bmatrix} \tag{4}$$

12.4.1.3.1.4 Calculate the Eigen Vectors and Eigen Values

Eigen vectors and Eigen values are computed from the covariance matrix in order to determine the principal components of the data. Principal components are new variables that are constructed as linear combinations of the variables. The combinations of Eigen vectors and Eigen values are done in such a way that they create new variables called principal components. The principal components are uncorrelated, and most of the information within the initial variables is squeezed or compressed into the first components. Generally, principal components reduce dimensionality without losing much information. This discards the low information components and maintains the remaining components as your new variables.

The Eigen value (λ) of the covariance matrix is calculated using the following formula:

$$del(\lambda I - C) = 0 \tag{5}$$

Where,
C—Covariance matrix
λ—Eigen values associated with the matrix
I—Identity matrix
del—determinant matrix

Eigen values are sorted by magnitude, after the computation of Eigen values. The highest Eigen values are maintained and discard the remaining. The Eigen values are used to form the basis vectors. The basis vectors based on Eigen values with the highest variance are used to reduce dimensionality with preserving the original information of the image. Then the Eigen vector is computed for a given Eigen value, as follows:

$$(\lambda_k I - C) \times V_k = 0 \tag{6}$$

Where,
λ_k—Highest Eigen value
I—Identity matrix

C—Covariance matrix
Vk—Eigen value to be calculated

12.4.1.3.1.5 Calculate the Basis Vectors

Form a Eigen vector matrix by assembling Eigen vectors (V1, V2, . . . VN) using the following formula:

$$EV = [V1, V2, \ldots VN] \tag{7}$$

Compute the basis vectors by multiplication of the mean adjusted matrix, using the Eigen vector matrix:

$$S_B = S_{mean} \times EV \tag{8}$$

Where,
S_B—Basic vector dimensions
S_{mean}—Mean adjusted matrix
EV—Eigen vector matrix

12.4.1.3.1.6 Represent a Linear Combination of Basis Vectors

Each sample can be represented as a linear combination of basis vectors using the following formula:

$$LC = \left(S_{sample} - \bar{S} \right)^T \times S_B \tag{9}$$

Where,
LC—Linear combination of basis vector
S_{sample}—Sample representation using basis vector

Finally, PCA transforms the image with a linear combination of basis vectors that are used for classification.

12.4.1.4 Classification

Classification is one of the most important processes in cancer diagnosis. In the modern world, various types of neural networks act as active and robust classifiers, which are used in different applications for classification. In a healthcare system, classification is mostly used for decision making. Classification helps to extract the feature from large datasets. In a healthcare system, artificial neural networks (ANN) are used to implement classification. The neural networks have the following features:

- Various self-adaptive techniques are available in neural networks, which adjust the distributional form of an image without any explicit specification.
- Accuracy of the classification is high because classification can be done by using functional relationships between group members and object attributes.
- Neural networks provide flexible modeling using nonlinear input–output mapping.
- Neural networks are able to perform statistical analysis.
- Neural networks handle large data sets.

Neural networks have the ability to extract the patterns using multidimensional nonlinear connections of data. In cancer diagnosis, ANN are for image classification. Numerous classification algorithms are available in AI. This section provides the most important advances of recurrent neural networks (RNN) in cancer diagnosis. RNN is one of the most important categories of neural networks, where the output from the previous step is fed as input to the current step [1]. It uses the output of the first step for the input of the second step which gives context to the data. RNNs are mainly used for sequence classification, sentiment classification, and video classification.

Figure 12.2 depicts the number of sequences for image or data classification are used in image analytics.

RNN uses number of sequences for image or data classification. RNNs are divided into number of units called long short-term memory (LSTM). This uses feedback loops to prioritize the data that is important information or non-important information. LSTM contains:

- **Memory cell:** It is used to control the input information flow.
- **Input gate:** It accepts the input data. It uses a built-in activation function to decide whether to let the input data in or erase the present state and to decide how the input will affect the output.
- **Output gate:** It is used to regulate and filter the output data.
- **Forget gate:** This allows disposal of information that is previously stored.

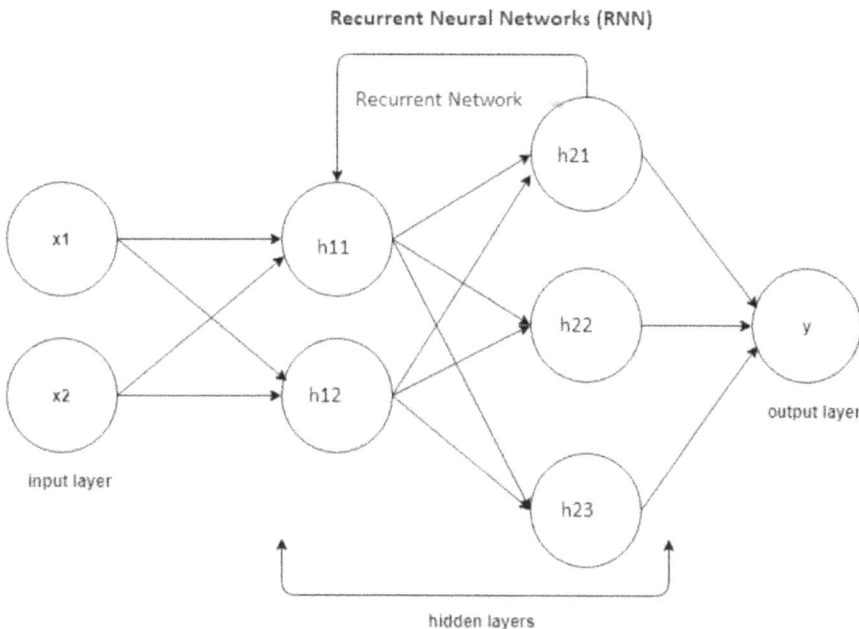

FIGURE 12.2   Recurrent neural network architecture.

LSTM is used to classify and identify output data based on a series of discrete-time input data. It uses gradient descent and back-propagation algorithms to reduce the error. In cancer image classification, RNN produces more accurate results.

12.5 FUTURE TRENDS OF AI AND ML IN A HEALTH CARE SYSTEM

In a healthcare system, AI and ML techniques can take more responsibility and reduce risk in diagnostic and treatment processes. The intention of modern technology is not to replace human clinicians but to enable a streamlined high-quality healthcare delivery process. AI and ML techniques play an important role in the medical automation process. These algorithms help to create intelligent robots, which act as the assistants to conduct surgeries, deliver medication, and monitor hospital patients. Robotic assistants remind doctors and patients of routine activities like medication intake, guidance, etc. Robots can act as a vessel for a silicon-based artificial brain. Generally, the imagined future development of artificial intelligence and machine learning can be used to develop hybrid human-artificial intelligence in healthcare systems.

12.6 CONCLUSION

Healthcare systems are facing various complex challenges. The most important complex challenge in healthcare is to handle voluminous data that is generated during the process of healthcare. The technology development of AI and ML can easily handle or solve this challenge using their intelligent architecture and learning and reasoning abilities. Healthcare systems use their techniques to solve the problem successfully. These technologies are helped to automate and speed up the disease diagnostic process. However, modern technologies cannot totally replace the human elements in patient care but can replace the expensive elements in patient care and treatment.

REFERENCES

[1] Chandra, B, et al., 2017. On Improving Recurrent Neural Network for Image Classification, *International Joint Conference on Neural Networks (IJCNN).* Anchorage, AK, US App. 1–10.
[2] Davenport, T; Kalakota, R; The Potential for Artificial Intelligence in Healthcare, *Future Healthcare Journal*, Vol. 6(2), pp. 94–98, 2019.
[3] Dhahri, H; Maghayreh, E; Mahmood, A; Elkilani, W; Nagi, M F; Automated Breast Cancer Diagnosis Based on Machine Learning Algorithms, *Journal of Healthcare Engineering*, Vol. 2019, pp. 1–11, 2019.
[4] Gardezi, S J S; Elazab, A; Lei, B; Wang, T; Breast Cancer Detection and Diagnosis Using Mammographic Data: Systematic Review, *Journal of Medical Internet Research*, Vol. 21(7), pp. 1–6, 2019.
[5] Kavita, R; Dubey, R; Chawda, R; Data Analysis and Its Importance in Health Care, *International Journal of Computer Trends and Technology (IJCTT)*, Vol. 48(4), pp. 176–180, 2017.

[6] Kunkle, S; Christie, G; Yach, D; El-Sayed, A M; The Importance of Computer Science for Public Health Training: An Opportunity and Call to Action, *JMIR Public Health and Surveillance*, Vol. 2(1), pp. 1–5, 2016.

[7] Mert, A; Kılıç, N; Bilgili, E; Akan, A; Breast Cancer Detection with Reduced Feature Set, *Computational and Mathematical Methods in Medicine*, Vol. 2015, pp. 1–11, 2019.

[8] Ramani, R; SuthanthiraVanitha, N; Valarmathy, S; The Pre-Processing Techniques for Breast Cancer Detection in Mammography Images, *International Journal of Image, Graphics and Signal Processing*, Vol. 5, pp. 47–54, 2013.

13 Big Data, Artificial Intelligence and IoT Enabled Smart Cities
Applications and Challenges

Aboobucker Ilmudeen

CONTENT

13.1 INTRODUCTION

In the era of "big data, artificial intelligence (AI), and Internet of things (IoT)", the recent developments in digital technologies have facilitated the rapid innovation and re-engineering of smart cities. Cities across the globe are always modernizing their digital infrastructure facilities to offer a comfortable lifestyle for their citizens. The digital infrastructure and technology are producing a digital revolution of lifestyle for citizens. Smart cities are constantly upgrading their state-of-the art digital infrastructure facilities to support the lifestyle of their citizens. Accordingly, the smart city can offer timely and quick responses to various stakeholders such as daily routine needs, peoples' living, safety, and security measures, citizen travelling and passage, health services, and commercial and industrial activities (Pramanik, Lau, Demirkan, & Azad, 2017).

The utmost objective of constructing a smart city is to experience and gain access to facilities with modern technologies and the latest structural arrangements of facilities to offer superior value to society and to encourage government service efficiency, public autonomy, and societal relationships (Sun & Zhang, 2020). However, cities have problems such as food and water safety, disaster recovery and prevention, environmental stewardship, facilities for shopping, tourism, and recreation, travelling facilities, energy efficiency, care of elderly people, and crime and accident prevention (Wu, Wu, & Wu, 2019). Smart cities are more and more moving to specific technologies to tackle problems that are linked to the environment, society, infrastructure, morphology, and many others. This rapid development and application of technology is possible when smart cities greatly implement the potentials of big data, IoT, sensor devices, and artificial intelligence.

Big data analytics via AI can significantly bring many economic prospects, superior management capabilities, and urban designing. The construction and planning of a smart city and urbanization must cope with the requirements of cities, technology, citizens, spaces, resources, and social responsibility (Sun & Zhang, 2020). Similarly, the applications of modern technologies in designing and maintaining the smart cities are continually advancing. Figure 13.1 shows the popularity of big data, AI, and IoT in Google Trends under the urban and regional planning category among academia, researchers, citizens, policy makers, engineers, designers, industrialists, etc.

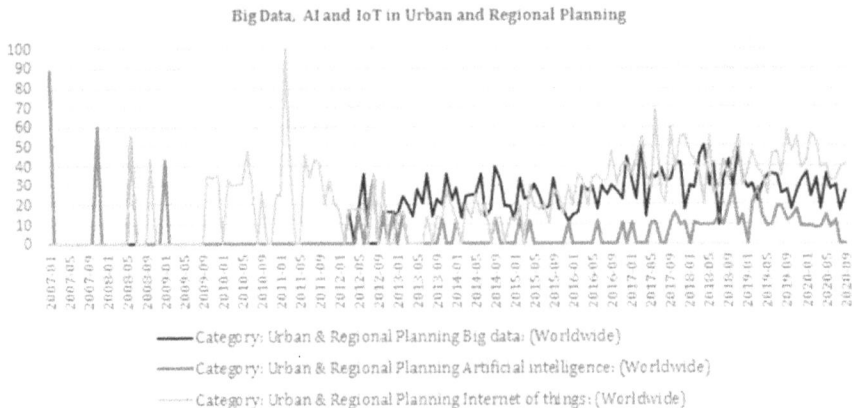

Big Data, AI and IoT in Urban and Regional Planning

——Category: Urban & Regional Planning Big data: (Worldwide)
——Category: Urban & Regional Planning Artificial intelligence: (Worldwide)
------ Category: Urban & Regional Planning Internet of things: (Worldwide)

FIGURE 13.1   Status of big data, AI and IoT 2007 to 2020.

Source: Google Trends, 15, September 2020

The smart city applies the data technology to advance people's living standards such as transportation of goods and services, renting houses, state sector services to the citizens, education, and health care services in an intelligent way (Sun & Zhang, 2020). Information and communication technology (ICT) is mainly employed in policy design, decision making, execution, and crucial useful services in smart city development. Rising urbanization, energy usage and consumption, retaining a green environment, improved economic and living standards of the citizens, and enriching the competencies of the citizen to proficiently handle the latest ICT are targeted issues in smart cities. Experts claim that improving manifold technologies to uplift the present infrastructures with supreme safety and security arrangements is essential for the future expansion of the smart city ecosystem (Al-Turjman & Malekloo, 2019). This chapter aims to fulfill the limitations that are highlighted in the literature. For instance, in the context of novelty, no prior study had been done on the IoT and big data related applications in an urbanized city's perspective (Bibri, 2018).

This chapter focuses systematic evaluation, synthesis, and critical applications and challenges of state-of-the-art application. In addition, this chapter will propose an applied conceptually designed prototype by incorporating the major technologies for the development of the smart city.

13.2 SMART CITIES

Smart cities refer to the deployment of modern technologies to sense, inspect, process, and incorporate massive amounts of valuable data and information from essential systems in functioning cities (Pramanik et al., 2017). Scholars outline the division of a smart city as people, living, citizen movement, economy, atmosphere, and administration (Nishant, Kennedy, & Corbett, 2020). Smart cities are advocating the prospect of linking sensors and big data via the IoT (Allam & Dhunny, 2019). Today, cities across the globe are moving toward digitalization by installing sensor devices, various communication technologies, and computing power. Hence, various sources such as open areas, land plots, buildings, and streets are interconnected, which can generate huge volumes of big data. This permits the smart city designers and policy makers to design and construct state-of-the-art smart cities.

Figure 13.2 depicts the required components of smart city development (Diary, 2018). Smart cities greatly improve the quality of citizen's lifestyle, such as smart health care, transportation, parking, governance and regulations, and so forth (Babar & Arif, 2017). The modern smart city is enabled by citizens' accessibility to the latest services, electronic governance, easy transportation, applications of modern ICT, energy and water supply, safety, security, etc. (Saravanan & Srinivasan, 2018). The smart city offers a number of inventive, smart services to meet residents' needs cautiously and powerfully, extending to individual and business environments (Kai, Li, Xu, Li, & Jiang, 2018). The notion of smart city advancement is evolving as a key retort to the quick expansion and socio-economic deficits encountered by cities across the globe (Israilidis, Odusanya, & Mazhar, 2019). In line with the smart city, the concept of a "green" smart city is also getting popular (e.g., Kaur, Tomar, & Singh, 2018; Li, Kisacikoglu, Liu, Singh, & Erol-Kantarci, 2017; Muhammad, Lloret, &

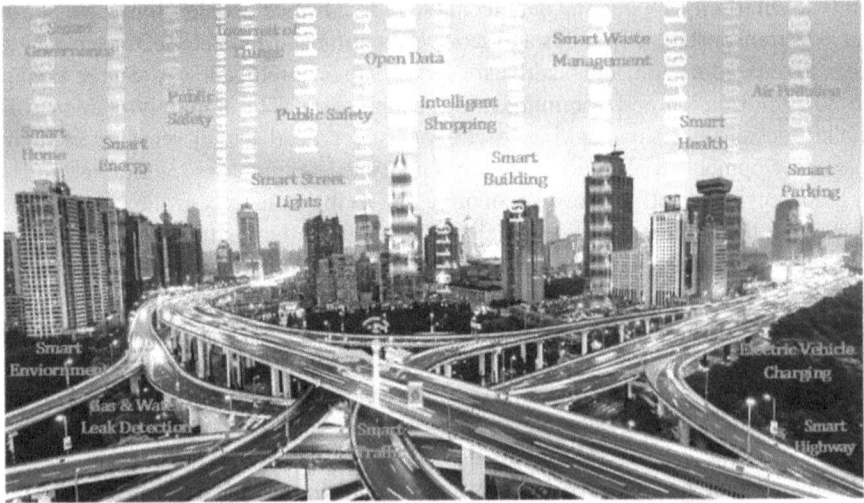

FIGURE 13.2   Smart city components.

Source: Diary, 2018

Baik, 2019; Sodhro, Pirbhulal, Luo, & de Albuquerque, 2019). The green smart city may consist of green smart homes, smart farming, smart heath, a smart water system, smart education system, smart retail, and smart transportation (Kaur et al., 2018).

According to Pramanik et al. (2017), there are three paradigm shift trends in smart health and smart cities. First, the move in health care is from traditional health care to ubiquitous health care and then intelligent health care. Second, in the city, the traditional town advances to digitalized city and then to smart city. Third, in data, database has shifted to data mining that is enabling big data. The smart city denotes a new future structure that mixes various ICT and IoT solutions' services in order to advance the citizens' lifestyle quality (Memos, Psannis, Ishibashi, Kim, & Gupta, 2018).

13.3   THE BIG DATA APPLICATIONS IN SMART CITIES

Today, big data is believed to be an essential technical foundation to construct the smart city (Sun & Zhang, 2020). Big data involves actions such as comprehensive data collection, storing data, and processing of data, and it recognizes the release, distribution, and recycling of data from the smart city in the Internet (Sun & Zhang, 2020). Researchers claimed that the large volume of data created by the IoT could be explored by AI to facilitate the administration, structure, and uptake of a smart city (Nishant et al., 2020). According to (Osman, 2019), mining the hidden insights and unseen associations from big data is a growing trend that can reveal better ideas for decision making to offer smart services to citizens. Gig data has various smart energy applications, for example, intelligent systems for energy management, forecasts of energy consumption, and utilization of IoT solutions (Marinakis et al., 2018). The applications of large volumes of data in smart cities are in various technological forms such as cloud computing, IoT, AI, mobile Internets, and other modern IT technologies to create the data decision centers (Sun & Zhang, 2020).

Researchers claim that crowdsensing and big data analytics are used to ensure the citizens' privacy when facial or motion recognition technologies are used (Laufs, Borrion, & Bradford, 2020). Big data offers ample inducement for the development of urban city creation, and it is the significant basis for smart city intelligence (Sun & Zhang, 2020). The large volume of data can be optimized to quickly extract and process the data, mining insights and knowledge, thus producing insights and value from it (Wu et al., 2019). The analytical power of bulk data via AI can significantly add to the metropolitan design, viability, and living comfort dimensions.

13.4   ARTIFICIAL INTELLIGENCE APPLICATIONS IN SMART CITIES

The term artificial intelligence (AI) refers to the computerized abilities and cognitive power of the machine to solve problems. AI can process the way computers mimic to think and simulate human behavior (Allam & Dhunny, 2019). AI is employed for synchronous data analysis, which allows for a quicker response. With the aid of interlinking through massive data and IoT, the AI can significantly simplify the urban governance process while offering much quicker data analysis to recognize the emerging, present, and future issues in a smart city. AI has penetrated various organizational processes, causing the fear that intelligence machines and devices will substitute many humans in decision making shortly (Jarrahi, 2018).

For instance, the technological evolution in cognitive machines offers unbelievable advances that unleash a surprisingly wide range of applications in AI. With a superior computing analyzing power and a methodical method, AI can lengthen humans' reasoning when dealing with complications. However, a human can give a complete, natural solution in addressing complexity in the process of decision making (Jarrahi, 2018). The purpose of using the AI is that the activities that the AI is handling are considerably difficult for humans (Casares, 2018). Some of the deployments of AI in a smart city, such as AI for renewable energy, are getting much more attention because it could alleviate effects of climate change. The AI is supporting for mapping, modelling, and planning energy sources for its improved effectiveness (Allam & Dhunny, 2019). Wide-ranging efforts are being made through AI, machine learning, and the ecosystem with the intention to offer solutions to the challenges in smart parking systems (Al-Turjman & Malekloo, 2019).

13.5   IOT APPLICATIONS IN SMART CITIES

The increasing level of Internet of Things (IoT) usage in various facets of human life is to alleviate resource expenditure, cost, and wastages in our modern era in order to improve human civilization. Various devices are employed in the network to produce large volumes of data while integrating with IoT. As a result, big data and IoT have great links with each other. The IoT devices are generating huge volumes of big data that are complex and hard to resolve using the most precise and efficient decisions. In this context, to maximize the decision making from the big data, the state-of-the art techniques such as AI, machine learning, and deep learning are highly suitable (Ullah, Al-Turjman, Mostarda, & Gagliardi, 2020). There are various usages of cloud enabled IoT applications in a smart city, for instance, smart transportation and parking facilities, automatic lighting system, advanced surveillance systems to safeguard public areas using sensors and cameras, conservation of national heritage, water supply, energy supply and maintenance, garbage management, advanced health care, and smart education (Saravanan & Srinivasan, 2018).

The smart city IoT is integrated with big data analytics to uncover insights through the process and analysis of large volumes of data (Babar & Arif, 2017). The IoT is linked with big data analytics, which enable the maximization of energy efficiency and reduction of environmental effects in many urban designs (Bibri, 2018). IoT can have a rich use of data and intelligence applications that can serve a key part in the construction of urban cities. A variety of big data are produced by cloud-centric IoT tools or sensors that play an important role in various uses of smart cities (Ullah et al., 2020). IoT devices offer real-time data that can be processed for appropriate learning and sense making. Various pioneering IoT-based solutions rely on chasing and sketching users' mobility and actions. Hence, the IoT has been adopted to uplift the productivity and achievement of urban cities (Allam & Dhunny, 2019).

13.6 PRIOR STUDIES IN BIG DATA, ARTIFICIAL INTELLIGENCE AND IOT APPLICATIONS IN SMART CITIES

There are several prior studies which highlight various state-of-the-art technological applications in smart city construction. This section discusses the prior studies that employed big data, AI, and IoT in detail. For instance, Ullah et al. (2020) reviewed the AI and machine learning applications in urban cities. Their in-depth, detailed review covered usage such as smart carrying systems, energy-efficient smart grids, cyber-security, effective use of driverless vehicles, modern communication services, and smart health care system. Laufs et al. (2020) conducted a systematic review to discover the latest literature about the smart city security technologies and investigated how it links with the existing security arrangements. Similarly, Zahmatkesh and Al-Turjman (2020) reviewed fog computing-based urbanized cities in the Internet of Things period. They focused on potential fog computing usages and possible supporting technologies for urban cities in which various techniques such as AI, ML, and unmanned aerial vehicles for collecting data from fog-computing-related IoT applications were broadly presented.

The study of Allam and Dhunny (2019) reviews the potentials of bulk data and artificial intelligence in urban cities which covers urban governance, culture, and livelihood that comply with UN Sustainable Development Goal 11 in urbanized cities. Israilidis et al. (2019) systematically reviewed the present smart city literature that linked with the knowledge management standpoints aiming to offer an outline for the directions of future smart city research. Kaur et al. (2018) proposed the design of green IoT architecture aiming to decrease energy usage and integrate the cloud-based system that decreases hardware deployment in the smart city. Some research admitted the possible applications of fog computing and possible supportive technologies in the urbanized cities considering the IoT atmospheres (Zahmatkesh & Al-Turjman, 2020). As there is an increasing use of modern technologies that are integrated with IoT applications, (Korczak & Kijewska, 2019) systematically reviewed the literature under the notion of smart logistics; accordingly, the virtual–physical systems, IoT, and stakeholders in the market are the core to create the smart logistics.

Anghel et al. (2015) discussed the innovative approaches that enable the green networked data centers to observe, regulate, reprocess, and improve energy consumption and creation, especially in renewable resources in smart grid cities. Bunders and Varró (2019) reviewed by questioning how urban and civil society participants problematize the difficulties and their issues that are linked to data-centric activities in Dutch cities to reveal their performance and practice. Nitoslawski, Galle, Van Den Bosch, and Steenberg (2019) researched by examining the present and evolving smart city advancements and technologies and emphasized the feasible deployment for smart city plantation and greenery field management. Krishnan Saravanan, Julie, and Robinson (2019) evaluated in detail the smart city development, architectonics, techniques, usages, and principles and threats. Table 13.1 shows the related studies in big data, AI and IoT in the literature.

13.7 MODERN TECHNOLOGIES, METHODS AND INFRASTRUCTURES IN DESIGNING SMART CITIES

Many prior studies applied the functions of AI and machine and deep learning to advance smart cities (Ullah et al., 2020). Numerous existing applications use big data, AI, and IoT in designing the urbanized city. For instance, the novel applications of big data, AI, and machine learning used in blockchain can significantly add to smart cities (Allam & Dhunny, 2019). Similarly, the scholars claimed that the integration and operation of the blockchain technique in building urbanized cities are unescapable (Sun & Zhang, 2020). The lighting system in today's modern city can connect through the nationwide grid or the usage or travelling of the vehicles on the street (Mukta et al., 2020).

The smart grid aims to meet goals of being effective, uninterrupted, and consistent in energy consumption to the smart city, which enhances the supply effectiveness, reduces processing costs, and offers improved connection with sustainable power supply so as to optimize electricity consumption (Zahmatkesh & Al-Turjman, 2020). As a result, it would help to elevate superior electricity systems and decrease the electricity charges in smart cities. The intelligent transportation network consists of connected street lights that can help reduce traffic jams, avoid accidents, and decrease noise and energy usage. For instance, in the health observing systems, the street cameras can detect the turned-on lights of the ambulance and can change the street lamp to send the ambulance during the traffic (Zahmatkesh & Al-Turjman, 2020). Table 13.2 shows smart city key indicators in prior studies.

TABLE 13.1

Big Data, Artificial Intelligence and IoT-Related Studies in Smart Cities

Author	Purpose of Study/System	Techniques/Platform Features	Data and System Design Features	Application Domain	Findings/Key Highlights
Sun and Zhang (2020)	Decentralized point-to-point secure system for smart city information sharing and exchanging	Blockchain big data platform for the decentralized system for low carbon release and non-polluting atmosphere	Urban city, block chain, big data, and IT implications are combined for the smart city big data platform	Creation of modern urban city	Carbon-less eco-friendly energy system on the creation of urbanized Hefei city
Mukta, Rahman, Asyhari, and Alam Bhuiyan (2020)	IoT enabled, cost-effective, sustainable street light system	Design and development of effective street lights focusing on outcome and ecological benefits	Design principles including advantages, disadvantages, and research challenges	Highway lighting systems	Advise scholars and administrators about the merits of enhanced energy saving for street light
Al-Turjman and Malekloo (2019)	IoT-based smart parking systems	The IoT, sensors, and supportive technologies	Smart parking while aiming on data interoperability and exchange	Smart parking	Smart parking systems in view of soft and hard design factors
Babar and Arif (2017)	Urban planning using big data analytics to link the compatibility of IoT	A generic solution using a range of data sets to validate the proposed architecture	Proposed architecture consist of three modules and analyzed datasets on Hadoop server	IoT device compatibility and Smart City Planning	Proves the proposed architecture is more valuable than the existing smart city architecture
Bibri (2018)	Sustainable smart cities using IoT and big data applications for eco-friendly viable solutions	Discusses big data handling resources and calculating models for urbanized viable cities	Review the modern sensing huge volume of data usage aided by IoT for ecologically viable	Designing of sustainable smart city	Conceptual, analytical, and supreme level big data and IoT innovative applications and their research directions
Kai et al. (2018)	Efficient communication for device-to-device connection in smart cities	The maximization of uplink subcarrier and power distribution in device-to-device essential mobile systems	Subcarrier assignment and energy distribution sub-issues and heuristic algorithm to assign subcarrier	Device-to-device transfer in cellular user devices in smart city	The algorithm's results prove notable advancement in power consumption

Reference					
Li et al. (2017)	Big data analytics for electronic automobile assimilation to smart sustainable cities	The electric vehicle-generated huge data from sensors to trip logs are studied by means of big data techniques	Developing smart charging algorithms, resolving energy efficiency problems, assessing power distribution systems to handle additional powering loads, and defining the market price for the services offered by electric automobiles	The electronic automobile assimilation in urban sustainable cities	Provided the direction for upcoming data analytics requirements and results for electric automobile assimilation with smart cities
Lau et al. (2019)	The big data merging in the smart city applications	Proposed multi-views classification to assess smart city selected applications	The multi-perspectives grouping to evaluate the urban city usages in the context of data fusion	Data fusion in the urban city	Development of multi-perspective and discussed future trends and difficulties for data fusion
Marinakis et al. (2018)	Intelligent energy management application using big data in energy sector	Big data architecture can support the construction, expansion, repairs, and utilization of smart energy for cross-domain data.	Based on the architecture, a web-enabled decision support system (DSS) is established that uses multi-sourced data for the design of energy administration plans	An application of smart energy management in the energy sector	The application of developed "data-driven" DSS supports for managing energy performance in smart cities
Memos et al. (2018)	IoT network-based architecture and its safety and confidentiality challenges in wireless sensor networks	An effective model for surveillance system in IoT network for smart city	Proposed algorithm is developed based on previous two algorithms such as WSN packet routing and security	A surveillance system to manage wireless sensor network security issues using IoT in a smart city	The proposed system is effective in privacy, media security, and sensor network memory requirements that are essential for future smart city development
Muhammad et al. (2019)	An energy-efficient framework is proposed for green smart cities by linking IoT, data prioritization, AI, and big data analytics	Highlightsthe key challenges of data prioritization, its future desires, and schemes for integration into green smart cities	An in-depth analysis of the recent approaches and trends of data prioritization for data of different natures, genres, and domains in green smart cities	Smart and energy-efficient data analyzing in green smart cities	Highlights major defies and sanctions for future study so as to develop the related smart services in smart cities

(Continued)

**TABLE 13.1
(Continued)**

Author	Purpose of Study/System	Techniques/Platform Features	Data and System Design Features	Application Domain	Findings/Key Highlights
Nicolas, Kim, and Chi (2020)	To categorize the enablers of smart cities and to measure their dynamic effects	Enabler clusters such as technological infrastructure, open governance, intelligent community, and innovative economy and performance objectives such as efficiency, sustainability, livability, and competitiveness	Structural equation modeling using the actual data of 50 smart cities	Smart city development	The non-technical enabler clusters and the technical drivers have significant impacts on the performances of smart cities with their highly interrelated, synergetic dynamics
(Pramanik et al., 2017)	Big data-based framework for health care in smart cities	Modern design and architecture of smart health care paradigm	Smart integration and technologies to design state of the art health care services	Electronic health record, biometric data, social media, and surveillance data	Combined big data and health care for smart services in the smart city

TABLE 13.2

Key Indicators of Smart City

Indicator	Zahmatkesh and Al-Turjman (2020)	Nitoslawski et al. (2019)	Memos et al. (2018)	Kai et al. (2018)	Ullah et al. (2020)	Sun and Zhang (2020)	Allam and Dhunny (2019)	Al-Turjman and Malekloo (2019)	Babar and Arif (2017)	Allam and Newman (2018)	Plageras, Psannis, Stergiou, Wang, and Gupta (2018)	Pramanik et al. (2017)	Sodhro et al. (2019)	Chaturvedi, Matheus, Nguyen, and Kolbe (2019)	Washburn et al. (2009)	Neirotti, De Marco, Cagliano, Mangano, and Scorrano (2014)
Smart citizens	✓	✓		✓			✓									✓
Smart living	✓	✓		✓		✓	✓					✓		✓	✓	✓
Smart education						✓	✓								✓	✓
Smart governance	✓							✓	✓		✓			✓	✓	✓
Smart economy	✓															✓
Smart mobility	✓			✓	✓	✓		✓	✓		✓				✓	✓
Smart infrastructure	✓		✓	✓	✓	✓		✓	✓		✓		✓	✓	✓	✓
Smart environment	✓	✓		✓	✓						✓					✓
Public safety											✓				✓	✓
Smart health care	✓					✓					✓	✓			✓	✓
Culture										✓						✓

Source: Adopted from Allam & Dhunny, 2019

TABLE 13.3

Smart Services by Big Data, AI, and IoT Applications in Smart Cities in the Existing Studies

Study	Domain/Smart Services									
	Governance/Policy	Environment	Water	Education	Energy	Health Care	Sustainability	Agriculture	Highway	Green Data
Nitoslawski et al. (2019)		✓					✓			
Mukta et al. (2020)					✓				✓	
Zahmatkesh and Al-Turjman (2020)					✓				✓	
Bibri (2018)	✓						✓			
Li et al. (2017)					✓					
Kai et al. (2018)					✓					
Marinakis et al. (2018)					✓					
Anghel et al. (2015);							✓			✓
Muhammad et al. (2019)						✓	✓			
Pramanik et al. (2017)	✓					✓	✓			
Silva, Khan, and Han (2018)	✓		✓				✓			✓
Neirotti et al. (2014)	✓	✓		✓		✓	✓			
Maye (2019)	✓		✓					✓		

In smart city construction, an elegant water management system would be able to check the water usage, water supply, and estimation of future water usage levels (Zahmatkesh & Al-Turjman, 2020). Hence, the smart city can offer an uninterrupted, hygienic—with detection of poisons or contaminants in real-time—efficient, and cost-effective water supply to the citizen and the business firms. The sensors embedded in the system will detect water losses and supply of smart water network system by exploring the sensor data. For this purpose, the cloud and fog computing system will include sophisticated wireless technologies and protocols so as to create a more viable, reliable, and efficient water supply network. Smart agriculture is the latest trend that would supply food in the smart city. For smart ag, the sensors' devices can be employed in the farming vehicles, which can gather data and information about the crops' growth and climate conditions from the cultivation field. When it comes to health care, big data, AI, and IoT are greatly used. For instance, health care experts and officers are helped with disease diagnosis, severity assessment, clinical observation, health monitoring, drug recommendations, and clinical treatment planning (Allam & Dhunny, 2019). Similarly, the applications of intelligent agents in health care are retrieving health information from big data, disease diagnostic decision support systems, scheduling and planning tasks for doctors, nurses, and patient, health care information sharing, medical image processing, automation, simulations, bioinformatics, medical data management, and health decision support systems (Pramanik et al., 2017).

Similarly, the cloud computing, fog computing, and machine learning infrastructure are also employed in designing the smart city. Thus, fog computing can be employed as an effective structure to decrease obstructions and improve insufficiencies in the system in the smart cities (Zahmatkesh & Al-Turjman, 2020). Machine learning is the computational techniques that are divided as three types, that is, administered, unmanaged, and supportive learning (Ullah et al., 2020). The ML models have been applied to economic, environmental, and social inquires such as smart cities and climate change that are the key artificial intelligence research areas (Nishant et al., 2020). Similarly, the ML possibly can be used for the distribution, production, and usages of energy, smart city functions and care, and citizens' ideas for sustainable power and predicting citizens' view from social media data, etc. The ML is applied using the techniques such as linear regression, artificial neural network, genetic algorithm, support vector machine, autoregressive moving average model, and the adaptive neuro-fuzzy inference system or network-based fuzzy inference system (Nishant et al., 2020).

13.8   DESIGN AND DEPLOYMENT OF BIG DATA, AI AND IOT APPLICATIONS IN SMART CITIES

Smart city designers keep in mind that the technology does not make the city smarter (Israilidis et al., 2019). Hence, the modern emerging technologies, for instance, cloud computing, IoT infrastructure, big data, artificial intelligence, and mobile Internet with urban planning, building, management, and operations are deeply incorporated in smart cities (Sun & Zhang, 2020). In today's smart city and IoT setting, smart parking and its innovative services have received vital attention among urban designers. The smart parking system that is implanted with sensor devices in vehicles and city services can reduce the blocks in parking issues and offer a better solution to citizens (Al-Turjman & Malekloo, 2019).

The application of blockchain technologies is ever increasing in the smart cities. Blockchain is the informationless interface that stores different types of data in blocks. It has features such as trust, shared ledger, shared data, and information traceability which can help increase efficacy for data sharing in smart cities (Sun & Zhang, 2020). Hence, researchers have said that the contribution of blockchain big data infrastructure in building the new urban cities is inevitable (Sun & Zhang, 2020). The green IoT spontaneously and cleverly makes the smart cities viable in a combined manner (Kaur et al., 2018). Researchers have discussed various storage techniques and the deployment of automated aerial vehicles and different AI and ML techniques in storing data for fog-enabled IoT systems (Zahmatkesh & Al-Turjman, 2020).

Machine learning is applied for processing the big data that are collected by IoT and reasoning AI. Similarly, knowledge-centered AI is also being employed in the urban city development (Nishant et al., 2020). For instance, smart city and urban designers apply satellite images and ML approaches to model the images to find the green spaces in their cities. Subsequently, the intelligent drones and RFID tags are used to more easily and more closely maintain and safeguard the plants' growth. Similarly, the architecture of the fog computing model enables delay-sensitive and simultaneous analysis of IoT systems. In the urban city construction, fog computing could be employed as an effective framework to decrease interruptions and increase cost-effectiveness of the system. The cloud-centric IoT devices that gather the updated details of passengers or street lamps and check the power parameter variables with an implanted system in order to ensure minimum energy are misused when the bulbs are idle (Mukta et al., 2020).

13.9 BIG DATA, ARTIFICIAL INTELLIGENCE AND IOT CHALLENGES IN SMART CITIES

The United Nation estimated that 68% of the global public will live in cities by 2050; thus, handling the present facilities and arrangement to satisfy a viable smart lifestyle for the increasing demands of the smart city public has become more challenging (Lau et al., 2019). Though modern technology has significantly increased comfortable living, there are various challenges that arise, namely unsecured devices, immorality, and privacy infringement, amongst others (e.g., Allam & Dhunny, 2019; Memos et al., 2018). Scholars claimed that the current short-range and self-centered cognitive systems intended to fulfill the water, energy, and food supply shortages (Nishant et al., 2020). In prior studies, various challenges have been highlighted. For instance, energy saving and keeping sustainable technology are the key aspects of urban city, hence, increasing intelligent street bulb systems is still a challenge for scholars (Mukta et al., 2020).

The heterogeneous and distributed IoT systems are surely creating challenges such as data complexity in sharing, software, and system proprietary issues, privacy and security issues, and resource scarcity (Al-Turjman & Malekloo, 2019). Babar and Arif (2017) highlighted that the use of dissimilar devices can make compatibility problems, and a heterogeneous environment increases the identification ambiguity of data retrieval sources in smart city. The challenges of a smart city can begin from its design stage to real functioning, for instance, costs for the plan and execution, selection of technologies, diversity of devices, the huge amount of data, issues in security and cyberspace, adoption, and speed of connection (Saravanan et al., 2019).

Though there is an increasing attention towards advancing the traditional cities to urbanized smart cities, studies highlight various hindering causes that can limit the expansion of the urban city. Hence, the industry experts and academic scholars have identified the following avenues which can be efficiently employed by using the AI, ML, and DL techniques to increase the efficiency of the urban cities (Ullah et al., 2020). Firstly, for a perfect and accurate management process, a massive amount of experimental data, such as automobile speed, location, distance between the vehicles, drivers' behavior, UAVs altitude, etc., are required to effectively test the ML and DL protocols. Secondly, the regularization of big data expansion in a smart grid connectivity protocols used during connection of different smart grids devices, and choice of the most effective AI, ML, and DL techniques can advance the smart grid's performance. Thirdly, in any platform the security issues and breaches are common and inevitable. Accordingly, modern technologies, end-user applications, and citizens' personal and confidential information will be used, and they are subject to be misused. Fourthly, urban citizens face numerous challenges such as traffic blocking, ecological deprivation, lack of safety, privacy breaches, lack of housing, smuggling, urban slump, and ineffective service delivery (Allam & Dhunny, 2019). The crowded travelling has led to serious traffic jams that cause waste of time and energy, air pollution, and parking scarcity.

13.10 PROPOSED CONCEPTUAL DESIGN ARCHITECTURE FOR BIG DATA, AI, AND IOT FOR SMART CITY

Prior studies considered various frameworks, models, and layers in design for the conceptual architecture of smart city. For instance, Pramanik et al., (2017) used three layers that outline the structure of the smart city, namely the perception, network, and application layer in which each handles, analyzes, and processes huge amounts of data. Kaur et al. (2018) proposed cloud-based green IoT in which the architecture consists of presentation, application, big data analytics, network, and sensor layers for the green smart city. This chapter proposes the conceptually designed architecture that includes four layers such as smart city data generation, data processing and analysis, design concepts and domains, and knowledge portal of smart city development. The following subsections describe each component.

13.10.1 SMART CITY DATA COLLECTION AND GENERATION

The proposed architecture's first layer is the data generation and collection. In today's smart city, the massive amount of data produced by machines, automobiles, tools, sensors, buildings, power grid, and IoT devices have augmented data transmission rates (Li et al., 2017). This layer involves various functions such as data sensing, data acquisition, and data collection from smart city components or domains. There are various sensors, devices, and tools that sense the big data. The sensed data involve detection, logging, and integration of data acquisition. The acquisition process is executed by diverse data acquisition techniques which convert analog data into digital. From the various smart city domains, the data are collected through transmission media or technologies.

13.10.2 SMART CITY DATA PROCESSING AND ANALYSIS

The second layer aims to process and analyze the data. It includes big data analytics tools and techniques, processing frameworks, and the output files. This involves the analysis and simplification of information gathered from various sources, turning it into meaningful output files using these analytics and techniques. There are sophisticated techniques, powerful algorithms, and recommended models to extract insights from the big data collected from the smart city.

13.10.3 SMART CITY KNOWLEDGE PORTAL OF SMART CITY DEVELOPMENT

The third layer is very much essential in this architecture and includes key elements. This layer is responsible to store data and information for prediction, forecasting, simulation, automation, visualization, estimation, planning, and alert generation. The structured and organized data are stored into the databases of various types of smart city decision making. Different artificial intelligence, neural networks, machine learning, deep learning agent systems, and modelling will be employed. This smart city knowledge portal can be linked with various applications and nodes for smooth accessibility.

13.10.4 SMART CITY DESIGN CONCEPTS AND DOMAINS

The final layer of this architecture includes the smart city design concepts and the domains. Various design concepts and domains formed the sophisticated smart city's construction. In each domain the citizen, enterprises, and other stakeholders' requirements are reflected. Hence, they are expecting smart services and innovative applications in each domain (see Table 13.3). The urban designers and architects must take into consider each domain's characteristics, development stage, risk factors, feasibility, future consequences, technical requirement, etc. In a well-developed smart city, these domains offer state-of-the-art services to the citizens and the enterprises.

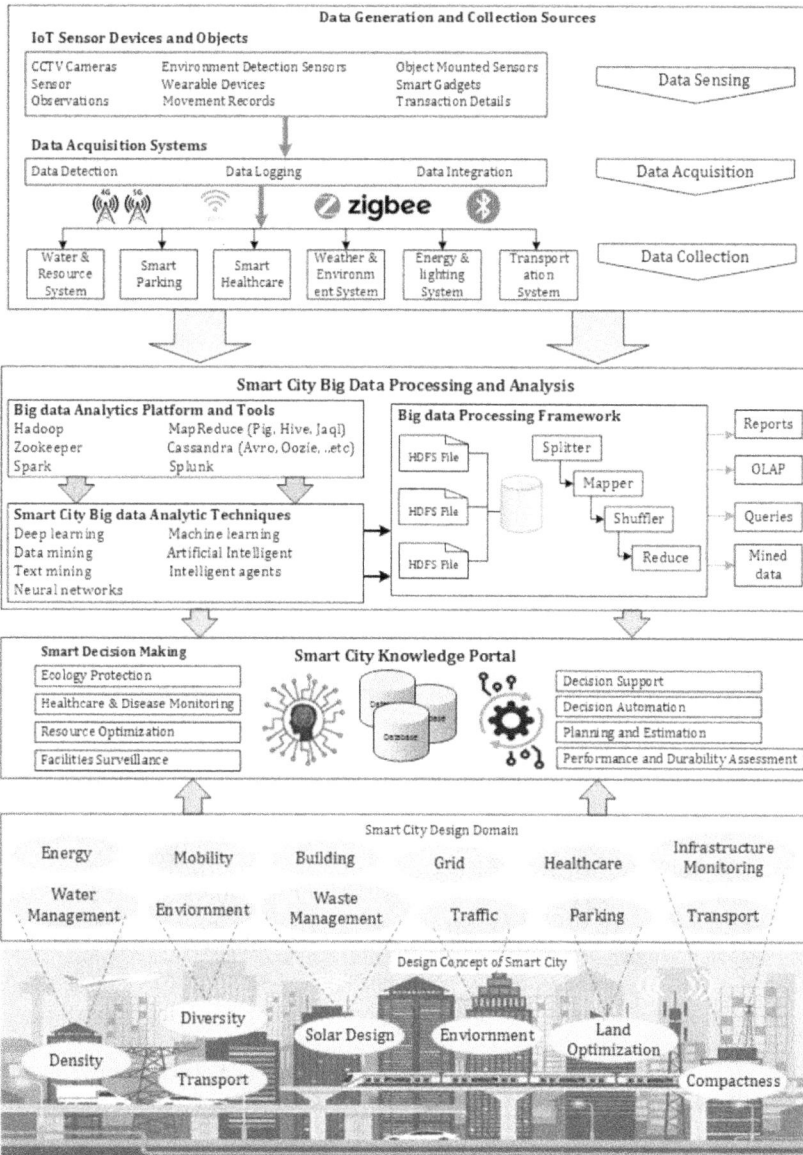

FIGURE 13.3 Proposed conceptually designed architecture for a smart city.

13.11 CONCLUSION

Today, smart cities are increasingly shifting towards sophisticated technologies to solve societal and environmental issues. The application of big data, AI, and IoT can greatly support assembling and cultivating the steps of environmentally viable growth in a smart city. Urban cities are expected to expand, optimize energy, sustain the viable atmosphere, advance the economic status and lifestyle of the people, and advance the people's competencies to professionally use and adopt the state-of-the-art IT applications. This chapter targets the urban designers, engineers, data scientists, and policy makers who are seeking to employ and integrate the big data, AI, and IoT in smart cities with the desire to enhance the people's comfort, living standard, modernization, and livability.

REFERENCES

Allam, Z., & Dhunny, Z. A. (2019). On big data, artificial intelligence and smart cities. *Cities*, *89*, 80–91. doi:10.1016/j.cities.2019.01.032

Allam, Z., & Newman, P. (2018). Redefining the smart city: Culture, metabolism and governance. *Smart Cities*, *1*(1), 4–25.

Al-Turjman, F., & Malekloo, A. (2019). Smart parking in IoT-enabled cities: A survey. *Sustainable Cities and Society*, *49*. doi:10.1016/j.scs.2019.101608

Anghel, I., Bertoncini, M., Cioara, T., Cupelli, M., Georgiadou, V., Jahangiri, P., . . . Velivassaki, T. (2015). GEYSER: Enabling green data centres in smart cities. *Energy Efficient Data Centers*, 71–86.

Babar, M., & Arif, F. (2017). Smart urban planning using big data analytics to contend with the interoperability in internet of things. *Future Generation Computer Systems*, *77*, 65–76.

Bibri, S. E. (2018). The IoT for smart sustainable cities of the future: An analytical framework for sensor-based big data applications for environmental sustainability. *Sustainable Cities and Society*, *38*, 230–253.

Bunders, D. J., & Varró, K. (2019). Problematizing data-driven urban practices: Insights from five Dutch 'smart cities'. *Cities*, *93*, 145–152. doi:10.1016/j.cities.2019.05.004

Casares, A. P. (2018). The brain of the future and the viability of democratic governance: The role of artificial intelligence, cognitive machines, and viable systems. *Futures*, *103*, 5–16. doi:10.1016/j.futures.2018.05.002

Chaturvedi, K., Matheus, A., Nguyen, S. H., & Kolbe, T. H. (2019). Securing spatial data infrastructures for distributed smart city applications and services. *Future Generation Computer Systems*, *101*, 723–736. doi:10.1016/j.future.2019.07.002

Diary, I. N. (2018, December 5). NEC wins Hubballi Dharwad Smart City Project. *English*. Retrieved from http://indiannewsdiary.blogspot.com/2018/12/nec-wins-hubballi-dharwad-smart-city.html

Israilidis, J., Odusanya, K., & Mazhar, M. U. (2019). Exploring knowledge management perspectives in smart city research: A review and future research agenda. *International Journal of Information Management*. doi:10.1016/j.ijinfomgt.2019.07.015

Jarrahi, M. H. (2018). Artificial intelligence and the future of work: Human-AI symbiosis in organizational decision making. *Business Horizons*, *61*(4), 577–586. doi:10.1016/j.bushor.2018.03.007

Kai, C., Li, H., Xu, L., Li, Y., & Jiang, T. (2018). Energy-efficient device-to-device communications for green smart cities. *IEEE Transactions on Industrial Informatics*, *14*(4), 1542–1551. doi:10.1109/tii.2017.2789304

Kaur, G., Tomar, P., & Singh, P. (2018). Design of cloud-based green IoT architecture for smart cities. In *Internet of things and big data analytics toward next-generation intelligence* (pp. 315–333). Cham: Springer.

Korczak, J., & Kijewska, K. (2019). Smart logistics in the development of smart cities. *Transportation Research Procedia*, *39*, 201–211.

Lau, B. P. L., Marakkalage, S. H., Zhou, Y., Hassan, N. U., Yuen, C., Zhang, M., & Tan, U. X. (2019). A survey of data fusion in smart city applications. *Information Fusion*, *52*, 357–374. doi:10.1016/j.inffus.2019.05.004

Laufs, J., Borrion, H., & Bradford, B. (2020). Security and the smart city: A systematic review. *Sustainable Cities and Society*, *55*. doi:10.1016/j.scs.2020.102023

Li, B., Kisacikoglu, M. C., Liu, C., Singh, N., & Erol-Kantarci, M. (2017). Big data analytics for electric vehicle integration in green smart cities. *IEEE Communications Magazine*, *55*(11), 19–25. doi:10.1109/mcom.2017.1700133

Marinakis, V., Doukas, H., Tsapelas, J., Mouzakitis, S., Sicilia, Á., Madrazo, L., & Sgouridis, S. (2018). From big data to smart energy services: An application for intelligent energy management. *Future Generation Computer Systems*. doi:10.1016/j.future.2018.04.062

Maye, D. (2019). 'Smart food city': Conceptual relations between smart city planning, urban food systems and innovation theory. *City, Culture and Society*, *16*, 18–24.

Memos, V. A., Psannis, K. E., Ishibashi, Y., Kim, B.-G., & Gupta, B. B. (2018). An Efficient Algorithm for Media-Based Surveillance System (EAMSuS) in IoT smart city framework. *Future Generation Computer Systems*, *83*, 619–628. doi:10.1016/j.future.2017.04.039

Muhammad, K., Lloret, J., & Baik, S. W. (2019). Intelligent and energy-efficient data prioritization in green smart cities: Current challenges and future directions. *IEEE Communications Magazine*, *57*(2), 60–65. doi:10.1109/mcom.2018.1800371

Mukta, M. Y., Rahman, M. A., Asyhari, A. T., & Alam Bhuiyan, M. Z. (2020). IoT for energy efficient green highway lighting systems: Challenges and issues. *Journal of Network and Computer Applications*, *158*. doi:10.1016/j.jnca.2020.102575

Neirotti, P., De Marco, A., Cagliano, A. C., Mangano, G., & Scorrano, F. (2014). Current trends in smart city initiatives: Some stylised facts. *Cities*, *38*, 25–36.

Nicolas, C., Kim, J., & Chi, S. (2020). Quantifying the dynamic effects of smart city development enablers using structural equation modeling. *Sustainable Cities and Society*, *53*. doi:10.1016/j.scs.2019.101916

Nishant, R., Kennedy, M., & Corbett, J. (2020). Artificial intelligence for sustainability: Challenges, opportunities, and a research agenda. *International Journal of Information Management*, *53*. doi:10.1016/j.ijinfomgt.2020.102104

Nitoslawski, S. A., Galle, N. J., Van Den Bosch, C. K., & Steenberg, J. W. N. (2019). Smarter ecosystems for smarter cities? A review of trends, technologies, and turning points for smart urban forestry. *Sustainable Cities and Society*, *51*. doi:10.1016/j.scs.2019.101770

Osman, A. M. S. (2019). A novel big data analytics framework for smart cities. *Future Generation Computer Systems*, *91*, 620–633.

Plageras, A. P., Psannis, K. E., Stergiou, C., Wang, H., & Gupta, B. B. (2018). Efficient IoT-based sensor big data collection: Processing and analysis in smart buildings. *Future Generation Computer Systems*, *82*, 349–357. doi:10.1016/j.future.2017.09.082

Pramanik, M. I., Lau, R. Y. K., Demirkan, H., & Azad, M. A. K. (2017). Smart health: Big data enabled health paradigm within smart cities. *Expert Systems with Applications*, *87*, 370–383. doi:10.1016/j.eswa.2017.06.027

Saravanan, K., Julie, E. G., & Robinson, Y. H. (2019). Smart cities & IoT: Evolution of applications, architectures & technologies, present scenarios & future dream. In *Internet of things and big data analytics for smart generation* (pp. 135–151). Cham: Springer.

Saravanan, K., & Srinivasan, P. (2018). Examining IoT's applications using cloud services. In *Examining cloud computing technologies through the internet of things* (pp. 147–163): Hershey, PA: IGI Global.

Silva, B. N., Khan, M., & Han, K. (2018). Towards sustainable smart cities: A review of trends, architectures, components, and open challenges in smart cities. *Sustainable Cities and Society, 38,* 697–713. doi:10.1016/j.scs.2018.01.053

Sodhro, A. H., Pirbhulal, S., Luo, Z., & de Albuquerque, V. H. C. (2019). Towards an optimal resource management for IoT based Green and sustainable smart cities. *Journal of Cleaner Production, 220,* 1167–1179. doi:10.1016/j.jclepro.2019.01.188

Sun, M., & Zhang, J. (2020). Research on the application of block chain big data platform in the construction of new smart city for low carbon emission and green environment. *Computer Communications, 149,* 332–342. doi:10.1016/j.comcom.2019.10.031

Ullah, Z., Al-Turjman, F., Mostarda, L., & Gagliardi, R. (2020). Applications of artificial intelligence and machine learning in smart cities. *Computer Communications, 154,* 313–323. doi:10.1016/j.comcom.2020.02.069

Washburn, D., Sindhu, U., Balaouras, S., Dines, R. A., Hayes, N., & Nelson, L. E. (2009). Helping CIOs understand 'smart city' initiatives. *Growth, 17*(2), 1–17.

Wu, Y. C., Wu, Y. J., & Wu, S. M. (2019). An outlook of a future smart city in Taiwan from post: Internet of things to artificial intelligence Internet of things. In *Smart cities: Issues and challenges* (pp. 263–282). Amsterdam: Elsevier.

Zahmatkesh, H., & Al-Turjman, F. (2020). Fog computing for sustainable smart cities in the IoT era: Caching techniques and enabling technologies: An overview. *Sustainable Cities and Society, 59.* doi:10.1016/j.scs.2020.102139

14 Application of Smart Technologies in Urban Water Supply System of Smart Cities

*J. Colins Johnny, G. Sakthinathan
and K. Saravanan*

CONTENT

14.1 INTRODUCTION

14.1.1 General

Countries around the world face difficulties in providing good water for their people due to severe water shortages where rainfall is low or due to other economic challenges (Mizuki et al., 2012). Globally, 51% of the population live in urban areas, where 96% of inhabitants get access to improved drinking water sources. Out of 96% in urban areas, 80% were provided with a piped water supply and get direct access to water in their premises ((JMP), 2012). With increased urbanization, the population increased in urban areas and the water supply systems were capable enough to supply for 97% of the population (WHO and UNICEF, 2019). About 30% of the population in the 1950s were residing in urban settlements, and with increasing urbanization, these areas are expected to house 68% of the world's population (UN DESA, 2019), which influences the gross domestic product of any country. Due to the high density of the population in urban areas, equitable water distribution and sustainable water management becomes more challenging. The efficiency of operation and management of water supply systems should be considered based on the growing population, cost optimization, shrinkage of water resources, shortages of supply and diversification of sources of supply. The holistic management of water supply network should include aspects of operation, monitoring, diagnostics and control.

14.1.2 Smart City

Smart cities are a new initiative that focuses on sustainable and inclusive development. It aims to develop the urban ecosystem and provides comprehensive development for the aspirations and needs of the citizens and urban planners. The development plans include four pillars such as institutional, physical, social and economic infrastructure. The process of building the infrastructure of a smart city could be a long-term goal and can be done incrementally. The main objective of the smart cities mission is to promote the development of core infrastructure to give a decent quality of life for citizens, a clean and sustainable environment and application of smart solutions (GOI and MoUD, 2015). The first and foremost core infrastructure element in the smart city is adequate water supply. The development goals along with goals for smart water and sanitation in the smart city mission are shown in Figure 14.1. A smart water supply concentrates on the smart technologies that can be included in the water supply system for the quantitative and qualitative management of water and identification of leakage and rectification of defects.

The smart city project is carried out through area-based development. The strategic component of area-based development is retrofitting that aims at city improvement and redevelopment that aims at city renewal. Enhancement of the existing water supply system in urban environments through smart solutions can help to achieve the UN's sixth goal of sustainable development. Smart metering can make a substantial contribution to better water management in the city. The mission outcomes of smart metering are measured through the percentage of non-revenue water. Cloud computing and Internet of Things (IoT) technologies play vital role in the implementation of smart city projects. Cloud IoT-based solutions (Saravanan and Srinivasan, 2017) are

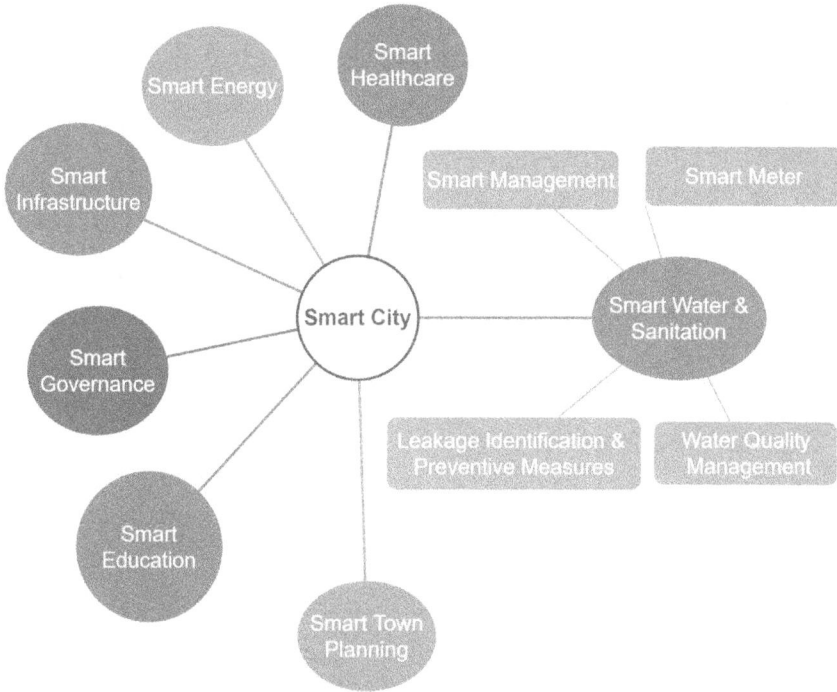

FIGURE 14.1 Smart water in a smart city.

proposed in all smart city verticals. Also, various smart city architectures are built based on cloud IoT (Saravanan et al., 2018b). Real-time monitoring of water supply and control of devices can be achieved by industrial IoT systems (Saravanan et al., 2018a). Both water quantity and quality can be measured remotely.

14.1.3 WATER DISTRIBUTION SYSTEM

The purpose of water distribution systems is to deliver water to the consumer with appropriate quality and quantity. The water distribution system shown in Figure 14.2 collectively includes facilities used to supply water from its source to point of usage. A typical system includes a source of water (river, lake, etc.), from which the water is extracted through the pumps and delivered to an overhead tank through a pump house that supplies water to the consumer. The distribution system should also consider the safety of the components in the system. The deterioration of the components could lead to abnormal pressure and poor water quality. The method of distribution adopted in a city depends on the elevation and water level of source, topography and other local conditions and also includes a gravity system, pumping system or combined gravity and pumping system. The basic layouts of water distribution networks are of two types, namely branched configuration and looped configuration. The principal design approaches of distribution system layouts are of four

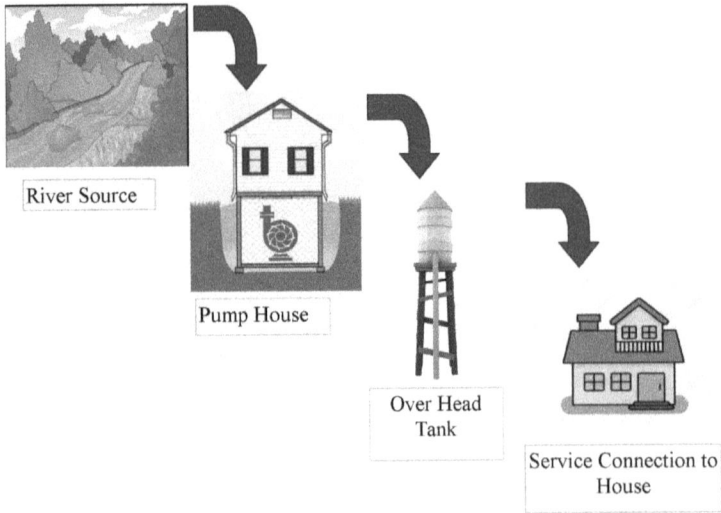

FIGURE 14.2 Water distribution system.

types, namely dead end or tree system, gridiron system, circular or ring system and radial system. The format of supply could be 24/7, otherwise called continuous supply of water, and discontinuous or intermittent supply of water. In global scenarios, a combined gravity and pumping system of distribution with a tree network of layout with intermittent supplies are found in most of the cities.

14.1.4 NEED FOR SMART WATER SUPPLY SYSTEM

The major direction in which the development of smart water supply systems moves is dependent on four needs:

i.   The need for decision support systems to control, monitor and detect faults in water distribution networks.
ii. The need for management support systems that help in maintenance of components of water distribution networks.
iii. The need for information systems for data analytics and forecasting.
iv. The need for the improvement of distribution efficiency, which focuses on the need for performance optimization of water supply networks.

14.2 SMART WATER SUPPLY SYSTEMS

14.2.1 SMART APPROACH

Conventionally, the operations of the water supply systems are managed by semi-skilled or unskilled workers, which requires a lot of work and training for human resources. The extempore management practices of operators typically rely on the experience of the field operators. The monitoring of flow of water and pressure in

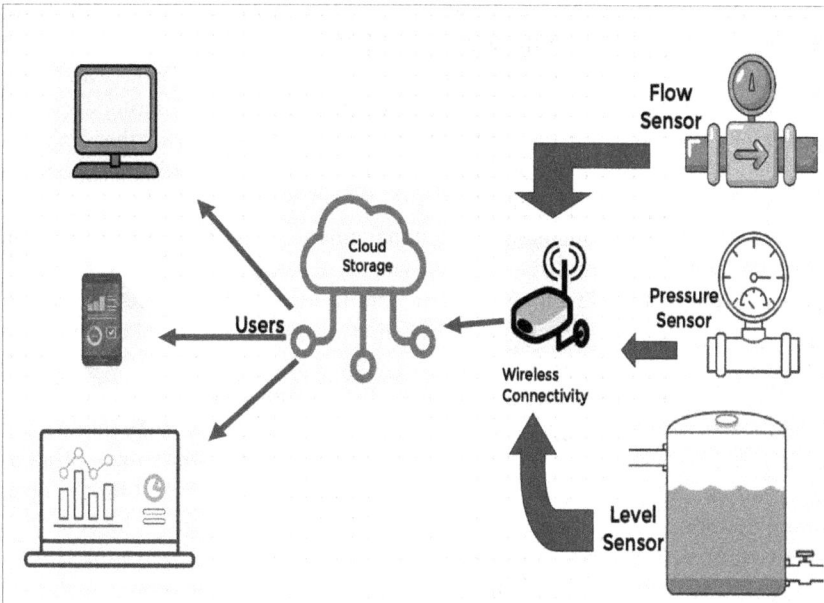

FIGURE 14.3 Smart water supply system.

the distribution network of the water supply system such as main pipes and distributaries in the consumer ends, water level monitoring in the reservoir, operation of valves in the distributaries, line and major cracks and bursts in pipes, detection of leakages, monitoring of water quality etc. is difficult. The difficulty further increases due to the need for monitoring continuously and simultaneously. Without adequate technological support, this monitoring is complex and almost impossible to achieve with the available resources. Technological support is essential to achieve the monitoring activities on a frequent time scale. In addition to all these challenges monitoring, operations and management have been unproductive with conventional practices and with the aging process in all the elements of the water supply infrastructure.

Implementation of smart technology in the water supply system can change the difficulties in the conventional water systems. The typical smart water supply system is shown in Figure 14.3. Smart technology provides smart instrumentation to detect the elements of water supply system, sense the activities in the water supply system and measure the values of the parameters related to quantity and quality of water in the water supply system. In a typical water supply system, the parameters such as flow, pressure and water level are measured. The measured values in the field are transmitted to the centralized server through wired or wireless networks and get recorded in a cloud or database management system. The recorded data can be viewed by the users such as operators and the managers of the interconnected networks. The back-end data can be processed for reports and graphs to help the managers and decision makers respond quickly to a situation and troubleshoot problems. Data analytics can help with the better-informed decisions such as optimizing cost of operation, energy conservation and time management.

By upgrading the existing water supply systems to a smart water supply system, we can ensure the quantity of the water to be delivered. It ensures a secure and flexible supply of water to the consumer and improves sustainability in the water supply by recording all the relevant data during the operation. The operating cost of the water supply system can be reduced by the data analytics platforms that are run on the big data storage. This technological advancement will help in the management of the water supply system and improve the cost-benefit ratio.

The main objectives of smart water management are to increase the efficiency in the supply of water and the productivity of the water supply system and to improve revenue. The developed system should also be cost effective, efficient and reliable, which ensures proper distribution of water for the community (Natividad and Palaoag, 2019). This management should balance the water supply demand by the reduction in water loss, that is, physical or commercial loss, and enables conservation of water. Smart water management also enhances the decision making which was usually experience based and sometimes blind and baseless to be more informed and intelligent. The energy required for the operation can be reduced and the revenue generated can be enhanced. The continuous monitoring can help the revenue collection system and can increase the end-user awareness that could lead to behavioral change. This improvement in the trust of the consumer toward the utility will progress the overall management of the water supply system.

14.2.2 SMART WATER MANAGEMENT

Smart water management is possible only through smart decisions. For a smart decision, information is required through data analytics. The analytics are possible only from data collected in the water stations (pump house, overhead tanks and booster stations). Particularly for operations and control, this data is needed in real time or at least near real time. There are various technological solutions that can be used as the elements of water supply systems. Some of them are listed here:

  i. **Smart meters and sensors**
     These are elements in a smart water distribution that can sense and measure the variables of a water distribution system and include water meters such as water level sensors, flow meters, pressure gauges and water quality sensors for real-time monitoring of water level, flow rate of water, supply pressure in the water distribution system and physico-chemical and biological quality characteristics of water.
  ii. **Smart communication**
     The data obtained in the water pumping stations has to be transferred to the dashboard where the decision is made on a real-time or near real-time basis. This is possible through smart communication using information and communication technology (ICT) tools. The ICT tools support transmission of data and communication of command through the Internet.
  iii. **Smart decision-making tools**

Decision making becomes easier when the data and information are available at a single place, that is, a dashboard. There may be uncertainties that can trouble the decision makers. These uncertainties can be analyzed and effectively be handled by soft-computing tools for decision making.

iv. **Smart operations and control systems**

The water distribution system can be automated, and the automated system can be monitored and controlled by electronic devises. The monitored data can be processed to identify anomalies and problems in the system, especially during emergency situations, to take decisions. Relays and actuators in the water distribution system can be controlled by critical decisions made by the system.

v. **Smart knowledge and information dissemination**

Data that gets stored in the database gets accumulated over the years and serves as an input for predicting the future. The prediction can be done effectively through data mining and data analytics. These data and knowledge can be built into a knowledge base, and the information can be disseminated to the managers and planners for policy decisions and regulations.

14.2.3   INFORMATION TECHNOLOGY INVOLVED IN SMART WATER SUPPLY SYSTEM

There are several information and communication technological tools that are available to support operations and management of water supply systems (Karwot et al., 2016). The tools are classified as three main groups:

i. **ERP systems**

Enterprise resource planning (ERP) is a type of system that is used to manage the daily activities such as accounting, procurement and supply chain operations of tools and components of the water distribution system. This tool can help in the registration of network components of a water supply system and the availability of standby components in the warehouse, such as tools available for repair work, etc.

ii. **SCADA systems**

Supervisory control and data acquisition (SCADA) system is used to supervise and acquire data. The collected data are stored and analyzed for anomalies and deviation in the system's behavior. Based on the deviation, critical decisions that control the distribution system can be attempted to solve the problem. This tool can help to automate the operation of the system and to monitor the condition of the water supply system.

iii. **GIS systems**

Geographic information system (GIS) is a tool that can store, retrieve, analyze and visualize spatial data. The details related to the water distribution system such as stations and its assets can be recorded with spatial significance. These spatial data can be used to analyze and manage the assets of the water distribution system.

vi. **Industrial Internet of Things**

IoT systems can connect millions of devices for sharing the data and decision making of the operations. These devices act as thin clients to generate the data and send the data to cloud server to aggregate and process for reporting, future trend analysis etc.

14.2.4 Software Used for Water Supply Systems

Software is used in different stages from planning to implementation of water supply system. To plan a water supply, project mapping of the source and the distribution area are needed, and this problem can be overcome through the mapping software such as AUTOCAD (Ramana et al., 2016), ARCGIS (Ramesh et al., 2012), QGIS, SAGA GIS, etc. The water availability at source can be analyzed through the hydrologic data and modeled using Hec-HMS, SWAT and other software. The water availability in the river source can be modeled using Hec-RAS. The groundwater sources and the water budget can be modeled with the help of MODFLOW software (Kirubakaran et al., 2018). The groundwater quality can be modeled with the help of the Modular 3-Dimension Transport (MT3D) Package (Colins Johnny et al., 2020). The assets in the water distribution network have to be recorded for effective planning, and therefore software, such as ARCGIS, QGIS, SAGA GIS etc., that can handle spatial and nonspatial data can be used. To design a water distribution network and analyze the input and output parameters under different conditions, EPANET software (Saminu et al., 2013) can be used that can model the hydraulic and water quality parameters in a water supply system.

14.2.5 Objectives of Smart Water Supply System

A smart water supply system should integrate different recent technologies that focus on the following objectives:

   i. Automatic detection of local failures and location of individual event with respect to geographical coordinates.
   ii. A decision support system that can provide solutions to remediate the fault in the water distribution networks.
   iii. A management support system that helps in the operation and maintenance of components of water distribution networks.
   iv. An information system that can perform data analytics and forecasting that aims for the improvement of distribution efficiency by optimization.

14.3 SURVEY AND ANALYSIS FOR UPGRADE TO A SMART WATER SUPPLY SYSTEM

Applications of GIS can improve the water supply system such as presentation of water assets, analysis of operations of actuators and other devices in distribution networks and their optimization, selection of optimal routes, localization and identification of leakages and locating faults. Design and analysis of distribution networks play an important role in the water supply system. Numeric hydraulic models can be used for the planning, design and analysis of small, medium and larger water supply projects.

14.3.1 POPULATION FORECASTING

The total requirement of water for a city is estimated by the product of the annual average daily draft in liters per day and the probable number of people who use the facility. Any water supply scheme should be designed with a provision for the estimated requirements of the future. This future period for which the provision is made is called a design period. The design period should not be too long or too short. The design period should not exceed the useful life of the component structure. The annual average demand of daily water requirements for each person, which includes domestic use, industrial use, commercial use, public use, wastes, thefts, etc., is the per capita demand. Under normal circumstances, a design period of 30 years is usually considered. It is necessary to estimate the future population depending upon the possibilities of future development and the trend of growth of population in the city. A typical representation of population forecasting is shown as a graph in Figure 14.4. The future population by the end of the design period is estimated (Punmia et al., 2005). There are various methods that are generally used to estimate population. Some of the popular methods used are the arithmetic increase method, geometric increase method, incremental increase method, decreasing rate method, simple graphical method, comparative graphical method, master plan method, apportionment /ratio method and logistic curve method (Punmia et al., 2005).

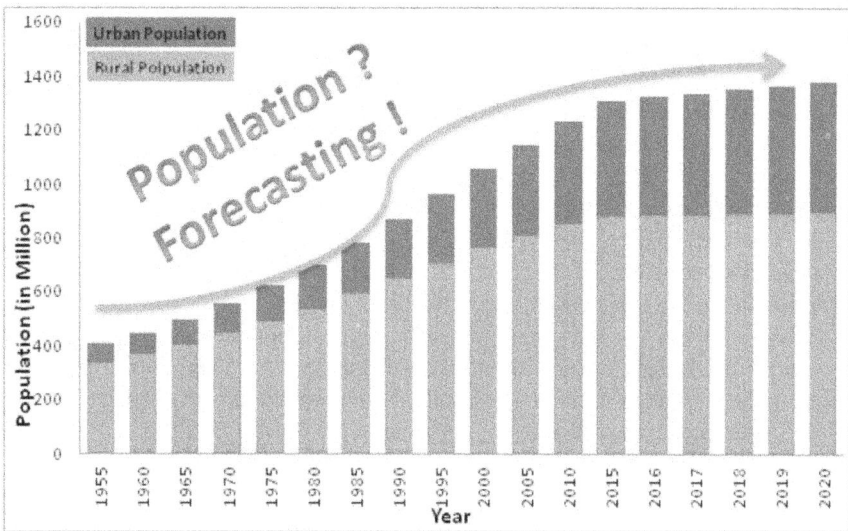

FIGURE 14.4 Population forecasting.

14.3.2 DEMAND FORECASTING

Water needed by a community for living is called water demand. Water demand estimation is the first step of any water supply scheme. The main object of this estimation is to quantify the water needed to feed the population of a city. Any water supply scheme has to be designed for a period by considering the need for the future. This period, called the design period, is usually considered to be 30 years for any project. Generally, schemes should ensure a reliable supply of water for the design period. The annual average draft of a city in liters per day (lpd) can be arrived at by the product of the probable number of people in the city at the end of design period and the annual average consumption. The water demand of the city is usually addressed in terms of liters per day or liters per year. This demand is estimated from the liters per capita per day (lpcd), arrived at from the standards (IS:1172, 1993). Various types of domestic demands considered for the water supply scheme of a city with a full flushing system are drinking (5 lpcd), cooking (5 lpcd), bathing (75 lpcd), washing of clothes (25 lpcd), washing of utensils (15 lpcd), washing and cleaning of houses and residence (15 lpcd), lawn watering and gardening (15 lpcd) and flushing of water closets and others (45 lpcd), which sums up to 200 lpcd. This demand is arrived at based on the average conditions and hence may also vary according to size of the city, climatic conditions, habits of people, quality of water supplies, pressure in the distribution, water costing and policies of metering. Water loss is also an important factor that needs to be considered during the estimation of water demand. This loss could be due to thefts, that is, water stolen by unauthorized water connections, and wastages and water lost in leakage due to bad plumbing and damaged meters.

14.4 SMART SENSING AND METERING DEVICES

14.4.1 SMART INSTRUMENTS IN WATER SUPPLY SYSTEMS

The conventional methods of water quality assessment are time consuming and labor intensive. Therefore, the need arises to monitor and protect the water with a real-time monitoring system with a view to reducing contamination by active measurements (Varsha and Wu, 2018). Smart sensing refers to detection of the status of pump, status of valves, detection of leaks, status of pipe cracks and bursts. Metering refers to the automatic measuring and recording of distribution network variables such as pressure, flow rate, reservoir water level, energy consumption and water quality parameters such as pH, total dissolved solids (TDS), conductivity, temperature, turbidity and ion-selective electrodes. Smart water meters are used to measure the consumption of water, monitoring pressure and flow that help in leak detection. Acoustic devices can also detect leakages in real time. Video cameras can help in asset management. These smart meters can be used in the pumping stations, booster stations and overhead tanks of the city.

14.4.2 INSTRUMENTATION IN STATIONS

Transducers play a vital role in the sensing and measurement of parameters. A typical transducer can manage electrical output from 4–20 mA. This electrical

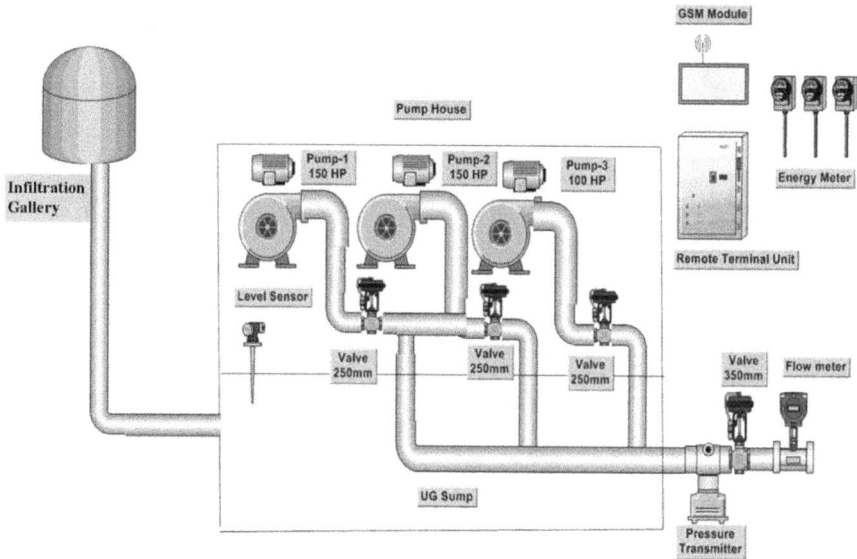

FIGURE 14.5 Components of SCADA in pumping station.

output has to be converted into engineering units such as m³/s or kg/cm². Though there are several sensors available, a typical instrumentation in a pumping station is shown in Figure 14.5. The water from the river could be collected through an infiltration gallery through pumps. There will be active and standby pumps for which energy consumption can be measured through an energy transmitter. The flow in the pipe can be measured through a flow transmitter. The pressure in the pipe can be measured through a pressure transmitter. Water level in the sump can be measured through a level transmitter. Placement of these devices should be such that all essential measurements are made with a minimum number of devices. The measured data needs to be transmitted to the remote terminal unit (RTU), from where data is transferred to the cloud. A typical booster station is shown in Figure 14.6; the instruments that are used in the pumping station can also be used here. The measurement related to pressure is given more importance, as these stations are established where there is a pressure drop in the distribution network. A typical instrumentation in an overhead tank is shown in Figure 14.7. The instruments that were used in the pumping and booster stations can be used in the overhead tanks. The water quality sensors such as the pH transmitter, TDS sensor turbidity sensor etc., can also be used in the overhead tank to ensure the quality of water delivered to the consumer. The main instruments that are very important for the distribution system are flow transmitters and pressure transmitters. Other transmitters can be added based on the requirements.

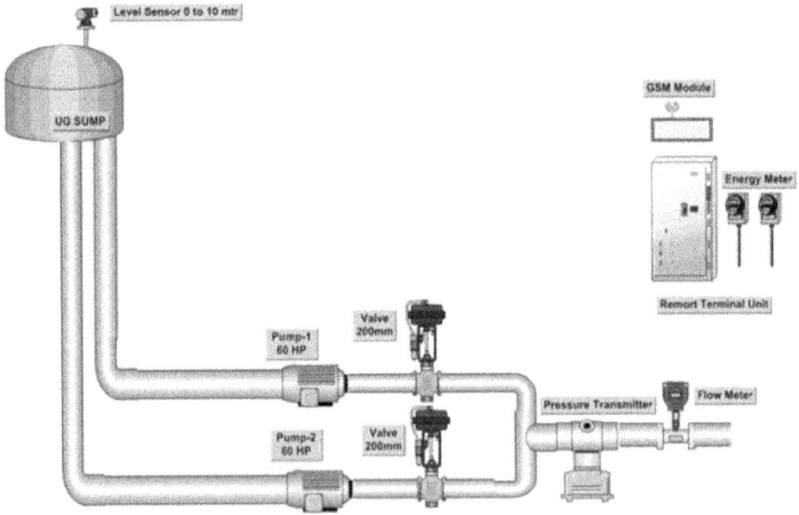

FIGURE 14.6   Components of SCADA in booster station.

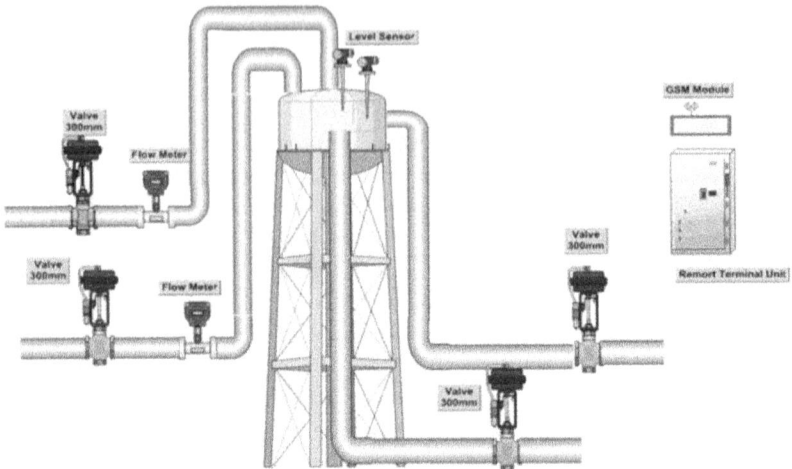

FIGURE 14.7   Components of SCADA in overhead tank.

14.4.3   FLOW TRANSMITTERS

A flow meter is a device used for the measurement of the volume of water discharged. It is usually measured in cubic meters per second (m^3/s). There are several kinds of flow meters that are classified based on the principle under which they operate.

   i. Differential pressure flow meters.
   ii. Positive displacement flow meters.

iii. Velocity flow meters.

iv. Mass flow meters.

Conventional data collection for discharge was made through water flow meters such as displacement water meters, otherwise called positive displacement meters, and velocity water flow meters, otherwise called internal capacity meters. Displacement water meters are generally used in residential applications and some commercial applications. The conventional measurement was replaced by electromagnetic and ultrasonic water flow meters. Electromagnetic meters are magnetic meters and are a variant of velocity water meters. These meters are based on Faraday's law and use electromagnetic properties such as induction to measure the velocity. The main advantage of these flow meters is the bidirectional measurement of flow. Ultrasonic water meters use ultrasonic sound waves to measure the velocity of flow. The measurements are accurate in case of substantial flow in the pipe. The ultrasonic transducers that are fitted on the pipe are allowed to send sound waves from the circumference of the pipe through the water; the waves get reflected from the other side of the pipe and come to the transducer in the same direction. This travel time is used to determine the velocity of water in the pipe. Compensation of resistance for the known meter associated with the construction and the impact of pipe should be allowed. Ultrasonic water meters, when installed externally, are easy to maintain, and the compensation for rate of flow through the pipe can be avoided.

14.4.4 PRESSURE TRANSMITTERS

A pressure transmitter is a type of transducer that converts pressure into an analog electrical signal. The main purpose of pressure transducers is to measure the pressure exerted in a water distribution system. There are four types of pressure sensors:

i. Strain gauge pressure transducers.

ii. Capacitance pressure transducers.

iii. Potentiometric pressure transducers.

iv. Resonant wire pressure transducers.

14.5 SMART WATER SUPPLY SYSTEM

14.5.1 SYSTEM ARCHITECTURE

Supervisory control and data acquisition (SCADA) is a system of software and hardware elements that allows monitoring and remote control of the water supply system. SCADA can be applied for the management of pressure in the water distribution system, to monitor levels in the reservoir and to control flow in distributaries. A typical SCADA architecture for a water distribution system is shown in Figure 14.8. The architecture of the SCADA system has different elements that are technically connected to supervise and control. The elements of the SCADA are programmable logic controllers (PLCs) or remote terminal units (RTUs), human machine interface (HMI), sensors, transmitters and other end devices that are interconnected. PLCs or

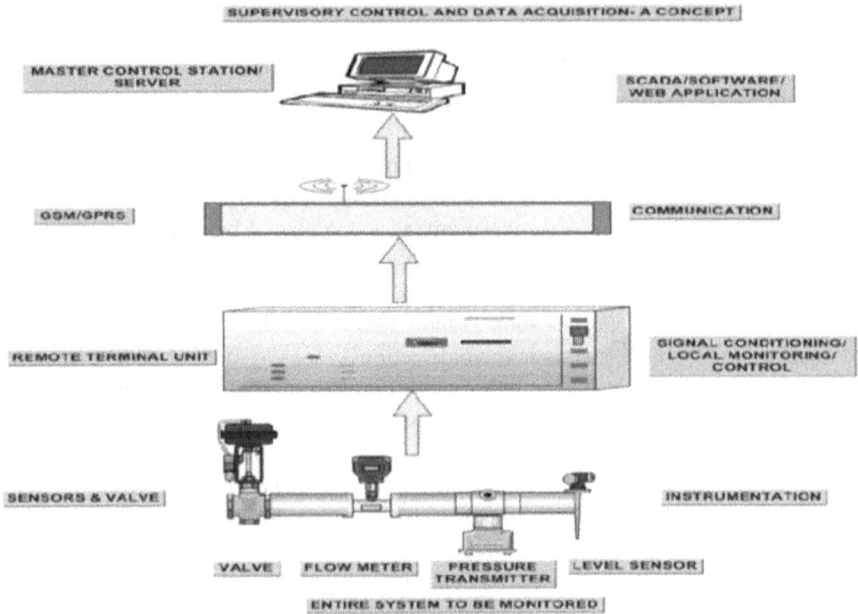

FIGURE 14.8 SCADA architecture for a water supply system.

RTUs are microprocessor-controlled electronic devices that communicate with an array of objects such as pumps, HMIs, sensors and other electromechanical devices. The data collected through the devices are then routed to computers mounted with SCADA software. The SCADA software will be capable to process and display the data in the HMI. The operators can visualize the data in the field and can take decisions based on the data. It helps the field operators in the field to interact with devices such as sensors, valves and pumps through HMI. The processed data are then transferred to the central server or cloud database server for storage. The stored data can be visualized by the stakeholders through software from the centralized server on a near real-time basis. These data can be analyzed through data analytics software. The results of the data analytics can be utilized for important policy decisions. SCADA systems are crucial for water distribution since they help to maintain the components and record, process and communicate data for smart decisions to reduce downtime and improve efficiency.

14.6 CHALLENGES IN MODERN WATER SUPPLY SYSTEMS

Limited new sources of water and the rapid population explosion has led to several challenges in modern water supply systems. Provision of upgradation of the existing and new conveyance system in accordance with the modern SCADA system will be the first challenge. The inclusion of a treatment system is unavoidable for the effective delivery of good quality water and the usage of reclaimed water. Adaptation of a present system for equitable water distribution and water conservation practices for a future demand is complicated in an urban agglomeration. Adaptation involving addition of linkages and components in a water supply system complicates the reaction of the system for growth. Water quality deterioration that takes place within the distribution system is difficult to monitor. Billing systems that are in place are capable to account for the billed authorized consumption and not for the nonrevenue water. Water loss in a system can be measured through the percentage water loss index (Ociepa et al., 2019). Advanced metering to account for nonrevenue water could not be economically justified and can cost customers severely.

14.7 CONCLUSION

The focus of smart water systems is to provide correct and real-time data to the central servers or cloud database. The data collected in the central server will enable the managers to make informed and systematic decisions. The smart meters and sensors will measure, store, display and transmit data to local area networks, metropolitan area networks or wide area networks using wired or wireless communication. The main advantage of smart metering is the reduction in the deployment of field staff for manual data collection and the facilitation of automated data collection. Data collected in the field are stored in a cloud and can be retrieved when needed through ICT technologies. The stored data can be analyzed through data analytics software, and a knowledge base can be built to enable effective decision making.

REFERENCES

Colins Johnny, J., Sashikkumar, M.C., Rajesh Banu, J., Kumar, Gopalakrishnan. 2020. "Chromium Transport Modelling in Tannery Effluent from a Surface Water Body to Groundwater Regime Case Study in Kodaganar Basin." *Journal of Hazardous, Toxic and Radioactive Waste* 25(2): 05020005-1-05020005-11. https://doi.org/10.1061/(ASCE)HZ.2153-5515.0000577

Government of India, and Ministry of urban Development. 2015. "Smart Cities." *International Encyclopedia of the Social & Behavioral Sciences: Second Edition*. https://doi.org/10.1016/B978-0-08-097086-8.74017-7

(JMP), WHO/UNICEF Joint Monitoring Programme for Water Supply and Sanitation. 2012. *Progress on Drinking Water and Sanitation 2012 Update.*

Karwot, Januz, Kazmierczak, Jan, Wyczolkowski, Ryszard, Paszkowski, Walldemar, Przystalka, Piotr. 2016. "Smart Water in Smart City: A Case Study." *16th International Multidisciplinary Scientific Conference SGEM2016, Book 3* 1 (SGEM2016 Conference Proceedings, ISBN 978-619-7105-61-2 / ISSN 1314–2704): 391–398. https://doi.org/10.5593/sgem2016B31

Kirubakaran, Muniraj, Jesudhas, Colins Johnny, Sisupalan, Samson. 2018. "Modflow Based Groundwater Budgeting Using GIS: A Case Study from Tirunelveli Taluk, Tirunelveli District, Tamil Nadu, India." *Journal of Indian Society of Remote Sensing* 46(11): 1–10. https://doi.org/10.1007/s12524-018-0761-7

Mizuki, Fumio, Mikawa, Kazuhiro, Kurisu, Hiromitsu. 2012. "Intelligent Water System for Smart Cities." *Hitachi Review* 61(3): 147–151.

Natividad, J. G., Palaoag, T. D. 2019. IOP Conf. Ser.: Mater. Sci. Eng. 482 012045.

Ociepa, Ewa, Mrowiec, Maciej, Deska, Iwona. 2019. "Analysis of Water Losses and Assessment of Initiatives Aimed at Their Reduction in Selected Water Supply Systems." *Water* 11(1037): 1–18. doi:10.3390/w11051037

Punmia, B. C., Jain, Ashok, Jain, Arun. 2005. *Water Spply Engineering.* New Delhi, India: Laxmi Publications. ISBN: 978-81-318-0703-3.

Ramana, Venkata G., Sudheer, V. S. S., Prasad, L. V. N. 2016. "Hydraulic Simulation of Existing Water Distribution System using EPANET at Dire Dawa City, Ethiopia." *Indian Journal of Science and Technology* 9(S(1)). doi:10.17485/ijst/2016/v9iS1/106859

Ramesh, H., Santhosh, L., Jagadeesh, C. 2012. "Simulation of Hydraulic Parameters in Water Distribution Network Using EPANET and GIS International Conference on Ecological." *Environmental and Biological Sciences*, Dubai.

Saminu, A., Abubakar, Nasiru, Sagir, L. 2013. "Design of NDA Water Distribution Network Using EPANET." *International Journal of Emerging Science and Engineering (IJESE)* 1(9): 5–9.

Saravanan, K., Anusuya, E., Kumar, Raghvendra, Son, Le Hoang. 2018a. "Real Time Water Quality Monitoring Using Internet of Things in SCADA." *Environmental Monitoring and Assessment* 190(9): 556. https://doi.org/10.1007/s10661-018-6914-x

Saravanan, K., Golden Julie, E., Herold Robinson, Y. 2018b. Smart Cities & IoT: Evolution of Applications, Architectures & Technologies, Present Scenarios & Future Dream. For the Upcoming Book Series, *Intel.Syst.Ref.Library*, Vol. 154, Valentina E. Balas et al. (Eds.): Internet of Things and Big Data Analytics for Smart Generation, 978-3-030-04202-8, 467407_1_En, (7). www.springer.com/us/book/9783030042028

Saravanan, K., Srinivasan, P. 2017. Examining IoT's Applications Using Cloud Services. In P. Tomar, G. Kaur (Eds.), *Examining Cloud Computing Technologies through the Internet of Things* (pp. 147–163). Hershey, PA: IGI Global. doi:10.4018/978-1-5225-3445-7.ch008

United Nations, Department of Economic and Social Affairs, Population Division. 2019. World Urbanization Prospects 2018: Highlights (ST/ESA/SER.A/421).

Varsha, Radhakrishnan, Wu, Wenyan. 2018. "IoT Technology for Smart Water System IEEE 20th International Conference on High Performance Computing and Communications." *IEEE 16th International Conference on Smart City; IEEE 4th Intl. Conference on Data Science and Systems.* doi:10.1109/HPCC/SmartCity/DSS.2018.00246

WHO, UNICEF. 2019. *Progress on Household Drinking Water, Sanitation and Hygiene 2000–2017: Special Focus on Inequalities.* New York: United Nations Children's Fund (UNICEF) and World Health Organization, 1–71.

15 An Artificial Intelligence-Based Evaluation of Soil Fertility

*A. Mohideen Pathumuthusabana
and S. Suja Priyadharsini*

CONTENTS

15.1   INTRODUCTION

Soil is formed from the method of rock weathering. Weathering is the breakdown of rocks into smaller particles upon contact with water, air or living organisms. As soils develop over time, horizons form a soil profile. Soil profile is a parallel layer of soil that allows us to inspect the structure of soil. Soil horizons are categorized into six patterns and are explained as follows:

O horizon: Thin in some soils, thick in others, and not present in all soils. Mostly organic materials are present in the O horizon.

A horizon: Most of the organic materials are present in the A horizon, and it looks darker than others. Rich in nutrients, organic matter and biological activity.

E horizon: This layer is mostly present in forestry land and classical soils. It contains discharge of slip, limestone and decomposed material, elements of quartz, etc.

B horizon: This is a clay-rich subsoil. This horizon soil has a lesser amount of nutrient substance and is rich in moisture content. This layer of soil is usually lighter in color and denser in soil consistency than the horizon A layer.

C horizon: Mainly tough rocky particles are found in the C horizon.

R horizon: Mass of rock that forms parent material for some soils. This is present under the C horizon [1].

Agriculture plays a key role in the development of human civilization. India ranks second in the world in farm outputs. Its economic contribution to the world's GDP (gross domestic product) is steadily declining with the country's broad-based economic growth [2].

Soil is the critical part of successful agriculture and is the original source of nutrients. The nutrients move from soil to plants. Nutrients are also a part of the food. Farmers use many practices to enrich the soil. There are many advanced technologies used to make modern agriculture more efficient. Precision agriculture is the study of management of practices to maximize food production and minimize environment impact. It is one of the many modern farm practices that make production more efficient [3].

Soil is a naturally occurring resource covering a major portion of the earth's land surface. Soil provides plants with essential minerals and nutrients. It protects plants from erosion and other destructive physical, biological and chemical activities. Soil holds water and maintains adequate aeration.

Soil can be classified into sand, clay, silt, peat and loam. Recognize the kind of soil facilitates increasing plant life.

Sandy soil: Due to its large percentage of sand and small amount of clayey particles, these types of soil are lighter in color. During the summer, sandy soils can easily get dried out, which results in shortage of nutrients. But we can improve the nutrient amount and moisture content by adding organic matter to the soil.

Clay soil: Rich in nutrient content due to its intense soil particles. This soil has 25% of clay, so it can hold a high amount of water.

Peat soil: Rich in organic compost and holds a huge quantity of moisture content. These kinds of soils are used to afford the finest soil layer for cultivation.

Chalk soil: These types of soil are rich in alkaline substances; because of this, these soils are not apt for growth of plants.

Loam soil: It is a mixture of sand, silt and clay soil and rich in nutrient content, which affords a fine drainage facility. It can be either sandy or clay loam based on its predominant composition [4].

Soil fertility is the ability of the soil to provide the necessary nutrients for plant nourishment. Fertile soils are rich in additional life-supporting nutrients for plants. Plants absorb water, nutrients and various minerals from the soil with their roots. Fertile soil provides plants with lots of additional food that supports plants life. The factors affecting soil fertility are parent materials, climate and vegetation, topography, inherent capacity of the soil to supply plant nutrients, physical condition of the soil, micro-organisms and soil fertility, availability of plant nutrients, soil erosion and the cropping system. The characteristics of fertile soil are as follows:

- Fertile soil is rich in nutrients, minerals, organic compounds and micro-organisms that support plant growth.
- It contains soil organic matter that helps to improve the efficacy of the soil. This allows the soil to hold more moisture and it is mostly found in topsoil (0–10 cm).
- The fertile soil has the pH from 6.0 to 6.8 [5].

Soil testing must be done before planting crops to find out whether there is the right amount of nutrients in the soil, so as to ensure there is maximum crop yield. Soil testing is a process that is essential before planting. Excess or lack of fertilizer has a significant effect on the crop yield. Crops also need well-distributed nutrients in the right amount to grow healthy. Applying chemical fertilizer to the soil for better yield might lead to soil damage. Careful soil sampling is essential for an accurate application of fertilizer to the soil. A sample must reflect the overall fertility of the field so an analysis accurately represents the fertility status of the soil. An accurate evaluation will result in more efficient use of fertilizer, which will exponentially increase the crop yield and reduce costs and environmental damage.

A smart city is a metropolitan spot that employs diverse categories of electronic techniques and sensors to gather information. The smart city concept incorporates information and communication technology (ICT) and different sensible tools linked to the IoT system to enhance the effectiveness of city functions and services. Smart city technology permits city executives to interrelate straightforwardly with both society and the metropolis framework and to supervise the city. ICT is used to improve value, interact with city aid, decrease expenses and material expenditure and to enlarge proximity between the general public and the administration. Smart city functions are evolving to control city traffic and permit immediate reactions. The main aim of the smart city is to construct proficient use of infrastructure through artificial intelligence (AI) and data analytics to sustain a tough and fit financial, communal and intellectual development. One developing technology for smart cities is smart agriculture. Smart farming is a plantation management idea using recent top tools to enhance the amount and worth of farming yields. In other words, information and IoT-based smart farming are empowering the potential of farming. With smart cultivation, cultivators can examine the ground situations from everywhere by their mobile devices. IoT-based smart farming is highly efficient. It makes cultivation precise and successful when contrasted with the traditional method. Smart farming based on IoT automation will help cultivators both to decrease waste and improve yield. Big data platforms directed by AI employ several real-time information, which provides for creating extra-informed food production decisions. The objective of this work is to making precision farming a reality, taking an essential step toward an additional sustainable food chain. In the future, pesticide and fertilizer use will drop. Overall efficiency will also be optimized. IoT technologies enable better traceability of food, which in turn leads to increased food safety. Through machine learning and data analysis, AI-assisted agricultural platforms will continue to enable long-term improvements to production through enhanced understanding of the whole agricultural process. In recent days, the deployment of smart farming is gradually expanding. As farming is a labor-intensive job, with most of it comprising repetitive and standardized tasks, it is an ideal area for automation and robotics. The most crucial aspects of crop maintenance include weeding and pest control. Developers are using machine learning to help robots identify weeds before pulling them. Furthermore, robots can be equipped with sensors, cameras and sprayers so as to identify the pests and application of pesticides [6].

In recent advanced agricultural technology, there are lots of tools available for the farmers to produce healthy crops. In order to improve the quality of the crops, soil must possess necessary nutrients. Soil organic carbon (SOC) and soil macronutrients (NPK) are the important factors for soil fertility. SOC is a measurable soil organic component. Soil organic matter (SOM) is the source and sink of soil organic carbon. SOC is usually reported as a percentage of topsoil (0–10cm). Erosion events remove topsoil, which contains the bulk of a soil's organic carbon.

Soil nutrients are classified as macronutrients and micronutrients. Macronutrients are those that are needed in large amounts, while those needed in small amounts are micronutrients. Nitrogen (N), phosphorus (P) and potassium (K) are the main macronutrients. Nitrogen (N) is the key element in plant growth. It is found in all plant cells, in plant proteins and hormones, and in chlorophyll. Atmospheric nitrogen is a source of soil nitrogen. Some soils are generally higher in nitrogen. Phosphorus is most commonly found in the soil in the form of polyprotic phosphoric acid (H_3PO_4) but is taken up most readily in the form of H_2PO_4. Only a limited amount of phosphorus is available in most of the soils because it is released very slowly from insoluble phosphates. Potassium controls the function of the stomata by a potassium ion pump. Stomata are important in water regulation. Potassium regulates water loss from the leaves and increases drought tolerance. The main goal of soil testing is to ensure efficient management of soil quality [7].

Soil testing can improve the soil and reduces the use of undesired fertilizers to be added to the soil. The traditional evaluation techniques are quite hard and difficult in practical applications, which results in an expensive and time-consuming process. Due to these methods of agricultural processes, the farmer faces many more problems in agricultural productivity than others. Also, the traditional soil testing program in laboratories takes a considerable amount of time to generate results. During the period from testing to getting results, the overall soil fertility changes often because of changing weather conditions. Time is a crucial factor in the determination of soil fertility, so there is a need to develop an automated evaluation technique for soil testing.

Various sensors such as color sensors and electrochemical, optical and radiometric sensors can be used in determining various physical and chemical properties of the soil. Various sensors have been developed to be used as on-the-go soil sensors, but among these sensors, electrical and electromagnetic sensors are widely used. In addition, ion-selective electrodes (ISEs) are used to determine the macronutrient level of the soil [8]. With the growth in world population, food consumption also grows rapidly. To meet the demand–supply ratio, change in the agriculture pattern is a necessity. In smart cities, smart agriculture is an emerging technology representing the application of modern ICT into agriculture, to increase the quantity and quality of products while optimizing the required human labor.

In this proposed research, deep learning (DL) models have been created for the classification of the SOC and NPK level in the soil. The soil samples are collected across three different regions, such as forest area, field area and seashore area, at three different depths (0–5cm, 5–10cm and 10–15cm). The soil is manually tested for ground truth values. Based on the results obtained from the laboratory, the soil samples are classified as organic high, organic medium and organic low and NPK as high, medium and low. For the evaluation of SOC and NPK level, the soil image is captured by using a mobile camera under controlled environmental condition. A total of 1700 soil images were taken. Theses soil images are the input to the DL classification algorithm. In the DL method of classification, three convolutional neural network (CNN) models have been created for the classification of SOC and NPK level in the soil for the evaluation of the soil fertility. The CNN models created are Lenet, AlexNet and Vgg16.

15.2 LITERATURE REVIEW

This section describes related work that provides the various soil testing methods employed in agricultural technology that has the capacity to enhance the productivity of the farmer. Peng-Tao Guo et al discussed the spatial distribution of SOM and its characteristics on a landscape scale [9]. This method can only predict the SOM without error by using Regression Kringing analysis. Antoine Denis et al (2009) improved SOC prediction accuracy using an Analytical Spectral Device (ASD) that gives better accuracy than the traditionally used methods [10]. Fu et al (2011) discussed an ANN modeling that demonstrated the abilities and accuracy of neural networks in predictive modeling of the spatial variation structure in specified riverside networks [11]. Nocita et al (2012) proposed a method to determine SOC content for moist samples with unknown moisture content. But this method needs improvement in SOC prediction for large-scale analysis [12]. Haiqing Yang et al (2012) and Peng Lu et al (2012) proposed a spectroscopic technique to predict organic carbon (OC) and inorganic carbon (IC) content in soil [13, 14]. This method has the potential benefits of mapping soil parameters and diagnosing the nutrient deficiency and stated that a MIR spectrophotometer is more accurate for predicting soil C content than a Vis-NIR spectrophotometer under laboratory conditions and is promising for portable or in-field measurement of C contents. Kennedy Were et al (2014) evaluated the performance of random forests (RF), support vector machines for regression (SVR) and artificial neural networks (ANN) models in predicting and mapping the variability of SOC stocks. The authors concluded that the ANN method is the best way to predict SOC stocks [15]. Deying Wang et al (2014) adopted the DNDC (denitrification decomposition) model for SOC change during the period 2010–2070. First, DNDC was calibrated using data from a long-term field experiment, then the modeling units were divided considering spatial continuity and variability of soil and climate parameters. And then the simulation sceneries were made by climate condition and crop types to simulate SOC change for the period 2010–2070 covering the entire study area [2]. Roxanne Stiglitz et al (2016) explain the commercially available color sensor to develop SOC prediction models for both dry and moist soils from the Piedmont region of South Carolina. In this method, the SOC prediction model was developed using regression analyses for dry and moist soils. This method suggests that soil color sensors have potential for rapid SOC determination, and soil depth and color are useful in predicting SOC content in soils [16]. Sudha R. et al (2017) describe a model based on the digital image processing technique where digital photographs of the soil samples were used for soil pH determination. Digital photographs were collected during sunlight, while photographs of the soil sample were taken in a dark room for the purity of digital value of the spectra [17].

John Carlo et al (2017) evaluate a method known as soil nutrient analyzer based on color recognition. This research employs the image processing and ANN technique to determine the soil pH and nutrient levels in soil via MATLAB. For image processing, this system contains 70% training, 15% testing and 15% for validation, and results will show the nutrient levels and pH of the soil. Gilbert Hinge et al (2018) predict digital mapping approach for the soil organic carbon distribution in Northeast India. Using climate data and satellite images, the random forest models have been trained to predict the parameters. Based on topographic factors and climate variables, this approach predicts and monitors the SOC in a most efficient way [18]. S. Panchamurthi et al (2019) analyses and predicts the suitable crop and fertilizer recommendation based on the processing of different parameters of soil. The farmers can get the soil testing services at the doorstep. This project replaces the primitive method of soil testing in an effective way, and so the farmers get to know about their soil quickly. The result provided by this project helps farmers to make a decision and prevents them from using unbalanced fertilizers [19]. The study from existing work indicates that the entire analysis of soil fertility was carried out in a manual method. Time becomes a critical factor for estimation of soil nutrients. From the period of testing to receiving results, the gross soil richness modifies in parallel with weather circumstances, so there is a requirement of a computerized method to calculate the soil fertility.

15.3 METHODOLOGY

The proposed research aims at classifying the soil organic carbon and soil macronutrients (NPK) using DL algorithms. Deep learning is a function of artificial intelligence that defines the workings of the human brain in data processing and creates patterns that are used in decision making. It helps to solve complex problems in unstructured and interconnected data. Convolution neural network (CNN) is the most popularly deployed deep learning model thatis used in analyzing visual imagery. CNN is one of the most widely deployed deep learning models. This model holds composite structure as it comprises a huge number of information-processing layers.

In this example, the enhanced deep convolution neural network models such as Lenet, AlexNet, VGG16 are modeled to improve the traditional soil testing methods with long convergence time and a large number of model parameters. In this proposed method, three CNN models have been created for the classification of soil nutrients in the soil. This proposed method includes five steps, namely soil testing, image capturing, image processing, training system and evaluation of obtained results. By adjusting the model parameters, the models are trained and tested with the soil images captured by mobile camera under a controlled environmental condition.

15.3.1 IMPLEMENTATION

Using a Jupyter notebook and Google Colab, the models have been created and trained based on the Keras framework. There are two types of models available

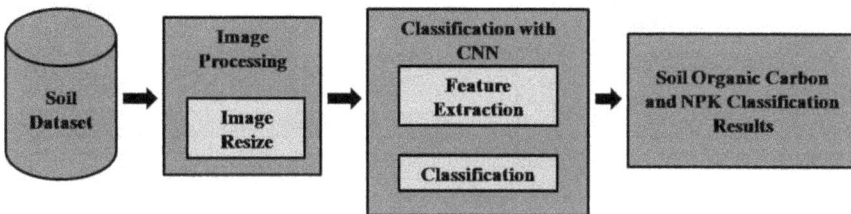

FIGURE 15.1   Deep learning framework.

in Keras: the sequential model and the model class used with the functional API (application programming interface). In this example, the sequential model has been implemented throughout. Keras allows for straightforward and quick prototyping of the neural system. Python coding is written for model creation.

In this example, for the classification of soil organic carbon and soil macronutrients, the soil samples are collected at three different depths (0–5 cm, 5–10 cm, 10–15cm) from three different areas: aforest area, field area and seashore area. The soil image was captured by using a mobile camera under controlled environmental conditions. A total of 1,700 images of soil samples are collected. The fertility of the soil varies with climatic factors such as humidity, temperature and moisture content present in the soil. Image annotation in a DL technique is an important step for object detection with more precision. The image annotation algorithm performs the frame selection and classifies the image based on training data obtained from the soil manual testing method by experts in the soil testing laboratory; the soil images are selected and labeled with the corresponding classes. The dataset that contains soil images have been annotated. The program will create the library for the group of soil images after the annotation step. The most important step in image annotation is the formatting of images, that is, resizing of images. All the images in soil datasets are automatically resized, which is computed by the Python program supported by Open CV. The CNN training network needs significant data. By using a huge amount of data, the CNN network can extract more features from the soil image. The overfitting problem in the training process of CNN can be controlled by means of data augmentation. After the completion of data augmentation techniques, the CNN model can study more unrelated prototypes via the training process, so that the overfitting problem can be avoided and yield better efficiency [20].

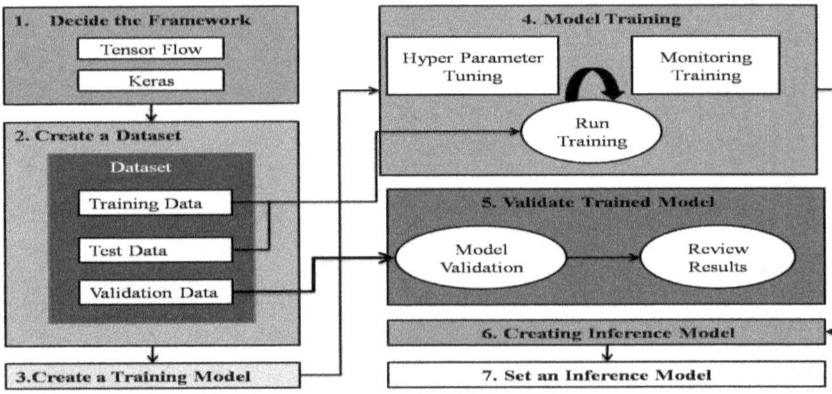

FIGURE 15.2   Flow diagram of proposed example.

15.3.2   Convolution Neural Network

A convolution neural network (CNN) is a type of DL model that is used in image recognition and processing because of its high accuracy. CNN has three layers, namely a convolution layer, pooling layer and fully connected layer. Each layer has different parameters and performs a different task on input data. It follows the hierarchal model and gives result as a fully connected layer in which all the neurons are connected to each other and the output is processed. It is specially designed to process pixel data. Convolution neural network is different from regular neural networks. It can automatically decide the important features without any human intervention [21].

In the proposed method, three CNN models are created for the evaluation of SOC and NPK levels in the soil.

15.3.2.1   Lenet Architecture

The Lenet model consists of two sets of convolution layer, a max pooling layer flattening convolution layer, two fully connected layers and softmax classifier.

15.3.2.1.1   First Layer

In the Lenet model, the first convolution layer contains six filters with stride of one. The input grayscale image size of $32 \times 32 \times 1$ is reduced to the output image of size $28 \times 28 \times 6$ by the first convolution layer.

15.3.2.1.2   Second Layer

The average pooling layer is the second layer of the Lenet architecture with a filter size 2×2 and a stride of two. The resulting output image has the dimension of 14×14×6.

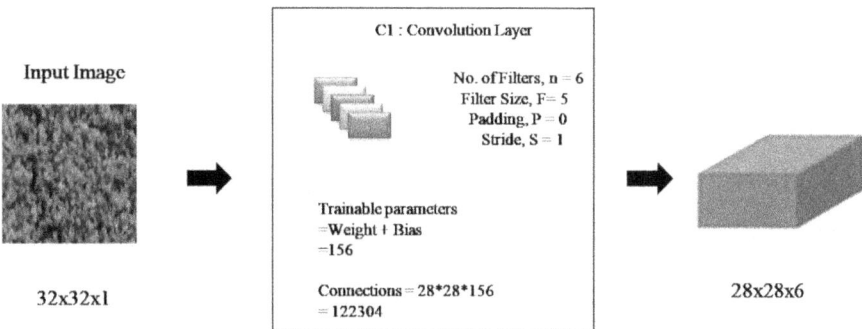

FIGURE 15.3   First layer of the Lenet model.

15.3.2.1.3 Third Layer

This is the second convolution layer having 5×5 size of 16 feature maps with a stride of 1. Only 10 feature maps are connected to 6 feature maps of the previous average pooling layer.

15.3.2.1.4 Fourth Layer

The fourth layer has the filter size 2×2 and a stride of 2, same as the second layer (S2).

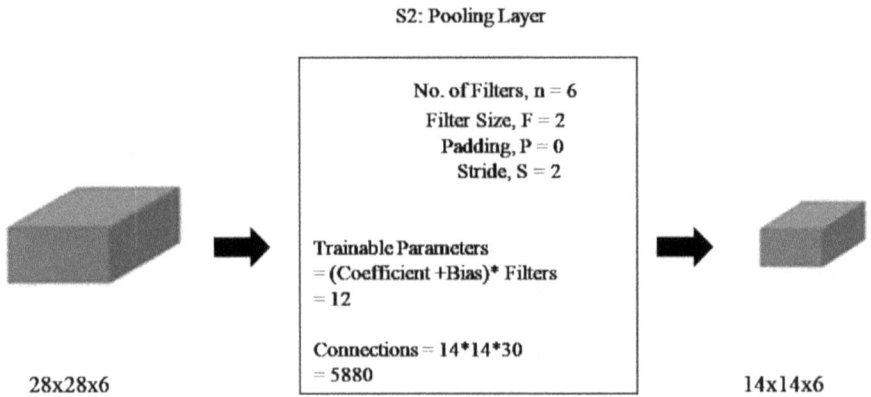

S2: Pooling Layer

No. of Filters, n = 6
Filter Size, F = 2
Padding, P = 0
Stride, S = 2

Trainable Parameters
= (Coefficient +Bias)* Filters
= 12

Connections = 14*14*30
= 5880

28x28x6

14x14x6

FIGURE 15.4 Second layer of Lenet model.

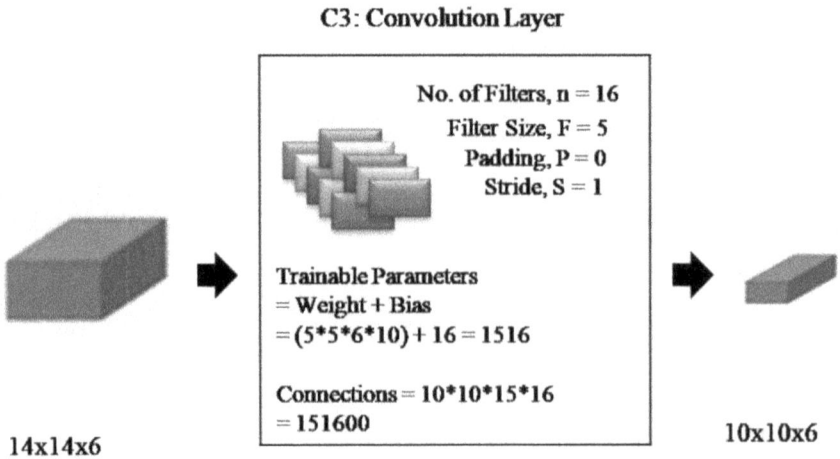

C3: Convolution Layer

No. of Filters, n = 16
Filter Size, F = 5
Padding, P = 0
Stride, S = 1

Trainable Parameters
= Weight + Bias
= (5*5*6*10) + 16 = 1516

Connections = 10*10*15*16
= 151600

14x14x6

10x10x6

FIGURE 15.5 Third layer of Lenet model.

15.3.2.1.5   Fifth Layer

This is the fully connected layer size of 11×1 with 120 feature maps. Each 120 units of fifth layer C5 is connected to 400 nodes in the fourth layer (S4).

15.3.2.1.6   Output Layer

The final layer in the Lenet architecture is output layer called softmax classifier. This layer has 10 possible values from 0 to 9.

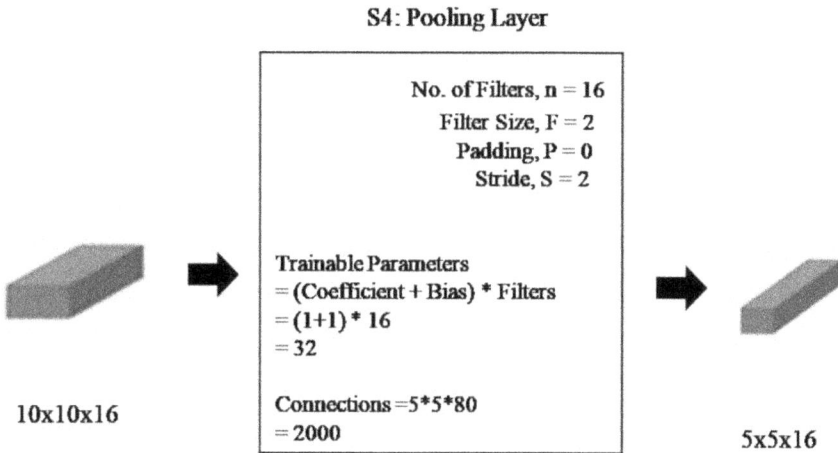

S4: Pooling Layer

No. of Filters, n = 16
Filter Size, F = 2
Padding, P = 0
Stride, S = 2

Trainable Parameters
= (Coefficient + Bias) * Filters
= (1+1) * 16
= 32

Connections = 5*5*80
= 2000

10x10x16

5x5x16

FIGURE 15.6   Fourth layer of Lenet model.

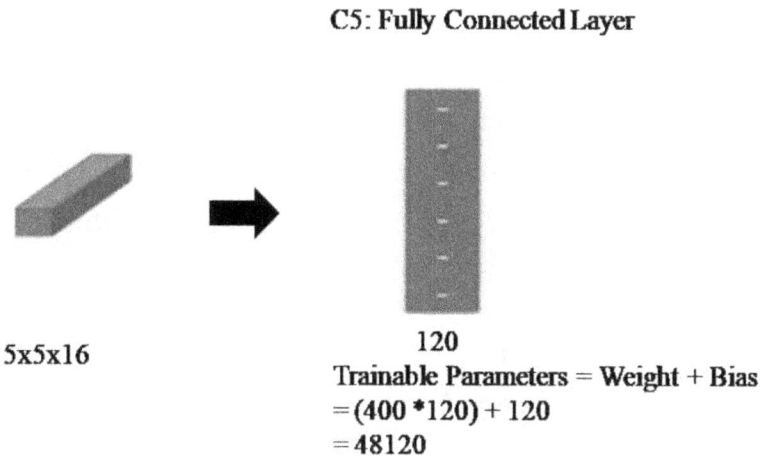

C5: Fully Connected Layer

5x5x16

120
Trainable Parameters = Weight + Bias
= (400 *120) + 120
= 48120

FIGURE 15.7   Fifth layer of Lenet model.

15.3.2.2   AlexNet Architecture

The AlexNet architecture has eight layers with five convolution layers and three fully connected layers. The overlapping max pooling layer is present after the fifth convolution layer. The output from max pooling layer feeds to a two fully connected layers. Finally, the softmax classifier with class labels has the input from the second fully connected layer.

C6: Fully Connected Layer   **Output**

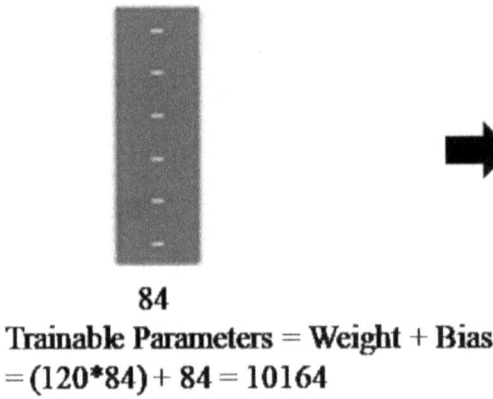

84
Trainable Parameters = Weight + Bias
= (120*84) + 84 = 10164

FIGURE 15.8   Output layer of Lenet model.

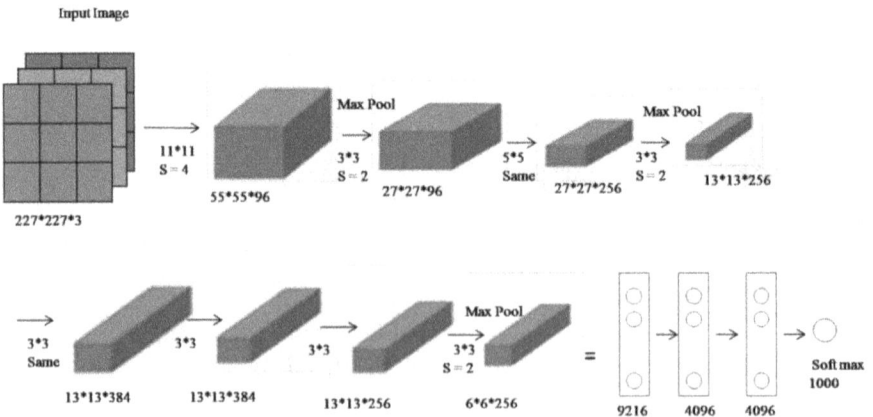

FIGURE 15.9   AlexNet architecture.

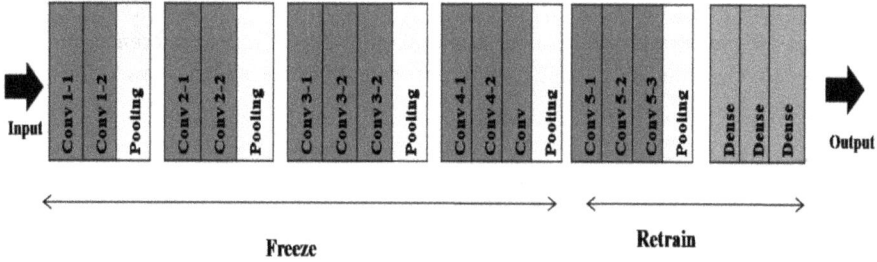

FIGURE 15.10 Vgg16 architecture.

15.3.2.2.1 *ReLU Nonlinearity*

An important feature of the AlexNet is the use of ReLU (rectified linear unit) nonlinearity. It has six times improvement in convergence from the tanh function. Hence, it avoids and rectifies the vanishing gradient problem. But its limitation is that it should only be used within a hidden layer. In ReLU, some gradients can be fragile during training and can die. It can cause a weight update, which will make it never activate on any data point again.

15.3.2.2.2 *Dropout*

A neuron with probability of 0.5 is dropped from the network. A dropped-out neuron has no forward or backward propagation. So each input goes through various network architecture. As a result, the learned weight parameters are more robust and do not get overfitted easily. Dropout increases the number of iterations.

15.3.2.3 VGG16 Architecture

VGG16 has 16 layers. It consists of 13 convolution layers, 5 max pooling layers and 3 dense layers which sum up to 21 layers but only 16 weight layers. There are 64 filters in Conv 1; Conv 2 has 128 filters; Conv 3 has 256 filters; and Conv 4 and Conv 5 have 512 filters. In VGG16, a neuron with a probability of 0.4 is dropped from the network.

15.4 RESULTS AND DISCUSSION

This approach employs a DL algorithm for classification of the SOC and NPK. Three CNN models are created for the classification of SOC and NPK. In this work, the models created are Lenet, AlexNet and Vgg16. For the classification of SOC using a DL algorithm, the number of training datasets is 1305, and the testing dataset is 141. For classification of NPK level in soil as high, medium or low, the number of training and testing datasets employed are 1114 and 280, respectively.

15.4.1 CLASSIFICATION OF SOC

Soil samples are collected from three different regions–forest area, field area and seashore area—at three different depths—0–5cm, 5–10cm and 10–15cm. The soil image is captured using a mobile camera under controlled environmental condition. The soil samples are tested manually in the laboratory. With the reference, the soil samples are classified as organic high, organic medium and organic low using a DL algorithm.

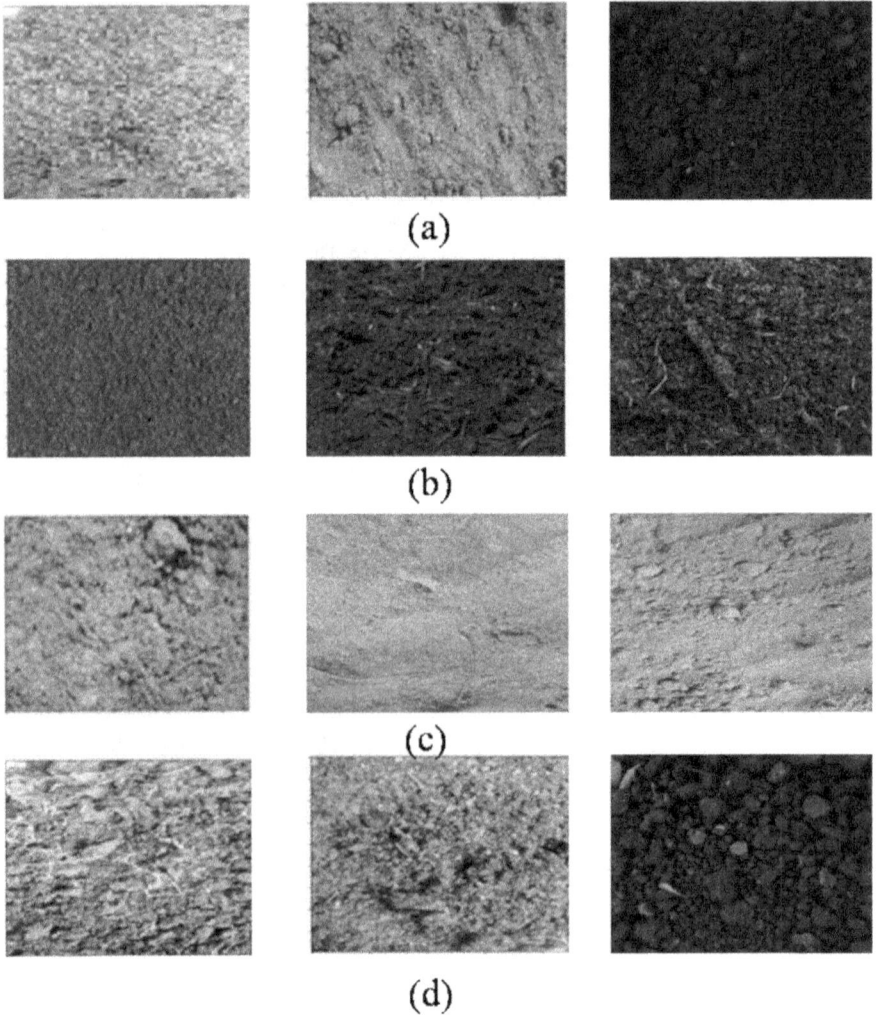

FIGURE 15.11   Sample images of (a) inorganic soil, (b) organic high soil, (c) organic low soil and (d) organic medium soil.

15.4.1.1 Lenet Model

The accuracy obtained from the Lenet model is 0.6446 and the loss is 0.3780. It is evident that the classification accuracy is low for a Lenet model. So further experimentation is carried out with the AlexNet model.

15.4.1.2 AlexNet Model

The classification accuracy obtained from the AlexNet model is 0.9061 and the loss is 0.2478. The classification accuracy was evaluated with equal number of images for each class. The AlexNet model yields slightly higher classification accuracy than prior model.

```
In [4]: model.summary()

Model: "sequential_1"
_________________________________________________________________
Layer (type)                 Output Shape              Param #
=================================================================
conv2d_1 (Conv2D)            (None, 62, 62, 32)        896
_________________________________________________________________
max_pooling2d_1 (MaxPooling2 (None, 31, 31, 32)        0
_________________________________________________________________
conv2d_2 (Conv2D)            (None, 29, 29, 32)        9248
_________________________________________________________________
max_pooling2d_2 (MaxPooling2 (None, 14, 14, 32)        0
_________________________________________________________________
flatten_1 (Flatten)          (None, 6272)              0
_________________________________________________________________
dense_1 (Dense)              (None, 128)               802944
_________________________________________________________________
dense_2 (Dense)              (None, 4)                 516
=================================================================
Total params: 813,604
Trainable params: 813,604
Non-trainable params: 0
_________________________________________________________________
```

FIGURE 15.12 Lenet model summary for SOC.

```
Model: "sequential_1"
_________________________________________________________________
Layer (type)                 Output Shape              Param #
=================================================================
conv2d_1 (Conv2D)            (None, 54, 54, 96)        34944
_________________________________________________________________
activation_1 (Activation)    (None, 54, 54, 96)        0
_________________________________________________________________
max_pooling2d_1 (MaxPooling2 (None, 27, 27, 96)        0
_________________________________________________________________
conv2d_2 (Conv2D)            (None, 17, 17, 256)       2973952
_________________________________________________________________
activation_2 (Activation)    (None, 17, 17, 256)       0
_________________________________________________________________
max_pooling2d_2 (MaxPooling2 (None, 8, 8, 256)         0
_________________________________________________________________
conv2d_3 (Conv2D)            (None, 6, 6, 384)         885120
_________________________________________________________________
activation_3 (Activation)    (None, 6, 6, 384)         0
_________________________________________________________________
conv2d_4 (Conv2D)            (None, 4, 4, 384)         1327488
_________________________________________________________________
activation_4 (Activation)    (None, 4, 4, 384)         0
_________________________________________________________________
conv2d_5 (Conv2D)            (None, 2, 2, 256)         884992
_________________________________________________________________
activation_5 (Activation)    (None, 2, 2, 256)         0
_________________________________________________________________
max_pooling2d_3 (MaxPooling2 (None, 1, 1, 256)         0
_________________________________________________________________
flatten_1 (Flatten)          (None, 256)               0
_________________________________________________________________
dense_1 (Dense)              (None, 4096)              1052672
_________________________________________________________________
activation_6 (Activation)    (None, 4096)              0
_________________________________________________________________
dropout_1 (Dropout)          (None, 4096)              0
_________________________________________________________________
dense_2 (Dense)              (None, 4096)              16781312
_________________________________________________________________
activation_7 (Activation)    (None, 4096)              0
_________________________________________________________________
dropout_2 (Dropout)          (None, 4096)              0
_________________________________________________________________
dense_3 (Dense)              (None, 1000)              4097000
_________________________________________________________________
activation_8 (Activation)    (None, 1000)              0
_________________________________________________________________
dropout_3 (Dropout)          (None, 1000)              0
_________________________________________________________________
dense_4 (Dense)              (None, 4)                 4004
_________________________________________________________________
activation_9 (Activation)    (None, 4)                 0
=================================================================
Total params: 28,041,484
Trainable params: 28,041,484
Non-trainable params: 0
_________________________________________________________________
```

FIGURE 15.13 AlexNet model summary for SOC.

15.4.1.3   Vgg16 Model

The accuracy obtained from the Vgg16 model is 0.9666 and the loss is 0.2882. The classification accuracy shows that the Vgg16 model outperforms when compared to both the Lenet model and the AlexNet model.

15.4.2   CLASSIFICATION OF SOIL MACRONUTRIENTS (NPK)

For the classification of NPK level in soil using a DL algorithm, three CNN models such as the Lenet model, the AlexNet model and the Vgg16 model have been created.

```
Model: "sequential_1"

Layer (type)                 Output Shape              Param #
=================================================================
vgg16 (Model)                (None, 7, 7, 512)         14714688

flatten_1 (Flatten)          (None, 25088)             0

dense_1 (Dense)              (None, 1024)              25691136

dense_2 (Dense)              (None, 512)               524800

dropout_1 (Dropout)          (None, 512)               0

dense_3 (Dense)              (None, 7)                 3591
=================================================================
Total params: 40,934,215
Trainable params: 33,298,951
Non-trainable params: 7,635,264
```

FIGURE 15.14   Vgg16 model summary for SOC.

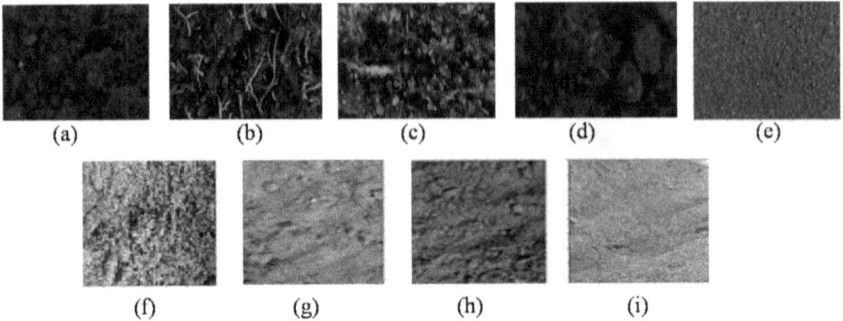

(a)   (b)   (c)   (d)   (e)

(f)   (g)   (h)   (i)

FIGURE 15.15   Sample images of (a) nitrogen-high soil, (b) nitrogen-low soil, (c) nitrogen-medium soil, (d) phosphorus-high soil, (e) phosphorus-low soil, (f) phosphorus-medium soil, (g) potassium-high soil, (h) potassium-low soil and (i) potassium-medium soil.

15.4.2.1 Lenet Model

The accuracy obtained from the Lenet model is 0.7747, and the loss is 0.3619. It is evident that the classification accuracy is low for the Lenet model, so further experimentation is carried out with AlexNet model.

15.4.2.2 AlexNet Model

The accuracy obtained from the AlexNet model is 0.8531 and the loss is 0.4074. The classification accuracy was evaluated with equal number of images for each class. The AlexNet model yields slightly higher classification accuracy than the previous model.

FIGURE 15.16 Lenet model summary for NPK.

FIGURE 15.17 AlexNet model summary for NPK.

15.4.2.3 VGG16 Model

The accuracy obtained from the Vgg16 model is 0.8738 and the loss is 0.5525. The classification accuracy shows that the Vgg16 model outperforms when compared to both Lenet and AlexNet.

15.5 PERFORMANCE MEASURES

15.5.1 Performance Evaluation for Classification of SOC Using Deep Learning

Table 15.1 shows the performance evaluation for the evaluation of soil organic carbon (SOC) using a deep learning algorithm.

```
⤷ Model: "sequential_1"

Layer (type)                  Output Shape              Param #
=================================================================
vgg16 (Model)                 (None, 7, 7, 512)         14714688

flatten_1 (Flatten)           (None, 25088)             0

dense_1 (Dense)               (None, 1024)              25691136

dense_2 (Dense)               (None, 512)               524800

dropout_1 (Dropout)           (None, 512)               0

dense_3 (Dense)               (None, 7)                 3591
=================================================================
Total params: 40,934,215
Trainable params: 33,298,951
Non-trainable params: 7,635,264
```

FIGURE 15.18 Vgg16 model summary for NPK.

TABLE 15.1
Performance Evaluation of SOC Classification Using Deep Learning

Model	No. of Epochs	Batch Size	Loss	Accuracy
Lenet	40	32	0.3780	**0.6446**
AlexNet	40	64	0.2478	**0.9061**
Vgg16	30	62	0.2882	**0.9666**

Note: No. of training samples = 1305; no. of testing samples = 141.

TABLE 15.2
Performance Evaluation of NPK Classification Using Deep Learning

Model	No. of Epochs	Batch Size	Loss	Accuracy
Lenet	40	32	0.3619	**0.7747**
AlexNet	40	64	0.4074	**0.8531**
Vgg16	30	62	0.5525	**0.8738**

Note: No. of training samples = 1114; no. of testing samples = 28.

15.5.2 PERFORMANCE EVALUATION FOR CLASSIFICATION OF NPK USING DEEP LEARNING

Table 15.2 shows the performance evaluation of NPK classification using deep learning.

From the models created, it is evident that the Vgg16 model outperforms the other models, with high classification accuracy, and gives a better solution to evaluate the fertility of the soil in a most efficient way.

15.6 FUTURE WORK

The proposed example concentrates only on the evaluation of soil macronutrients (NPK) in the soil. In the future, this study will extend to detect soil micronutrients as well as minerals in soil. In deep learning, it can also create other CNN models with a larger dataset for higher accuracy. This example will extend to develop a mobile app for rapid analysis of soil fertility.

15.7 CONCLUSION

In this chapter, a DL algorithm has been applied for the evaluation of SOC and NPK levels in the soil collected from forest area, field area and seashore area. For classification of SOC and NPK levels, the soil image has been captured using a mobile camera under controlled environmental condition. Lenet, AlexNet and Vgg16 are the models that have been created for the evaluation of soil fertility and executed in Jupyter notebook and Google Colab. For the classification of SOC, the classification accuracy yielded by the models are 64.46%, 90.61%, 96.66% by Lenet, AlexNet and Vgg16 models, respectively. For the classification of NPK levels in soil as high, medium and low, the classification accuracy obtained is 77.47%, 85.31%, 87.38% by Lenet, AlexNet and Vgg16 models, respectively. From the results obtained from the models, it is observed that the Vgg16 model outperforms the other models, with high classification accuracy. In smart cities, smart agriculture is an emerging technology. Therefore, the method proposed here helps to evaluate soil fertility in a most efficient way. In practical applications, CNN models can be used for classification of soil organic carbon and soil macronutrients.

REFERENCES

[1] www.soils4teachers.org/soil-horizons
[2] Wang, Deying, Yanmin Yao, Haiqing Si, Wenju Zhang, Huajun Tang. "Simulation and Prediction of Soil Organic Carbon Spatial Change in Arable Lands Based on DNDC Model." In 2014 The Third International Conference on Agro-Geoinformatics, Beijing, China, pp. 1–5. IEEE, 2014.
[3] Masrie Marianah, Mohamed Syamim Aizuddin Rosman. "Detection of Nitrogen, Phosphorus and Potassium (NPK) Nutrients of Soil Using Optical Transducer." In 2017 IEEE 4th International Conference on Smart Instrumentation, Measurement and Application (ICSIMA), Malaysia-Kuala Lumpur, pp. 1–4. IEEE, 2017.
[4] www.boughton.co.uk/products/topsoils/soil-types
[5] https://en.wikipedia.org/wiki/Soil_fertility
[6] www.isaf-forum.com/related-news-details.html
[7] https://en.wikipedia.org/wiki/Plant_nutrition
[8] Puno, John Carlo, Edwin Sybingco, Elmer Dadios, Ira Valenzuela, Joel Cuello. "Determination of Soil Nutrients and pH Level Using Image Processing and Artificial Neural Network." In 2017 IEEE 9th International Conference on Humanoid, Nanotechnology, Information Technology, Communication and Control, Environment and Management (HNICEM), Philippines, pp. 1–6. IEEE, 2017.
[9] Guo, Peng Tao, Hong-Bin Liu. "Spatial Prediction of Soil Organic Matter Using Terrain Attributes in a Hilly Area." In 2009 International Conference on Environmental Science and Information Application Technology, Wuhan, China, vol. 3, pp. 759–762. IEEE, 2009.
[10] Denis, Antoine, Bernard Tychon. "Improving Soil Organic Carbon Prediction by Field Spectrometry in Bare Cropland by Reducing the Disturbing Effect of Soil Roughness." In 2009 IEEE International Geoscience and Remote Sensing Symposium, Cape Town, South Africa, vol. 5, pp. V-351. IEEE, 2009.
[11] Fu, Yingchun, Xiantie Zeng, Xueyu Lu. "Spatial Prediction of Dissolved Organic Carbon Using GIS and ANN Modeling in River Networks." In 2011 Seventh International Conference on Computational Intelligence and Security, Sanya, Hainan China, pp. 401–406. IEEE, 2011.
[12] Nocita, Marco, Antoine Stevens, Carole Noon, Bas van Wesemael. "Prediction of Soil Organic Carbon for Different Levels of Soil Moisture Using Vis-NIR Spectroscopy." *Geoderma* 199, pp. 37–42, 2013.
[13] Lu, Peng, Zheng Niu. "Prediction of Soil Organic Carbon by Hyper Spectral Remote Sensing Imagery." In 2012 Third Global Congress on Intelligent Systems. Wuhan, China, pp. 291–293. IEEE, 2012.
[14] Yang, Haiqing, Weiqiang Luo. "Prediction of Organic and Inorganic Carbon Contents in Soil: VIS NIR Vs MIR Spectroscopy." In 2012 2nd International Conference on Consumer Electronics, Communications and Networks, China, pp. 1175–1178. IEEE, 2012.
[15] Were, Kennedy, Dieu Tien Bui, Øystein B. Dick, Bal Ram Singh. "A Comparative Assessment of Support Vector Regression, Artificial Neural Networks, and Random Forests for Predicting and Mapping Soil Organic Carbon Stocks Across an Afromontane Landscape." *Ecological Indicators* 52, pp. 394–403, 2015.
[16] Stiglitz, Roxanne, Elena Mikhailova, Christopher Post, Mark Schlautman, Julia Sharp. "Using an Inexpensive Color Sensor for Rapid Assessment of Soil Organic Carbon." *Geoderma* 286, pp. 98–103, 2017.
[17] Sudha, R., S. Aarti. "Determination of Soil PH and Nutrient Using Image Processing." *International Journal of Computer Trends and Technology*, pp. 58–61, 2017.

[18] Rao, Gilbert Hinge, Y. Surampalli. "Prediction of Soil Organic Carbon Stock Using Digital Mapping Approach in Humid India". *Environmental Earth Sciences*, 77, no. 5, pp. 1–10, 2018.

[19] Panchamurthi, S., M.D. Perarulalan. "Soil Analysis and Prediction of Suitable Crop for Agriculture Using Machine Learning." *International Journal for Research in Applied Science & Engineering Technology*, 7, no. 3, pp. 2328–2335, 2019.

[20] https://medium.com/supahands-techblog

[21] https://en.wikipedia.org/wiki/Convolutional_neural_network

KEY TERMS AND DEFINITIONS

Artificial Intelligence: The study of computer systems that attempts to model and apply the intelligence of the human mind. It is the capability of machine to imitate intelligent human behavior.

Deep Learning: Deep learning is a class of machine learning algorithm in the form of a neural network that uses a cascade of layers of processing units to extract features from data and make predictions about new data.

Convolution Neural Network (CNN): CNN is a specialized kind of neural network for processing data that has a known grid like topology such as time series (1D grid), image data (2D grid) etc. It is a supervised deep learning algorithm used in speech recognition, image retrieval and face recognition.

Soil Fertility: The status of the soil with respect to its ability to supply elements essential for plant growth without a toxic concentration of any element.

Soil Organic Carbon (SOC): SOC is the major constituent of soil organic matter (SOM), which plays a critical role in soil productivity and a wide array of ecosystem processes. This improves soil aeration (oxygen in the soil) and water drainage and retention and reduces the risk of erosion and nutrient leaching. SOC is also important to chemical composition and biological productivity, including fertility and nutrient holding capacity of a field.

Soil Nutrients: Nutrients are the chemical components that are essential for the growth and fertility. The major nutrients necessary for plant nourishments are nitrogen (N), phosphorus (P) and potassium (K).

16 Query Auto-Completion Using Knowledge Graph to Minimize Energy Usage

Vidya S. Dandagi and Nandini Sidnal

CONTENTS

16.1 INTRODUCTION

The Internet has become an integral part of modern life and economy. Information retrieval (IR) is the artificial intelligence (AI) field that extracts appropriate information from a data directory from a specific user based on a precise context. The success of the Internet has brought tremendous change for publishing and spreading data through the web. The voluminous data on the Internet has advantages and as well as disadvantages. With the increase in the data, the propagation of information and sharing becomes faster. Information retrieval is among the top of user queries that play a vital role in the web search engine. The main disadvantage is that because the information available is unprocessed and tons of data is available, and retrieval of relevant data becomes extremely difficult. The Internet is estimated to use 10% of global energy consumption [3].

The World Wide Web came into existence when there was a demand for automated information-sharing. The Web 1.0 was first implemented in 1989 by Tim Berners Lee; it was defined as web of data connections. This was known as a read-only web with little action, where the user could exchange the data, but it was not possible to associate with the website. Web 1.0 pages did not have content that the machines could understand; it was only readable by humans.

Web 2.0 is the second generation, which was considered as read-write web. Web 2.0 included social networks like Facebook and twitter. Web pages could be read and comprehend by humans, whereas computers could read and interpret but could not examine on their own. HTML (Hypertext Markup Language) is the representation of web pages in the World Wide Web. It is a language that makes visual presentation useful. This comprises a collection of "markup" symbols to be displayed on the web, usually intended for human use. The search engines present were very reliable but have insignificant results that are returned. Retrieving the precise data from the search outcome is a huge challenge because it does not take place in real time, and the need of the day is to collect relevant data. The Semantic Web is the potential solution to the aforementioned problem. Web 3.0 is the third generation Internet, Semantic Web. It includes semantic tagging of content. It is a read-write and executable web. The Semantic Web is a network of data that represents meanings and prepares models for spreading information across applications using the logic of business rules. The Semantic Web constructs relevant information. The key component that creates it is ontology. Storing the data in a machine-processable format in ontology helps in faster data recovery. Online searching is a very quick and impactful evolution of human experience. It is becoming a key technology that people rely on every day to get information about almost everything. Searching is typically performed with a common purpose underlying the query. If the user does not know the knowledge of the keywords to be searched, then the user spends more time to frame the query. The search may not contain the user's intended answer [29]. Semantics presents the rules for the syntax interpretation of a phrase that does not express the meaning directly but restricts the potential interpretations for what it has been stated [14]. Semantic search improves the search accuracy by understanding user intent and contextual meaning of terms as they appear in searchable data space, whether on the Web or within a system, to generate more significant results [2].

In green computing, the word 'green' means considering the computational climate and the ecosystem. Computing can be hardware fabrication, application creation, operating system functionality, and energy consumption reduction.

Green computing emphasizes reducing the energy consumption by the computing systems, thereby limiting their impact on the environment. Search engines are the interface between users and computers. An average query search on Google uses about 0.0003 KWh of energy, translating to 0.2 grams of CO_2, that is, equal to burning a 60 W bulb for 17 seconds. Google uses around 0.013 percent of global energy consumption. The shorter the search time, the lesser the energy consumption. The search time is reduced through query auto-completion. It is a process of suggesting words, and it is the foremost step of the search engine. The relevant suggestions recommended for the completion of the query are computed by the time taken for the search process. The present web content has a scarcity of the structure regarding the representation of information. The drawbacks of the current web are ambiguity in the information, which results in poor interconnection, not able to deal with the content of trust at all levels. Machines are not capable of understanding the given data since there is an absence of universal format. To overcome these drawbacks, the Semantic Web came into existence.

16.1.1 Impact of Smart Search Engines in Green Cities

Smart cities generate large volumes of data with great variation, range, velocity, and scale, and that data is complex. Data diversity and complexity can be attributable to the existence of several different formats. Smart city data grows very fast, such as the location of the bus, status of the people, or location of the garbage collectors every day and week. RDF stores accumulate the triples, and SPARQL is used to query them.

The RDF store has few characteristics that maintain precise rules that help the people of a smart city to take up the services. The RDF store of a smart city maintains the following: (1) Spatial indexing: a smart city RDF store should provide information near to a given geographical point. (2) Should have full text indexing: triples that contain subjects and objects have precise text space such as desc_ription, street_name, location_names. (3) Temporal indexing: in smart cities, a lot of knowledge and features shift over time, for instance, the weather condition and its associated prediction, the flow of traffic identified by the traffic sensors, the position of buses and actions taking place inside the city. For this reason, it is very important that the RDF store must carry out provisional searching so that information based on time is easily retrieved. (4) Huge number of queries: clustering is needed since it is difficult to deal with a huge RDF store as many users query the data. The information cannot be handled by a single server; when information is distributed among different servers, then scale-out clustering is implemented [51].

16.1.2 Query Auto-Completion

In the new search engine, the most important thing is query auto-completion (QAC). It is the method for measuring and recommending a set of words or sentences in real-time for each keystroke. The important goal of a QAC system is to decrease the user's effort that is needed in entering a text query. The difficulties namely include typing long texts, typo mistakes to be corrected, and finding the most relevant query words [13]. Traditionally, a web search engine is keyword-based on user-related information [5]. The features of the query auto-completion are as follows:

- Assists in locating the precise terms to use.
- Minimizes typo mistakes.
- Helps speedier contact.
- Predicts the desired query and enhances the efficiency of the search.

- Helps to quickly answer goal questions and reduce uncertainty.
- Intends to list the credentials before executing a query.

16.2   SEMANTIC WEB

The Semantic Web is the creative dynamic force in information management [9]. The Semantic Web is a smart web and an expansion to the current World Wide Web (WWW). The Semantic Internet uses traditional W3C standards. Search is carried out in the Semantic Web depending on the meaning of the terms. An artificial intelligence application is created by Semantic Web with the help of ontology [8]. The Semantic Web allows interconnection between information where the data is straightforwardly accepted and processed by machines. The Semantic Web facilitates the use of connected open data to alter the conceptual data dynamically.

16.2.1   AIMS OF SEMANTIC WEB

Berners-Lee et al. put forth the important goals of the Semantic Web:

- To build a method that conveys information and builds vocabularies and rules [15].
- To order the web pages according to meaningful information.
- To imbibe semantics into the data [14].

The potential approaches and key disputes in building a Semantic Web that not only minimize the space for storing and the search time but also return search results that are oriented and important [47] are:

1. Availability of content
   One of the greatest problems is developing content for the Semantic Web. For the web content, annotations are generated, but the tools only annotate static pages and focus on creating new content. Dynamically generated content is not taken into account, and existing content is excluded from the Semantic Web.
   Possible solution: Annotation services can be created of static and dynamic web documents that include multimedia and web services. Static material annotations can be applied through the use of wrapping technology. Dynamic content annotations can be rendered by annotating the query that will retrieve the information. Using NLP techniques, information extraction from the textual explanation of images, automatic speech recognition transcripts, and event reports in video sequences is done.

2. Ontology availability, development, and evolution
   All domains can use constructions of the kernel ontologies. The process of development of ontology that provides methodological and technological support includes knowledge acquisition and conceptual modelling. Configuration management tools can implement the development of ontologies and their relation to the annotated data.

Possible solution: The IEEE Standard Upper Ontology Team aims to create ontologies for the kernels. Machine learning techniques are of great benefit for the acquisition of knowledge and conceptual modeling. The semantic Web for ontology coding uses the Resource Description Framework Scheme (RDFS), Web Ontology Language (OWL).

3. Organizing, storing and mining the information
   Semantic Web content can be organized and grouped through semantic indexes. A peer to peer architecture may investigate the mining of content. Semantic subject routing can be accomplished by using agents and negotiation strategies, such as WordNet's semantic gap.

4. Multilinguality
   Semantic Web implements access information in multiple languages. Multilingualism plays an important role in the ontology and annotation levels.
   Possible solution: To personalize data access based on the user's resident language, strategies such as internationalization and localization are used. In order to allow providers to annotate data in their native languages, data annotations can be rendered in different languages. EuroWordNet can be explored to support multilingual existing services such as WordNet.
   Techniques of internationalization and localization can be used to personalize the access to information based on the user's native language. Content annotations can be executed in different languages, allowing providers to annotate content in their native languages. EuroWordNet and WordNet can be explored to support multilingual existing services.

5. Visualization
   Demand for clear and meaningful content is becoming increasingly necessary for intuitive visualization of information. Intuitive visualization can allow the user to quickly reorganize relevant information. Using semantic indexes and routers to store, organize and locate information can help with visualization. Semantic relationships can be explored using Java3D, Shockwave3D, and 3D real-time graphic representation, and a three-dimensional interface can be created automatically. This approach requires less space and lets users effectively communicate with the web.

6. Semantic Web languages standardization
   Possible solution: Construction of the Semantic Web resources is important. The tools rely on the language of the Semantic Web. The tool that standardizes the Semantic Web language is the W3C Semantic Web Activity.

16.2.2 ONTOLOGY

Ontology is able to give good acknowledgment of concept and the knowledge of understanding the relationship between entities, query understanding, and intention of ranking. With the help of ontology information, retrieval has gained better results. Ontology can be used for indexing and ranking processes and also query reformulation and expansion purposes. Ontologies are used to describe the relations between various forms of semantic knowledge information. Ontology is used to devise search strategies for data. Ontology languages have been introduced, namely XML (Extensible Markup Language), which is machine interpretable. Web Ontology Language (OWL) have OWL-Lite, OWL-DL, and OWL-FULL flavors. OWL was developed on top of RDF and DAML+OIL [1]. A major benefit of using the ontology of the domain is the ability to describe a semantic data model together with the associated domain information [10].

The structure of an ontology.

```
<!--www.semanticweb.org/vsd/ontologies/2019/11/vsd#isa-->
  <owl:ObjectProperty rdf:about="&vsd;isa">
    <rdfs:subPropertyOf rdf:resource="&vsd;isLocatedIN"/>
  </owl:ObjectProperty>
<!--www.semanticweb.org/vsd/ontologies/2019/11/vsd#isLocatedIN -->
  <owl:ObjectProperty rdf:about="&vsd;isLocatedIN">
    <rdfs:domain rdf:resource="&vsd;TajMahal"/>
    <rdfs:range rdf:resource="&vsd;Agra"/>
  </owl:ObjectProperty>
  <!--www.semanticweb.org/vsd/ontologies/2019/11/vsd#TajMahal-->
  <owl:NamedIndividual rdf:about="&vsd;TajMahal">
    <isLocatedIN rdf:resource="&vsd;Agra"/>
  </owl:NamedIndividual>
```

Figure 16.1 depicts ontology for the smart city domain. It consists of classes like a FunctionCell, which has subclasses CityService and CityGovernance. TouristService and BusService are the two subclasses of the CityService. CityCorporation_Service is a subclass of CityGovernance. TouristPlace is a subclass of TouristService. TajMahal is an instance of TouristPlace class.

16.2.3 RESOURCE DESCRIPTION FRAMEWORK

Resource Description Framework (RDF) allows explanation of the resources using statements that consist of subject, predicate, and object called triples.

16.2.4 SPARQL

[1] The data in the RDF format is retrieved and manipulated by using a query language known as SPARQL. A SPARQL query is shown as follows:

```
SELECT {?subject ?object ?predicate}
```

Ontologies and semantic reasoners give a better structure for representation and improve the data interpretation and processing. One may assume ontological information to obtain reasoning statements [16].

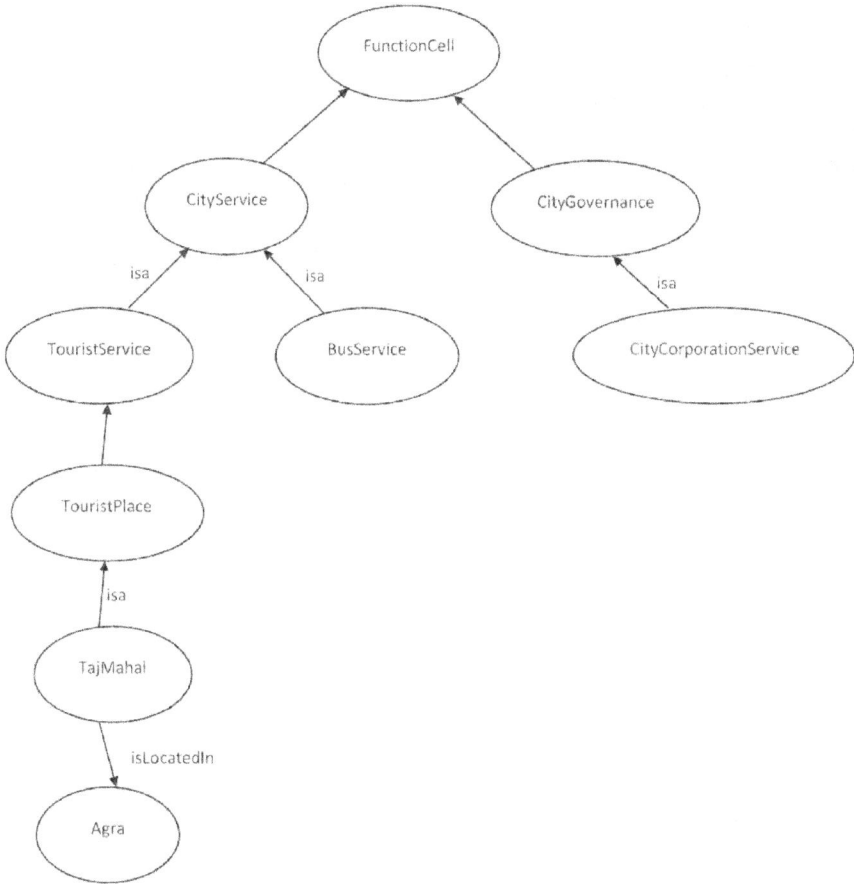

FIGURE 16.1 Ontology for the smart city domain.

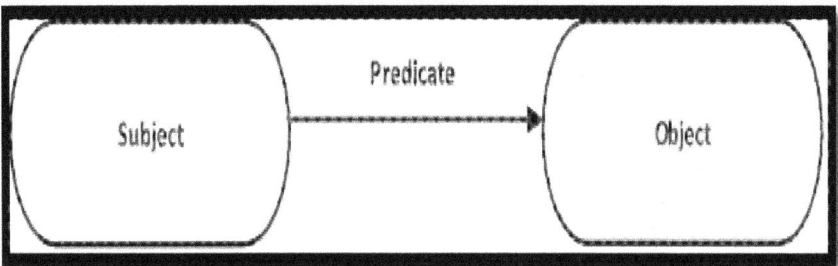

FIGURE 16.2 A RDF statement.

16.2.5 SEMANTIC SEARCH ADVANTAGES IN SMART CITY APPLICATIONS

Semantic search engines are used by the World Wide Web to access related infor-mation. Conventional search engines like Google, Yahoo, and Bing offer infor-mation based on keywords mapped into the content of the site. Developers of web content used these keywords to publicize their services and encourage users to visit their websites. Consequently, irrelevant information is recovered. Users are spending more time finding insignificant findings from the infinite URLs. Semantic search engines surmount the drawback of conventional search engines. Ontologies have semantic information that the semantic search engines use. Semantic engines require earlier annotation of the web content for better search precision. Semantic annotation is made using the methods of natural language processing (NLP) [50].

The integration and use of emerging physical infrastructures, electronic technol-ogy systems, and institutional frameworks for knowledge exchange and innovation distinguishes smart cities [48].

Semantic network innovations have an incredibly high potential and functional impact by developing new e-services within city ecosystems. Semantic Web tech-nologies work with Internet of Things embedded systems and offer new prospects for collaborating with social media and shared intelligence [49].

Ontology for a smart city that includes services such as tourist services and bus services in the city has been implemented.

16.2.6 ARCHITECTURE OF SEMANTIC QUERY AUTO-COMPLETION

Figure 16.3 depicts the proposed architecture of semantic query auto-completion. It consists of six layers, starting at the bottom:

1. The resource layer represents the data in the format using the Unified Resource Identifier (URI).
2. The transformed data is stored in the semantic framework layer in an RDF database called the triple store. Data representation models that are used to represent ontologies use OWL (Web Ontology Language).
3. The third layer is composed of an Application Programming Interface and a web service called SPARQL end point. The purpose of the SPARQL query is to enable application to retrieve data from the triple store, and the resultant will be subject, predicate, and object. Pellet reasoner is used for inferring the facts.
4. In the security service layer, the proposed architecture gives data security with authentication and authorization.
5. The knowledge graph layer is where the combination of knowledge base and reasoning creates a graph model using nodes and edges.
6. Creation of embeddings of nodes and applying a machine learning algo-rithm for prediction makes up the sixth layer.

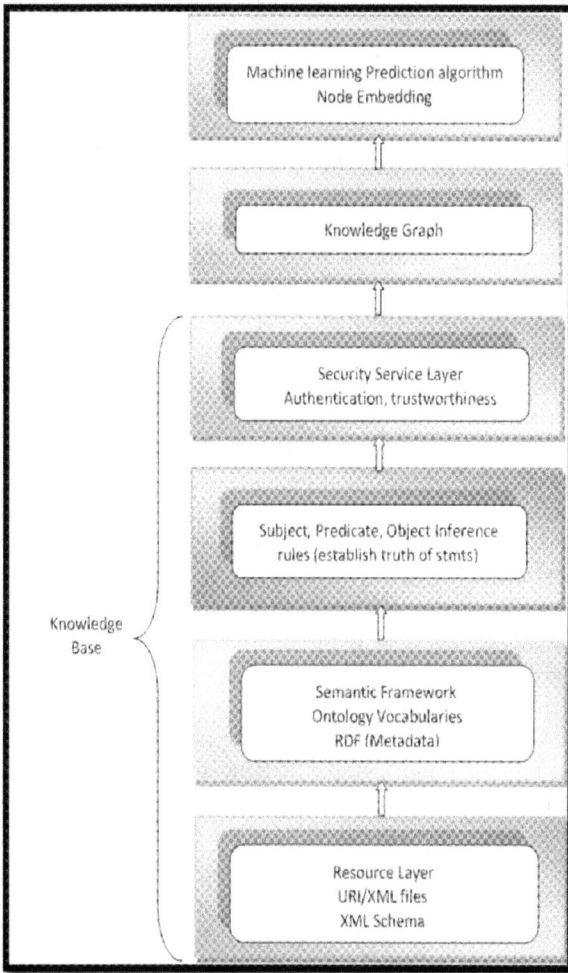

FIGURE 16.3   Architecture of semantic query auto-completion.

16.2.7   LITERATURE REVIEW

The literature survey reveals that query auto-completion uses multifaceted search technique for the retrieval of information based on content. Depending on the previous history, prefix, and past clicks, queries are ranked from the list of suggested queries. The generation of queries is through the use of neural network models. Reasoning is essential to build up the knowledge graphs. Different methods such as tensor factorization and simplE are put forward to improve the accuracy of relation prediction. These methods use semantic characteristics to understand the relationships between entities. The approach known as probabilistic soft logic is used as a tool of reasoning.

The method of tensor factorization also aids in the problem of relation prediction. There are numerous graph embedding models, including RESCAL, the first model based on matrix factorization, and TRANSE—this model uses embedding based on translation. TRANSH, TRANSR, and TRANSD are TRANSE's extended variants, which project the embedding vectors of entities into different spaces.

Ontology is a better method to solve the problem of differences in data heterogeneity to simplify the search process. Integration and mapping find interrelationships between concepts of two or more ontologies. Ontology is a model for establishing semantic concepts used by various sources. It has been shown that ontology makes the distribution of data become easier without reducing the semantic meaning [7].

Keyword expansion and knowledge storage are the two options for information retrieval. Expansion of keywords increases recall by increasing semantically relevant keywords that are generated from the initial keyword. Knowledge storage is where the semantic tool contains the data that the user needs to access [15].

Semantic Web technologies make possible the sharing and linking of the distributed information source. Conversion of green economy data into RDF improves the accessibility to more people as Linked Open Data [9].

A multi-label deep neural network model highlights relation prediction. The purpose of this model is to predict the relationships among the entities in a knowledge graph. Link prediction among entities is significant for creating huge ontologies and for knowledge graph completion. If the relation is predicted accurately, it can be augmented to a given ontology [19].

Ontology carries out the reasoning by using a collection of stated facts and axioms to elaborate logical conclusions. Reasoning on the relation of ontology brings new conclusions. Temporal and spatial reasoning are separate information-rationing strategies. Time interval-based reasoning is known as temporal reasoning. Spatial reasoning produces a general understanding of the object and the spatial relationship between the entities [30].

Knowledge graphs are important tools for many artificial intelligence projects, such as question and answering, semantic search, and semantic analysis. Embedding the knowledge graph is the main step of representing knowledge. Knowledge graphs have crypto graphed the relationship information into organized data. A knowledge graph represents the triples denoted as (h, r, t) [31]. A knowledge graph that covers the entire network is defined as a Semantic Web. A knowledge graph retrieves and incorporates data into ontology and makes use of a reasoner to extract new knowledge. Capabilities for reasoning are important features for understanding new knowledge [32].

Knowledge graphs are composed of the Semantic Web language known as the Resource Description Framework. A new method for embedding a knowledge graph is through tensors that are really valued. That tensor-based embedding retrieves the information graph relationship. The model of tensor decomposition works well when the mean degree of any term is elevated in a graph. This further enhances the estimation of new information through the knowledge graph [33].

Many embedding techniques, such as TransE and PTransE, are available. TransE and TransH are simple and efficient. They build embedding entities and relationships by taking the translation from the head and tail entity into account. TransE successfully performs in 1-to-1 relationship. It has problems in modeling relationships

1-to-N, N-to-1 and N-to-N. TransH and TransR let an entity have different depictions for different relationships. A cascading embedding mechanism is implemented in which semantic graph features are removed. Such features are taken from the embedding of a knowledge graph [34].

A simple, bilinear model for completion of the knowledge graph is proposed. Knowing the existing connections between the entities, estimation of connections helps to predict new links to a knowledge graph. The method of factorizing the tensor is useful for computing the connection between entities. Canonica Polyadic is the first tensor factorization tool; this method calculates two embedding vectors for each entity. SimpleE is Canonical Polyadic enhancement in which the two entities are trained dependently, one embedding vector for each relationship and two vectors for each entity [35].

Completion of the knowledge graph includes time data as one of the attribute in the embedding approach. The long short-term memory (LSTM) model is used to learn the facts inferred by the knowledge graph in which time is added [36]. Knowledge graphs are created by adding the triples and deleting them; they are dynamic. Static information graphs are targeted at embedding models. Dynamically generated knowledge graph embedding fails to remember. Dynamic knowledge graph embedding consists of two modes of representation, embedding information and embedding contextual dimension. Knowledge graph embedding is learned from scratch by an online algorithm that provides greater precision. This model is more robust and flexible to test embedding online [37].

Statistical relational learning is defined as the development of the statistical model for relative data. There are two models; the first depends on the latent model of the attribute, such as factorization of the tensor and multiway neural networks. The second model is based on extracting the patterns observed in the graph. A variation of this has improved computational capacity at low cost. Models of mathematical graph methods and text-based methods of knowledge extraction can be used in building knowledge graphs. RESCAL is a related latent feature model that triples the latent features through pair-wise interaction [38].

A system is proposed that integrates the combination of learning and rule-based methods. Embedding is tested for removal of the rules from current triples and old triples. Rules are learned by t-norm fuzzy logic by injection of axioms. Embedding entities are represented as vectors and their relationship as matrices. It is done through a linear map [39]. A new approach for producing negative samples to train the embedding information graph is called adversarial learning. This system is based on the policy gradient of generator preparation, which generates negative triples. The discriminator model can learn better by generating the negative samples. A single step REINFORCE method is used to allow the error to propagate back [40].

The relation predictor in this model assists in the classification process and captures the groups, the labeled and the unlabeled data. An algorithm called Classification Using Link Prediction (CULP) uses a novel format called Label Embedded Graph and link predictor to look for the unlabeled class of data. As predictors for CULP, a relation predictor known as the compatibility score is used. CULP uses a graph called Labeled Edge Graph enabling the use of link predictors [41].

A new approach called Entity Link Prediction for Knowledge Graph (ELPKG) is proposed to improve the accuracy of the relation prediction. The prediction algorithm for entity connections is used to know the relationship between entities. The method adopted is known as the soft logic based probabilistic reasoning method. This approach adds paths and semantic dependent characteristics to understand the relationship between entities. The Euclidean distance is computed to calculate the similarity of tuples. When the gap between them is the shortest, the two tuples are identical. This ELPKG tackles the lack of knowledge relations in the knowledge graph [42].

A relational learning agent named MINERVA (Meandering In Entities Networks to Reach Verisimilar Answers) is being developed that takes perfect steps of a decision to select the edges of the relationship to get the correct answer. This model is solid and has a long chain of reasoning to understand [43]. A single high volume recurrent neural network (RNN) model is implemented, which enables reasoning chains over various types of relationships. The approach of neural attention focuses on using several paths to reason. The pooling is achieved to increase rapidity and precision across multiple paths [45].

A novel method is put forward that combines the convolutional neural network and long short-term bidirectional memory. A combination of specific triples and corrupted triples grouped together form the training data. Then a path-ranking algorithm is adopted to classify the relationship paths appropriate to each training instance. On the whole graph random walks are performed to understand what relationship links the source entity to its target. A vector representation is generated to explain the semantic co-relationship between two entities and the direction between them. The focus cycle takes in the paths between the entities. This model performs multi-step reasoning in an embedding space over a representation of the path. Path encoder is effective in extracting path functions from large graphs [44].

For a given object, this model predicts all the triples. To obtain triples as a sequence in a knowledge graph, a new paradigm is proposed to use a multilayer RNN unique to the KG. A beam search method with a large window size is implemented to predict the triples [46].

This is proposed to assist users without using the query log to complete queries of the question type. An algorithm is based on ranking and answering the query, using Wikipedia page titles to create dictionary entities, and using dictionary methods to detect phrases. The association rules (word chains) are intended to recommend a keyword for the candidate to complete [22].

Semantic similarity and frequency of query enhances efficiency in predicting query purposes and helps formulate queries. To improve the accuracy frequency of the queries, semantic similarity of the query helps in predicting the intended query and also formulates the query. Word2vec method is used to compute the semantic similarity [20]. To create personalized query completion, a neural network language model is used. To make predictions for users not seen during training, an adaptive model is used [21].

Formulating the query online is proposed. Completion of the query depends on the click the user performed dynamically. The QAC must represent the time and adjust as changes are dynamically occurring. Completion of the query that is not for the particular purpose of the user is given more priority and is more likely to be clicked. Using multi-armed bandits, the lower ranked completion which are more important are placed in the higher place [23].

Online question and answer proposals will revisit successful program examples posted by the software developer for the programming questions. The code search engines do not handle linguistic queries and will return accurate code only when the query has the variables of class-name or method name. To identify semantically related terms in a given linguistics, a neural network model is used. Continuous bag-of-words model is used to represent the class-names as vectors. The class-name present in the Application Programming Interface and the user's query are used to calculate the semantic distance. A weighted-sum ranking schema progresses the searching of the code [24].

Synonyms and abbreviations are used on the large-scale databases for auto query completion. There are three data structures that endorse varying complexities in space and time. The topmost K synonyms are used to complete the description. Twin Tries (TT) is a lightweight information system. Expansion Tries (ET) is the one that checks for interchangeable and string rules. ET is quicker compared to TT and takes up more space in producing top-K completions. Hybrid Tries (HT) decreases the speed of finding and the cost of the space. Short queries infer that they take less memory space and take less time for a small data set [25].

The two secular popularity query structures in the hybrid QAC model, which is the normal shift in the query and the sudden leap in the popularity of the query, is called burst. Discrete Fourier transform is used to learn the periodic shifts in the importance of queries. This also predicts future query by the customer and future success for auto-query completion [26].

Generative auto-completion of large-scale queries does not rely on log information. It is used in confidential applications or to supplement a query-based approach. This approach reacts to the appropriate time frame for the interactive user interface. The terms in the n-gram model are pre-pruned using the search terms for the index. The search index is verified either partial or full in terms of the query prefix. The language model n-gram helps to detect popularity [27].

Queries are selected from the storage depending upon accurate feature like popularity. If no match is found in the current storage there would be no recommendations for a question. To generate fluctuating length suggestion for unseen queries for any starting prefix, a neural language model is built up with RNN [28].

16.3   KNOWLEDGE GRAPHS

When knowledge represented as a graph, the information is augmented with context. Entities are represented as nodes; edges represent the relationship between two entities [14]. A knowledge graph (KG) represents a semantic network and is a graph data structure. It consists of node-edge–node structure. The node is an entity and the edge is the link connecting two entities. A knowledge graph is used to explain the concepts, entities, and rich association among them. They are many fields where KGs can be implemented such as finance, medical, and semantic search [12]. The KG covers a huge number of domains and provides connection between entities without any constraints [14]. The basic concept of KG completion is instinctively deducing missing triples by taking into consideration triples that exist. This is attained by carrying out the link prediction on the KG. Prediction of the link calculates the likelihood of an entity connecting to another entity through a relationship in the graph. For example, given a question (Sanman, Isa,?), it could predict that Hotel is the missing entity [6].

16.4   GRAPH

Graphs are universal data structure used extensively throughout the computer science field. Graphs are used in social networks, recommender systems; these systems capture interaction between nodes and edges. Graphs play a major role in the current models of machine learning.

16.4.1   NODE EMBEDDINGS AND METHODS

Node embedding is the summary of the graph position and organization of the graph neighbor nodes that is mapped to the low-dimensional vectors. Embedding nodes in graphs is a method that is to transform nodes, edges, and their features into vector space. The different methods for node embeddings are encoder-decoder. Encoder maps each node to an embedding. The decoder decodes the structural data from the learned embeddings. Encoding embedding algorithms are matrix factorization and random walk. The matrix factorization method is the first method for studying representation for nodes. Laplacian Eigenmap is the most popular method that is viewed in the encoder-decoder framework [17].

The encoder-decoder squeezes the information about the graph structure into a low dimensional space. The encoder maps the node v_j, to a low-dimensional vector embedding z_j. The decoder digs up the user-specified information from the low-dimensional embedding.

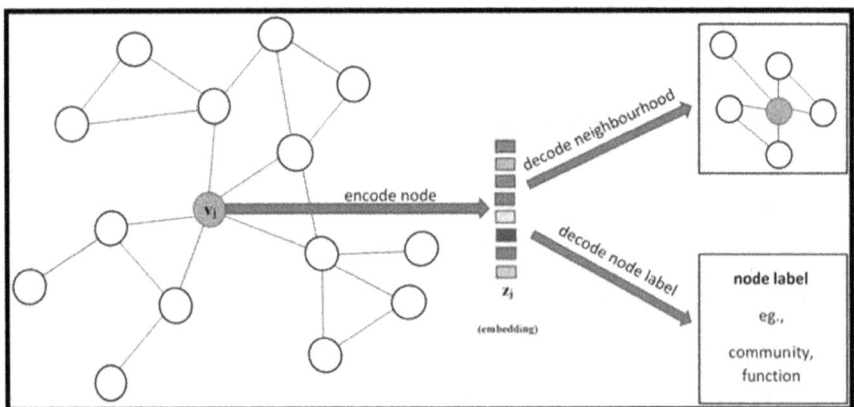

FIGURE 16.4   Encoder-decoder approach.

The decoder is a function that understands a group of node embeddings and decodes the graph specified by the user details from these embeddings.

DEC: $R^d \times R^d \rightarrow R^+$

Node2vec is a framework consists of an algorithm to learn the feature of nodes in the network; p and q are the two parameters that redirect the random walk. Node embeddings are derived by word-embedding algorithm Word2vec. Learning embeddings is done by analyzing the relationship between the entities. Node2vec uses a sampling strategy that takes in four arguments:

a. Number of walks: Random walk created from individual node.
b. Walk length: Number of nodes in individual random walks.
c. p: return hyperparameter.
d. q: InOut hyperparameter.

For random walk generation, the algorithm transits to each node in the graph and creates random walk and walk length.

Figure 16.5 depicts the random walk procedure in node2vec. The probability of movement from <v1> to any one of its neighbors is <edgeweight> × <α> (normalized), where <α> depends on the hyper parameters. Constraint p reins the probability

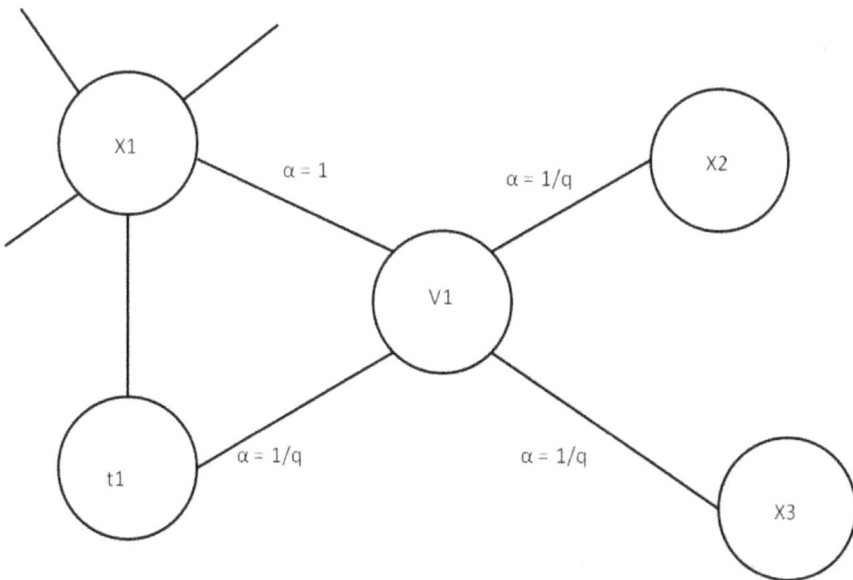

FIGURE 16.5   Random walks procedure in node2vec.

to return to <tl> through the <vl>; q reins the probability to move and investigate undiscovered parts of the graph [4].

Large-scale embedding of the information network (LINE) is not based on random walks but rather a very flourishing direct encoding approach. It combines two functions, one for first order proximity and another for second order proximity. HARP is basically a graph-preprocessing step that simplifies the graph to make for faster training. Coarsening methods on the graph are used that subside the related nodes together into supernodes and then node2vec, LINE are used to execute the coarsened graph.

16.4.2 RECURRENT NEURAL NETWORK

Recurrent neural networks (RNNs) are the class of neural networks. To process the sequence of inputs, RNNs use internal hidden states. In an RNN, the data from the previous texts are represented as low dimensional vector. Figure 16.6 depicts the architecture of the RNN.

A component of the neural network, h, occurs at the input x and the value y is generated. A loop allows the passage of information from one step of the network to the next. The chain-like nature indicates RNNs are linked to list sequences. RNNs cannot link the knowledge [11].

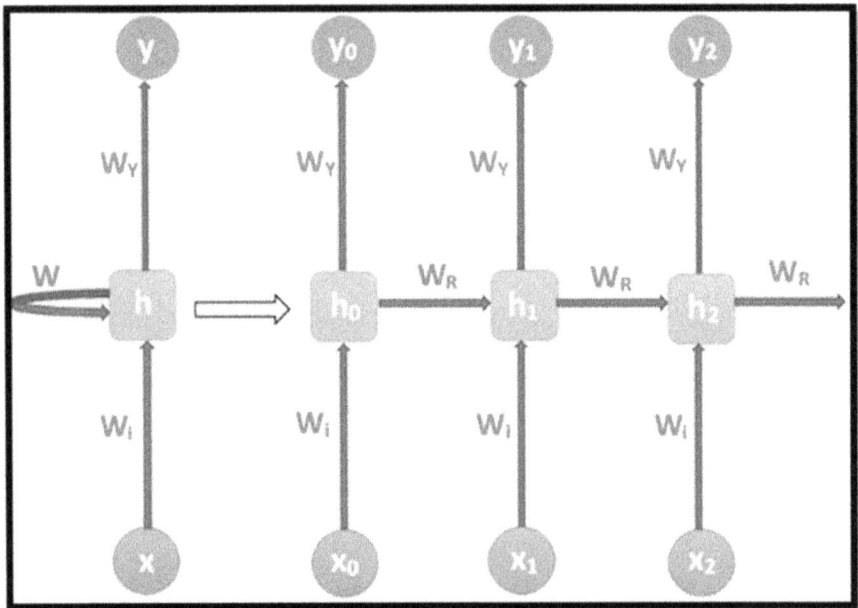

FIGURE 16.6   Architecture of recurrent neural network.

16.4.3 Long Short-Term Memory Model

The long short-term memory model (LSTM) is a special form of RNN that can learn long-term dependencies. The proposed example uses LSTM with sequential data (Figure 16.7).

Cell state is the key element; it behaves like a conveyor belt of LSTM that is represented as horizontal line. Adding and removing of data to the cell state is controlled by the gates. There are three gates, input gate, forget gate, and output gate. The forget gate rules the cell-state data. This decision, either 0 or 1, is made by the sigmoid layer; 1 means to hold, while a 0 represents removal. Input gate is used to revise the cell state. The sigmoid output determines what information the tanh output is to retain. The output gate rules the hidden state; it includes previous input information. Predictions are done through the hidden state.

16.4.4 Results

Queries relating to the city domain are represented as a graph and embedded using node2vec. There are five layers of Embedding, Dropout, LSTM, Dense layer for activation by Relu, and sigmoid. The LSTM model is to predict the next word given the subject and predicate. One hot code is used to represent the subject and predicate. The model is trained for 250 epochs, with the categorical cross-entropy as the loss function. The accuracy of the model is 89.4%; it infers that the relevant terms given by the random walk help in auto-completing. The MRR score for prediction of the object is 0.891.

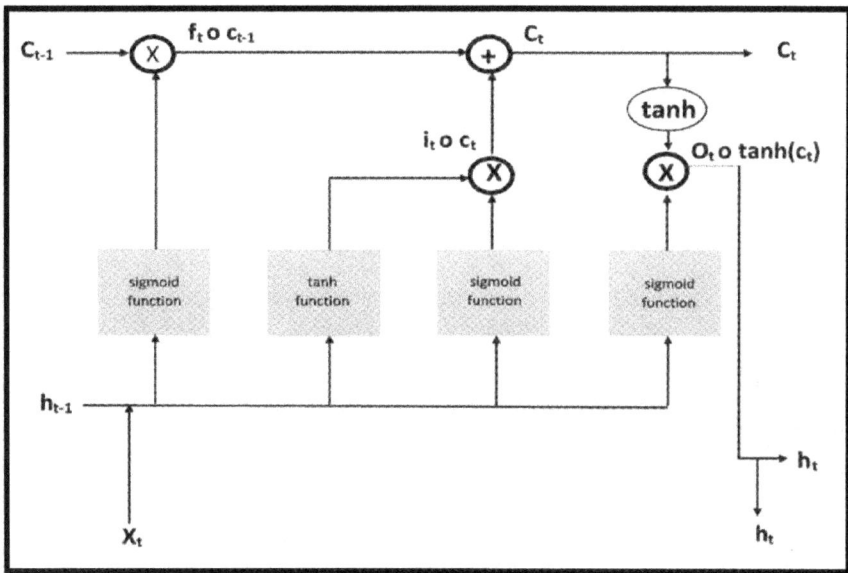

FIGURE 16.7 LSTM cell.

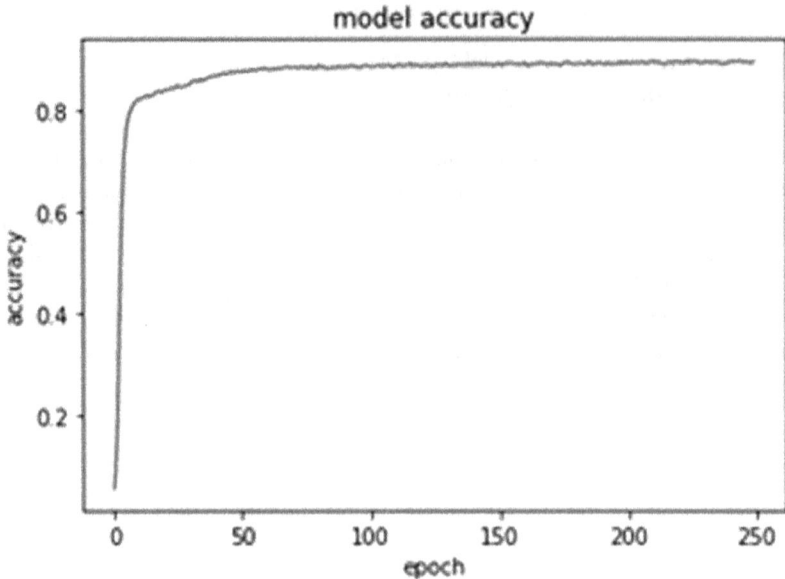

FIGURE 16.8   Model accuracy.

Figure 16.8 shows the graph of epoch versus accuracy. As the number of epochs increases the accuracy of the model improves, and also relevant suggestions are suggested.

16.5   CONCLUSION

In this chapter, creation of ontology was described that makes the auto-completion process more efficient and effective. The user will take less time to write the query since the terms that are relevant are suggested, since QAC is the most important event for the search engine. Implementing QAC using a knowledge graph and using the proposed model LSTM, the time taken for predicting the next word (completion) is 193 ms. QAC terms represented semantically takes less time for completion, so the search is done faster and hence contributes to smart city computing.

REFERENCES

[1]   Azizan, Azilawati, Zainab Abu Bakar, and Shahrul Azman Noah, "Analysis of retrieval result on ontology-based query reformulation." International Conference on Computer, Communications, and Control Technology (I4CT), Langkawi, Malaysia, pp. 244–248, IEEE, 2014.
[2]   Chauhan, Rashmi, Atul, Sharma, "Domain ontology based semantic search for efficient information retrieval through automatic query expansion." 2013 International Conference on Intelligent Systems and Signal Processing (ISSP), Vallabh Vidyanagar, India, pp. 397–402, IEEE, 2013.

[3] Gan, Kim Soon, et al., "A comparison of distance methods effectiveness in retrieving relevant articles in agricultural domain." 2014 International Conference on Computational Science and Technology (ICCST), Kota Kinabalu, Malaysia, pp. 1–7, IEEE, 2014.

[4] Grover, Aditya, and Jure Leskovec, "Node2vec: Scalable feature learning for networks." Proceedings of the 22nd ACM SIGKDD International Conference on Knowledge Discovery and Data Mining, San Francisco, CA, USA, pp. 855–864, ACM, 2016.

[5] Jadhav, Payal A., Prashant N. Chatur, and Kishor P. Wagh, "Integrating performance of web search engine with machine learning approach." 2nd International Conference on Advances in Electrical, Electronics, Information, Communication and Bio-Informatics (AEEICB), Chennai, India, pp. 519–524, IEEE, 2016.

[6] Jagvaral, Batselem, et al., "Path-based reasoning approach for knowledge graph completion using CNN-BiLSTM with attention mechanism." *Expert Systems with Applications*, vol—142, p. 112960, 2019.

[7] Jayadianti, H., L. E. Nugroho, P. I. Santosa, C. A. B. S. Pinto, and W. Widayat, "Semantic interrelation in distributed system through green computing ontology." 2013 International Conference on Information Technology and Electrical Engineering (ICITEE), Yogyakarta, pp. 216–220, 2013.

[8] Khamparia, A., B. Pandey, and V. Pardesi, "Performance analysis on agriculture ontology using SPARQL query system." 2014 International Conference on Data Mining and Intelligent Computing (ICDMIC), New Delhi, pp. 1–5, 2014.

[9] Moyo, S., and J. V. Fonou-Dombeu, "Architecture of a semantic repository of green economy data in South Africa." 2019 International Conference on Advances in Big Data, Computing and Data Communication Systems (icABCD), Winterton, South Africa, pp. 1–6, 2019.

[10] Munir, Kamran, and M. Sheraz Anjum, "The use of ontologies for effective knowledge modelling and information retrieval." *Applied Computing and Informatics*, vol—14.2, pp. 116–126, 2018.

[11] Park, Dae Hoon, and Rikio Chiba, "A neural language model for query auto-completion." Proceedings of the 40th International ACM SIGIR Conference on Research and Development in Information Retrieval, Shinjuku, Tokyo, Japan, pp. 1189–1192, 2017.

[12] Wang, R., B. Li, S. Hu, W. Du, and M. Zhang, "Knowledge graph embedding via graph attenuated attention networks." *IEEE Access*, vol—8, pp. 5212–5224, 2020.

[13] Wang, Y., H. Ouyang, H. Deng, and Y. Chang, "Learning online trends for interactive query auto-completion." *IEEE Transactions on Knowledge and Data Engineering*, vol—29.11, pp. 2442–2454, 2017.

[14] Lampropoulos, Georgios, Euclid Keramopoulos, and Konstantinos Diamantaras, "Enhancing the functionality of augmented reality using deep learning, semantic web and knowledge graphs: A review." *Visual Informatics*, vol—4.1, pp. 32–42, 2020.

[15] Drury, Brett, et al. "A survey of semantic web technology for agriculture." *Information Processing in Agriculture*, vol—6.4, pp. 487–501, 2019.

[16] Dadkhah, Mahboubeh, Saeed Araban, Samad Paydar, "A systematic literature review on semantic web enabled software testing." *Journals of Systems and Software*, vol—162, 2020.

[17] Hamilton, William L., Rex Ying, and Jure Leskovec, "Representation learning on graphs: Methods and applications." *Bulletin of the IEEE Computer Society Technical Committee on Data Engineering Cornell University*, vol—1, pp. 1–24, 2017.

[18] Futia, Giuseppe, Antonio Vetrò, and Juan Carlos De Martin, "SeMi: A semantic modeling machine to build knowledge graphs with graph neural networks." *SoftwareX*, vol—12, ISSN 2352-7110, pp. 1–10, 2020.

[19] Onuki, Yohei, et al., "Relation prediction in knowledge graph by multi-label deep neural network." *Applied Network Science*, vol—4.1, article no—20, 2019.

[20] Shao, Taihua, Honghui Chen, and Wanyu Chen, "Query auto-completion based on word2vec semantic similarity." *Journal of Physics: Conference Series*, vol—1004.1, IOP Publishing, 2018.

[21] Jaech, Aaron, and Mari Ostendorf, "Personalized language model for query auto-completion." arXiv preprint arXiv:1804.09661, 2018.

[22] Mao, Xian-Ling, Yi-Jing Hao Wang, and Heyan Huang, "Query completion in community-based question answering search." *Neurocomputing*, vol—274, 2018.

[23] Wang, Yingfei, Hua Ouyang, Hongbo Deng, and Yi Chang, "Learning online trends for interactive query auto-completion." *IEEE Transactions on Knowledge and Data Engineering*, vol—29.11, p. 24422454, 2017.

[24] Zhang, Feng, Haoran Niu, Iman Keivanloo, Ying Zou, "Expanding queries for code search using semantically related API class-name." IEEE Transactions on Software Engineering, 2017.

[25] Xu, Pengfe, and Jiaheng Lu, "Top-k string auto-completion with synonyms." International Conference on Database Systems for Advanced Applications, Springer, Cham, 2017.

[26] Jiang, Danyang, Honghui Chen, and Fei Cai, "Exploiting query's temporal patterns for query autocompletion." Mathematical Problems in Engineering, 2017.

[27] Maxwell, David, Peter Bailey, and David Hawking, "Large-scale generative query auto-completion." Proceedings of the 22nd Australasian Document Computing Symposium, Brisbane, QLD, Australia, ACM, 2017.

[28] Hoon Park, Dae, and Rikio Chiba, "A neural language model for query auto-completion." Proceedings of the 40th International ACM SIGIR Conference on Research and Development in Information Retrieval, Shinjuku, Tokyo, Japan, ACM, 2017.

[29] Al-Chalabi, Hani, Santosh Ray, and Khaled Shaalan. "Semantic based query expansion for Arabic question answering systems." 2015 First International Conference on Arabic Computational Linguistics (ACLing), Cairo, Egypt, IEEE, 2015.

[30] Gayathri, R., and V. Uma, "Ontology based knowledge representation technique, domain modeling languages and planners for robotic path planning: A survey." *ICT Express*, vol—4.2, pp. 69–74, 2018.

[31] Guan, Niannian, Dandan Song, and Lejian Liao, "Knowledge graph embedding with concepts." *Knowledge-Based Systems*, vol—164, pp. 38–44, 2019.

[32] Ehrlinger, Lisa, and Wolfram Wöß, "Towards a definition of knowledge graphs." SEMANTiCS, Posters and Demos Track, Leipzig, Germany, September 13–14, 2016.

[33] Padia, Ankur, et al., "Knowledge graph fact prediction via knowledge-enriched tensor factorization." *Journal of Web Semantics*, vol—59, p. 100497, 2019.

[34] Li, Daifeng, and Andrew Madden, "Cascade embedding model for knowledge graph inference and retrieval." *Information Processing & Management*, vol—56.6, p. 102093, November 2019.

[35] Kazemi, Seyed Mehran, and David Poole, "Simple embedding for link prediction in knowledge graphs." Advances in Neural Information Processing Systems 32, Canada, 2018.

[36] García-Durán, Alberto, Sebastijan Dumančić, and Mathias Niepert, "Learning sequence encoders for temporal knowledge graph completion." Proceedings of Conference on Empirical Methods in Natural Language Processing, Association for Computational Linguistics, Brussels, Belgium, pp. 4816–4821, October 31–November 4, 2018.

[37] Wu, Tianxing, et al., "Efficiently embedding dynamic knowledge graphs." arXiv,1910. 06708v1 [cs.DB], October 15, 2019.

[38] Nickel, Maximilian, et al., "A review of relational machine learning for knowledge graphs." *Proceedings of the IEEE*, vol—104.1, p. 1133, 2015.

[39] Zhang, Wen, et al., "Iteratively learning embeddings and rules for knowledge graph reasoning." Association for Computing Machinery, Washington, DC, USA, 2019.

[40] Cai, Liwei, and William Yang Wang, "Kbgan: Adversarial learning for knowledge graph embeddings." Proceedings of NAACL-HLT, Association for Computational Linguistic, New Orleans, Louisiana pp. 1470–1480, June 1–6, 2018.

[41] Fadaee, Seyed Amin, and Maryam Amir Haeri, "Classification using link prediction." *Neurocomputing*, vol—359, pp. 395–407, 2019.

[42] Ma, Jiangtao, et al., "ELPKG: A high-accuracy link prediction approach for knowledge graph completion." *Symmetry*, vol—11.9, p. 1096, 2019.

[43] Das, Rajarshi, et al., "Go for a walk and arrive at the answer: Reasoning over paths in knowledge bases using reinforcement learning." arXiv preprint arXiv:1711.05851, 2017.

[44] Jagvaral, Batselem, et al., "Path-based reasoning approach for knowledge graph completion using CNN-BiLSTM with attention mechanism." *Expert Systems with Applications*, vol—142, p. 112960, 2020.

[45] Das, Rajarshi, et al., "Chains of reasoning over entities, relations, and text using recurrent neural networks." Proceedings of the 15th Conference of the European Chapter of the Association for Computational Linguistics: Volume 1, Long Papers, pp. 132–141, Valencia, Spain, April 3–7, 2017.

[46] Das, Rajarshi, et al., "Chains of reasoning over entities, relations, and text using recurrent neural networks." Proceedings of the 15th Conference of the European Chapter of the Association for Computational Linguistics: Volume 1, Long Papers, pp. 132–141, Valencia, Spain, April 3–7, 2017.

[47] Benjamins, Richard, et al. "The six challenges of the Semantic Web," First International Semantic Web Conference, ISWC2002, Cerdeña, Italia, ISBN 978-3-540-43760-4, June 2002.

[48] Consoli, Sergio, et al., "Producing linked data for smart cities: The case of Catania." *Big Data Research*, vol—7, pp. 1–15, 2017.

[49] Abid, Tarek, et al., "Towards smart city ontology." 2016 IEEE/ACS 13th International Conference of Computer Systems and Applications (AICCSA), IEEE, Agadir, Morocco, 2016.

[50] Saravanan, K., and A. Radhakrishnan, "Dynamic search engine platform for cloud service level agreements using semantic annotation." *International Journal on Semantic Web and Information Systems (IJSWIS)*, vol—14.3, pp. 70–98, 2018.

[51] Bellini, Pierfrancesco, and Paolo Nesi, "Performance assessment of RDF graph databases for smart city services." *Journal of Visual Languages & Computing*, vol—45, pp. 24–38, 2018.

17 Privacy and Security Issues in Green Smart Cities

S. Porkodi and Kesavaraja Duraipandy

CONTENTS

17.1   INTRODUCTION

17.1.1   INTERNET OF THINGS IN DIGITAL ERA

The Internet of Things (IoT) is mainly billions of sensors and devices throughout the world connected with the Internet that are used to acquire and share data. From a small light bulb or door locks to big airplane or satellite, anything can be part of IoT [24]. The IoT devices can be used to transfer real-time data without any human involvement. The data from various IoT sources are merged to make a device that is smart, automated and leading to a better world. In the era of digital technology, the adoption of IoT is increasing in various fields. At present there are more than 20 billion IoT devices connected to the Internet, most of which are smart watches, medical equipment, speakers, sensors, etc. [16]. In the next five years, it is predicted that there will be more than 40 billion IoT devices connected to the Internet, most of which will be wearable and smart home devices. The IoT also plays an important role in building a green smart city [33].

17.1.2   GREEN SMART CITY

Green smart cities are basically developed urban areas that use various sensors and IoT devices to acquire data that are processed and analyzed to get the knowledge that is used to efficiently manage resources, assets and various services [25]. The acquired data are mainly used to improve the safety and development of the city. The data is acquired from IoT devices, citizens, assets and building that are used to manage and monitor various services of the green smart city, such as water supply, power distribution, crime detection, hospitals, education, traffic management, waste management, etc. [26]. In the past two decades, the green smart city gained lots of interest, and it is actively implemented and growing in various developed and developing countries. These cities consist of machines and objects connected to the Internet; the data are transmitted via wireless technology, and further data can be stored and processed in the cloud [34]. The citizens of the green smart city can access or interact with the IoT objects with the help of their smart phones. The green smart city solves the majority of the problems with modern technologies, reduces overall costs, improves sustainability, reduces pollution level, improves energy usage, efficiently maintains the economy of the city and improves the security and quality of life by making better decisions.

17.1.3   VARIOUS NEW TECHNOLOGIES IN THE GREEN SMART CITY

The development of smart applications also leads to lots of privacy and security issues in the green smart city. The acquired data from a citizen consists of medical history, travel history, current location coordinates, credit card numbers, etc., which are highly sensitive and are vulnerable to attacks such as unauthorized access. If the data is changed, then the false data would also initiate various serious problems. For example, if health data is forged, then the medicine prescription might go wrong, even leading the patient to death. Unauthorized access to the GPS data of the citizen may lead to intrusion and burglary. In 2015, hackers attacked Ukraine's power grid system, and electricity was disconnected for citizens for a long time.

The traditional cybersecurity protection techniques cannot be directly applied in the applications of the green smart city due to the dynamic, scalability and heterogeneity characteristics of IoT devices. The lack of protection or inefficient protection measures lead to serious privacy and security threats to the whole system. Advanced smart applications need more advanced protection mechanisms. There are various new technologies that can be used as protection mechanisms such as biometrics, cryptography, data mining and machine learning, game theory, blockchain and nontechnical supplements. These technologies can be used to ensure privacy and security in green smart cities.

There are many surveys based on security to the IoT ecosystem in the recent past. Zhang et al. [32] suggest some security solutions such as biometrics and cryptography for green smart city applications. Sicari et al. [27] describe the present challenges and issues in the IoT devices such as privacy, security and trust; some solutions such as biometrics, cryptography and data mining are also explained. Eckhoff et al. [6] studied about nine different privacy protection technologies that include biometrics, cryptography and data mining, based on which a survey is also taken at a contest about smart cities. Nia et al. [21] discuss various security issues in IoT, edge computing, and the solutions for the security issues such as cryptography, data mining and nontechnical supplements are also discussed. There are only few surveys on the privacy and security issues in smart cities. Gharaibeh et al. [10] in 2017 present the smart city achievements, and the existing issues in privacy and security are discussed from a data-centric perspective. Solutions such as cryptography, machine learning and blockchain are highlighted as the solutions to the security issues. The protection mechanisms to ensure privacy and security in smart city of various examples in this chapter are listed and compared in Table 17.1.

In this chapter, an overview of the green smart city is presented along with its characteristics and applications. The privacy and security issues in the green smart city are identified and explained. The corresponding requirements that are needed to build a secure and stable green smart city are discussed. Finally, existing technologies to ensure privacy and security in the green smart city are explained, and the challenges and further direction of research and development are identified.

TABLE 17.1
Comparison of Technologies in Various Works to Ensure Privacy and Security in a Smart City

	Zhang et al. [32]	Sicari et al. [27]	Eckhoff et al. [6]	Nia et al. [21]	Gharaibeh et al [10]	This Chapter
Biometrics	Yes	Yes	Yes	No	No	Yes
Cryptography	Yes	Yes	Yes	Yes	Yes	Yes
Data Mining	No	Yes	Yes	Yes	No	Yes
Machine Learning	No	No	No	No	Yes	Yes
Game Theory	No	No	No	No	No	Yes
Blockchain	No	No	No	No	Yes	Yes
Nontechnical Supplements	No	No	No	Yes	No	Yes

17.2   OVERVIEW OF GREEN SMART CITY

17.2.1   GREEN SMART CITY ARCHITECTURE

There are various architectures evolving due to the continuous development in the green smart cities. The basic four layers of the green smart city are perception layer, network layer, support layer and application layer as shown in Figure 17.1.

Perception Layer: First layer of architecture is the perception layer; it can also be expressed as the data acquisition, data recognition and data sensing layer. It collects data from all the sensors, actuators and devices in the world, and the collected data are transmitted through the network layer.

Network Layer: The second layer of the architecture is the core layer, called network layer, that consist of wireless and wired networks such as communication networks, wireless sensor networks (WSNs) and Internet. It transmits acquired data from perception layer to servers or smart things.

Support Layer: The third layer of the architecture is the support layer; it mainly supports the application layer with all necessary requirements with techniques of intelligent computing such as edge computing, cloud computing or fog computing.

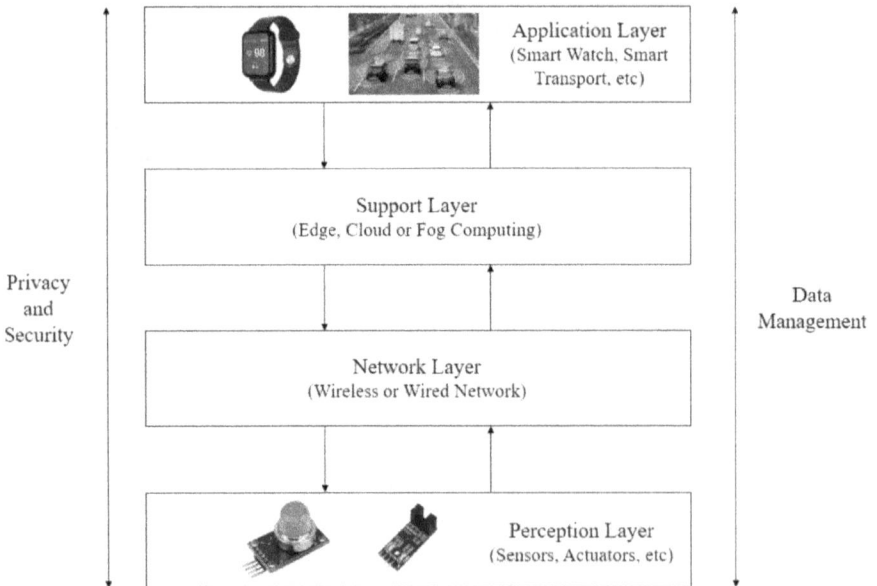

FIGURE 17.1   Green smart city architecture.

Application Layer: The top layer of the architecture is the application layer; it provides real-time services or intelligence to the users according to the personalized requirements of the user.

17.2.2 Characteristics of the Green Smart City

There are various characteristics to be considered in the green smart city while developing privacy and security solutions. A few characteristics of the green smart city are listed in Figure 17.2.

Infrastructure Development: The physical infrastructure (roads, buildings, other assets) should be efficiently used for the development of the city by using data analytics and artificial intelligence. The infrastructure should be built in such a way that it must have the ability to adapt according to environmental changes of the city. Example, smart homes, bus stops with chargeable ports for electric vehicles, smart traffic lights, Vehicular ad hoc Network (VANET), buildings can be built with solar panels at the top to generate green energy, etc.

Connectivity, Availability and Scalability: Any device and every device are connected to each other or to the Internet based on the necessity of the smart application. Connectivity is a basic feature for building a smart world. Smart devices mostly operate on real-time data, and they cannot

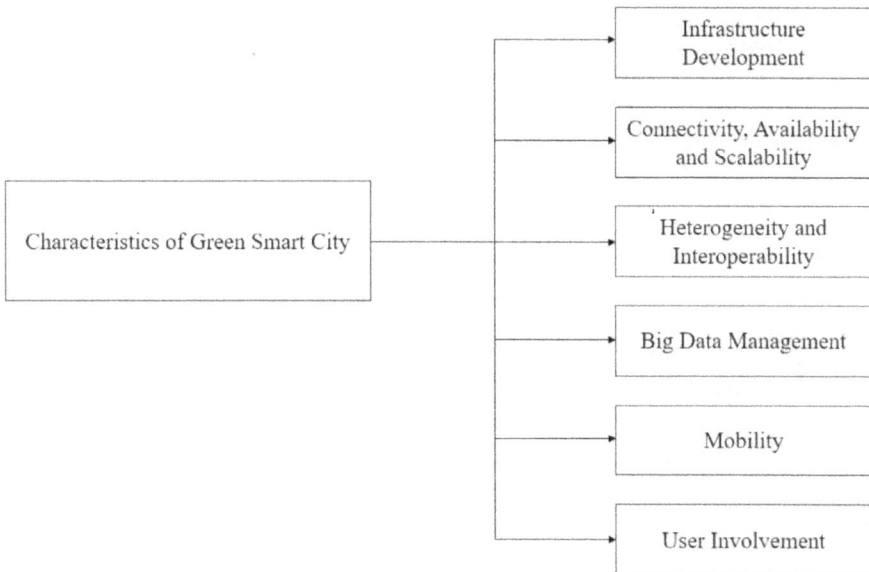

FIGURE 17.2   Characteristics of Green Smart City.

work properly if there is downtime, so availability of data is highly required for the active functioning of the smart devices in critical or emergency situations. The volume of data is growing huge, and the network traffic should be maintained; thus, scalability of the system is very important.

Heterogeneity and Interoperability: The systems in the green smart city are distributed, independent and used by various users for different needs. Thus, heterogeneity is important to handle different IoT nodes, diverse technologies, various communication protocols, platforms, applications, etc. The green smart city should also maintain interoperability, which is to exchange data from various sources to obtain better usage of information.

Big Data Management: As huge volumes of data are generated in real time from various sources with different velocity, an efficient technology is important to manage all the big data. Massive amounts of data are generated, acquired, transmitted, processed and stored in the green smart city. But all these data are highly useful when information or knowledge is gained from them, which can be used for better decision making and to manage assets of the green smart city.

Mobility: The movement of vehicles within the city and traffic control monitoring on a real-time basis comes under mobility. An instant reaction or a quick solution must be provided in case of any problems or emergency situations. For example, in the case of a traffic jam or an accident, the route should be changed automatically and all the vehicles in that route should be given an alert and directions quickly. Mobility is customized with the help of better development in the communication infrastructure.

User Involvement: The green smart city is built to improve the quality of the citizen's life, so user involvement is needed to improve the smart applications and to build new applications. User involvement requires education, creativity and continuous learning of new technologies. Better user understanding and identifying user needs will efficiently lead to creating better applications.

17.2.3 GREEN SMART CITY APPLICATIONS

There are lots and lots of green smart city applications to improve the quality of citizens' lives and to provide benefits to citizens in different aspects. A few green smart city applications are listed in Figure 17.3, and there are many other green smart city applications that are used in different smart cities around the world [1].

Smart Health Care: The IoT sensors and various health devices and mobile health applications collect the user data and send it to hospital servers via the Internet. The collected data in the server are then analyzed at the hospital, and the results or prescription are sent to the user/patient from the doctor. The doctor can monitor the real-time health of the remote patient and can meet the patient virtually. If there arises any emergency

Smart Healthcare		Smart Parking
Smart Power Meter		Smart Transportation System
Smart Water Distribution		Smart Surveillance
Smart Energy Management	Green Smart City Applications	Smart Street Lighting
Smart Waste Management		Smart Home
E - Governance		Disaster Deduction System
Smart Road		Smart Education

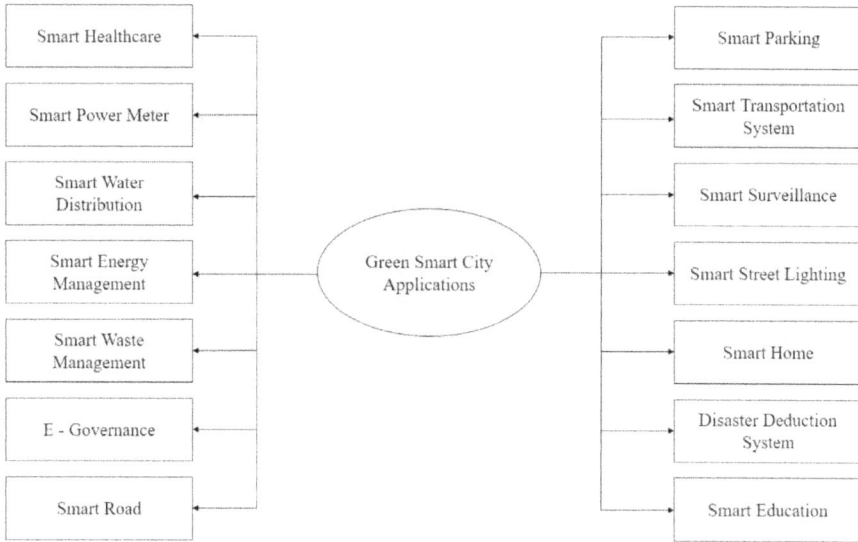

FIGURE 17.3 Green smart city applications.

situation, an immediate alert is sent to the ambulance along with the user location.

Smart Transportation System: All the vehicles are connected to the Internet, where traffic flow and mobility can be managed with smart traffic lights [28]. Speed limits of the vehicles can be monitored, and alerts can be sent if the limit is exceeded. The shortest routes to the destination can be found through a map application available in smart phones; also, alerts regarding any accidents can be sent immediately to avoid traffic jams. Smart parking can be used to predict the nearest free space and allocate it for the vehicle based on the given destination [17]. Driverless vehicles can be developed to function automatically based on processing and analyzing various data from the surroundings. Smart roads can be constructed that deliver warning and diversion details based on the climatic conditions to avoid accidents [1].

Smart Water Distribution: Smart water distribution consists of valves that are electrically actuated, which controls the water flow in homes, street-wise and area-wise. The speed of the pump is managed by the electrical valve; water availability can be monitored with water level sensors. The water leaks can be identified based on water pressure so that alerts to the repair team can be sent, along with the specific coordinates of the leakage. Quality of the water is maintained regularly. Smart water meters are placed at every home to quickly calculate the bill for water usage every month [15].

Smart Home: The smart home consists of various smart applications, from smart fans and smart lighting systems to energy or waste management systems. The smart energy management system is used to monitor and manage the power that is used by the user; this ensures reducing power consumption efficiently. The smart energy meter calculates the bill for power consumption every month [29]. The smart waste management at home consist of sensors that can alert the city collection service when a certain level of waste is reached. Fill patterns and short routes to the destinations can be studied to minimize operational costs [12].

E-Governance: All the government related services comes under e-governance, to provide efficient services, increase data transparency and improve communication between people and government. For example, people can demand the services they need, online complaints can be registered to the government, online bills can be paid to the government, notifications and any information can be sent to citizens directly from the government, etc.

Smart Surveillance: Smart surveillance can mainly be used for crime detection and prevention and to provide security. The whole green smart city can be monitored with AI-enabled video surveillance connected to the Internet. This system can also be used for various other purposes such as vehicle number plate recognition systems, smart tolls and monitoring low-emission zones for reducing air pollution [8].

Smart Education: A new generation of digital learning model that is interactive with students and maximizes student engagement is smart education. In 2020, since all the traditional educational schools and colleges are closed, most students were using smart phones to learn in virtual classrooms through video conferencing and writing online tests, which comes under smart education [11].

17.3   PRIVACY AND SECURITY ISSUES IN THE GREEN SMART CITY

All the smart applications are developed only for the development of society and to improve the quality of citizens' lives, but all these smart applications and services are vulnerable to cyberattacks and hacking. The cyberattacks include eavesdropping, collision, background knowledge attacks, spamming, denial of service attack, Sybil attacks, identity attacks, forgeries, etc. The smart application also handles sensitive data of citizens; for example, the smart health care system contains all the medical details of the user, and thus any inappropriate change of data may end up costing the life of the citizen. The user pattern can also be studied by accessing various smart application databases, and the details can be used against the user. Some lack of privacy and security issues in the new technologies are listed next.

17.3.1   PRIVACY ISSUES BY BOTNET ACTIVITIES

Botnet activities and attacks are increasing in IoT devices. Botnet is just a group of devices connected to the Internet that are in control of a central server. But the term botnet is often used to refer to malicious hijacking such as distributed denial of service (DDoS). Numerous computers are accessed by the botnet owner using a Trojan virus and attacking the security of the systems. Then all these computers can be commanded at the same time to perform a malicious activity. The botnet attack also includes account takeovers, which means the leaked credentials can be validated to take over the account of someone else, an attacker can use a botnet to gain control over the device and the network on which the device is connected, online web applications are attacked for stealing data, etc. [3, 18]. Some examples are listed here.

- In 2016, the Mirai botnet was used to shut down a huge portion of Internet including Netflix, Twitter and many banks in Russia [20].
- In 2017, a DDoS attack was performed using 75,000 bots against a customer at Akamai [31].
- In 2020, an Emotet botnet was used by cybercriminals to send malicious emails and infect various users with DDoS and ransomware [5].

17.3.2 LACK OF PRIVACY IN VIRTUAL REALITY AND AUGMENTED REALITY

Virtual reality (VR) is simulating a virtual environment where a user can interact with the environment virtually. The virtual reality technology is already adapted by aerospace and the defense industry for training soldiers and pilots in flight simulation. This technology is evolving in terms of efficiency, and at an affordable cost, in the future it can be used in gaming, hospitals and educational fields. Augmented reality (AR) adds digital elements in the physical real world. Some examples of AR are filters in Instagram and Snapchat and games like Pokémon Go. Both VR and AR have many security issues; for example, the VR sensors can be compromised, the system can be hacked or the user can be tricked to crash into the wall or fall from a location. User authentication issues, DDoS attacks and data breaches may happen in the VR and AR technology. When collaborating with IoT devices and systems in the VR and AR, the system can be hacked with ransomware, and money can be demanded to decrypt the hacked system [9, 14, 22].

17.3.3 SECURITY RISKS IN ARTIFICIAL INTELLIGENCE

Artificial intelligence (AI) is used in many smart applications to make the decision-making process faster. When more training is given, AI performs much smarter and more quickly and provides better solutions than humans. However, it can also backfire. Hackers can create new programs or modify the existing program by altering inputs and training AI with wrong data sets that can allow the hacker into the system, crossing all the security measures. The hacker can do any malicious process on entering the system and can change the way that the system works and responds. Once the system is modified by the hacker, it is very hard to return to the normal state. Also, AI is basically designed to study about a user or a situation and learn to take decisions according to the situations. Thus, when a hacker gains control over a system, the sensitive data about the users are leaked, and the hacker can use those details to target the user. The attackers with AI knowledge are becoming smarter, and thus a strong security mechanism and technology is required to avoid security risks in AI [2, 23]. In 2018, Taskrabbit, which is an online marketplace, encountered a cybersecurity breach on a massive level by affecting 145 million users. The data of the users that included details of bank accounts and Social Security numbers was compromised [36].

17.3.4　Security Issues in Driverless Cars

Tesla and Uber are testing their driverless cars, which will be used by many people in the future green smart cities. Google's self-driving car project is carried out by Waymo-00 LLC, which is a company under Alphabet Inc. The road testing in 25 cities and six states was already completed in 2018, and in June 2020 Waymo collaborated with Volvo for integrating Waymo's driverless car technology in Volvo's vehicles. The driverless cars or vehicles minimize human errors due to distractions, high speed of the vehicle, and long trips at night, and that would reduce thousands of accidents happening every year. Also, driverless cars are highly useful for people who do not know to drive, people with disabilities and for older citizens. Driverless cars will reduce the collision of vehicles by automatically sensing obstacles. In driverless cars, not only life safety but also cybersecurity is a concern. Criminals and hackers can hack and access the passenger data and steal it and the operations of the car can be manipulated or disturbed (for example, brakes could be controlled remotely). Criminals can trap the passenger within the car and demand a ransom. Robbery or assault can be planned by the attacker hacking into the destination details of the car. Hijackers can hack and collapse the whole automated transportation system [4, 7, 19]. Waymo's driverless cars and vans were attacked physically by some stones and sharp objects.

17.3.5　Lack of Security in Water Management

A major cyberattack on the water infrastructure at a smart city in Israel was attempted in April 2020 [35]. The attack was on the control system and controls of the sewers, pumping station and wastewater treatment plants. The measures that Israel could take were changing the Internet passwords, updating software of the control system and even taking the system offline until some measures were taken for improving their system security. These attacks are likely to be caused by weak or default passwords, open ports and known vulnerabilities. If the system has weak security measures, the intruder can change the level of chlorine in the water, which can affect the whole smart city, and mishandling sewer water can also add problems for the people of the smart city. So it is highly important to provide the water management system with good cryptographic security, and also blockchain can be used to monitor all the activities in the water management system to avoid any intruders into the system.

17.3.6　Cyber Security and Privacy Issues in Smart Bulbs

The smart bulbs can be hacked easily by hacker just with the help of laptop and antenna within a 100 meter radius; also, the hacker can plant malware or take over the network [37]. The Zigbee protocol is used in the IoT smart devices present inside the smart home such as Samsung Smart-Things, Phillips smart bulb, Amazon Echo Plus, etc. In the Phillips smart bulb, the brightness or color of the light bulb can be controlled by the hacker, diverting the user to misunderstand it as a glitch in the bulb. These smart bulbs or devices are controlled by the application in the user smart phone. So, as the glitch occurs, the user tries to reset the application and again connect to the smart bulb. At the same time, the hacker can use the vulnerabilities in the Zigbee protocol to trigger a buffer overflow based on heap in controlling the bridge, which is done by sending a huge volume of data. The hacker uses the data to control the bridge and install malware that can be used to gain control over the home network or target business. The hacker can also spread spyware, ransomware and even hijack the smart home via this technique. So it is highly dangerous to citizens and the society. In 2019, the researchers of Tel Aviv University took control of the Phillips smart bulb present on a targeted network, and they also successfully installed the malware on it [38]. More security should be provided to the smart city and smart home applications with cryptographic techniques, biometric authentication and other new technologies to avoid unauthorized intruders and hackers to take over the system.

17.3.7 SECURITY RISKS IN SMART GRID

Smart power girds are used to manage the power supply throughout the smart city. These smart grids are vulnerable to various cyberattacks that can lead the entire smart city into a complete mess. Consider that if there is a power blackout due to a cyberattack at the smart grid, many smart city applications that are functioning on power will also stop functioning and hospitals and other infrastructure that needs power to function will collapse, which leads to lots of economic loss, and citizens' lives are also in danger. In 2015, hackers attacked Ukraine's power grid system, and electricity was disconnected for the citizens for a span of six hours, affecting 230000 citizens in Ukraine [39]. This attack was likely to be carried out with various different techniques, including BlackEnergy malware attached to a phishing email, the control system supervisory control and data acquisition (SCADA) was seized and turned off remotely, some components of the IT infrastructure were destroyed or disabled, the files that were stored on workstations and servers were destroyed with KillDisk malware, and to deny information regarding the blackout to the consumers, a denial of services (DoS) attack was used. Thus, blockchain should be used to manage and control activities in the smart grid network to avoid these types of attacks the chance to take over the system, causing blackouts and trouble to the citizens.

17.4 REQUIREMENTS OF SECURED GREEN SMART CITY

To build a green smart city secured from hackers, cyber criminals, hijackers and intruders, there are some precautionary requirements to be taken, such as, integrity, availability, confidentiality, authentication, privacy protection and intrusion prevention.

17.4.1 INTEGRITY AND AVAILABILITY

Integrity and availability of data are so important for green smart cities. The green smart city applications should deliver accurate and reliable data; thus, integrity is needed. Smart device should provide real-time data continuously and real-time services whenever the user requires; thus, availability is needed. Even if the system is attacked, the data integrity and availability should be maintained, else the green smart city will be at huge chaos. The IoT devices are connected to the network where the data are transmitted between the IoT device and storage like the cloud. The data tampering might happen in between the transactions of data, so a strong cryptographic protection mechanism is necessary. Basic protocols and firewalls cannot ensure the security at the end devices, since the computational power is low. Also, the smart system should be capable of detecting abnormal activities in its systems, since the smart devices are vulnerable to attacks and intrusions. A strong protection should be developed for the smart systems that can learn and adapt to the growing intelligent attacks and produce countermeasures when a smart application is under any attack.

17.4.2 CONFIDENTIALITY AND AUTHENTICATION

Confidentiality is to keep all the sensitive information of the user private and secure from the hackers and attackers. When using IoT devices, the devices or communication can be subjected to eavesdropping attacks. Encryption can be used for protecting the data confidentiality to maintain a secure storage system and reliable communication network. The authentication is used for user identity verification to avoid letting unauthorized users into the system. Green smart city applications can be subjected to attacks such as a replay attack, insider attack, user impersonation attack, stolen smart card attack, denial of service attack and password guessing in an offline attack. The protocols of user authentications are used for stopping malicious users from gaining access to the system, and only the authorized user should be given permission to access the system. Also, advanced technologies should be used to protect the system from malicious users.

17.4.3 PRIVACY PROTECTION

There are different types of private data that need to be protected in green smart city applications. They are user actions and behaviors, location, user media files, social life and state of mind and body. User actions and behaviors are mainly collected to study about the user pattern of buying products and to show targeted advertisements. This data can be used to analyze much more about the user. Recently, in smart phones, users can cancel the option of collecting user data to show targeted advertisement. But many are unaware of this option, and in IoT devices there is no such option, and also, only essential data should be collected by the corresponding IoT devices. Location details are necessary for some IoT devices such as showing routes in maps, driverless cars, etc. But not all applications and devices need locations, and cyber criminals can use these details against the user by planning burglaries if, for example, the home and workplace of the user are revealed, etc. IoT devices have access to media and storage devices to store the acquired data, but accessing other media in the devices such as CCTV footage, photos and videos of the user is violation of privacy. The metadata from social media can reveal who the user talks to and how long, or what political thoughts the user has. These details can be used against the user. For example, false articles or negative articles about a political leader can be shown as an ad to change the mind of a user who supports that particular political leader. This can even reflect changes in voting results. As the biometrics are always connected to the user, the state of mind and body of the user can be studied, such as health, mental state, etc. This can be used by the insurance company and employers to discriminate against users; thus, the third party should be eliminated and a strong security system should be used.

17.4.4 INTRUSION PREVENTION

An intrusion prevention system is a much needed requirement of the green smart city, as network and IoT devices are highly vulnerable to intrusion. The smart system should detect abnormal activities or a change in network traffic and prevent the intrusion from happening. An intrusion detection system (IDS) is used mainly to detect an anomaly in usage, application misuse and specification-based intrusion detection. But this traditional IDS is not efficient for complex and heterogeneous green smart city applications. So an efficient intrusion prevention technology is needed. That system should prevent the intruder even before intruding into the system, for that intrusion should be predicted and then prevented. Many intrusion prevention system have failed in the past, failing to detect the upcoming threats and attacks, so intelligence should be added to the system, and it should be trained to predict and prevent intrusion from happening.

17.5 TECHNOLOGIES TO ENSURE PRIVACY AND SECURITY IN THE GREEN SMART CITY

Apart from traditional privacy protection mechanisms, there are some new technologies to ensure privacy and security in green smart cities; they include biometrics, cryptography, data mining and machine learning, game theory, blockchain and non-technical supplements.

17.5.1 BIOMETRICS

In the IoT applications and devices of green smart city, biometrics can be used for user authentication. A person's unique attribute or behavior like fingerprints, retina, facial recognition, handwriting, signature, etc., can automatically recognize and allow the user to gain access to the smart applications. Leading technology has initiated authentication based on brain waves that has a high level of accuracy in authentication. Mutual or two-way authentication or key exchange protocols can be used to protect all the confidential data of the user when storing in a database or storage device. These protocols defeat attacks in the security system efficiently and maintain affordable cost for communication. While using biometrics, any unique details of a user are highly confidential and sensitive data that need to be protected.

Biometric devices Attached to User's Smart Phone	Smart Home Applications (Smart Light, Smart AC, Smart TV, Smart Home Security System, Etc)	Storage Device (Hard disk, Cloud, etc)

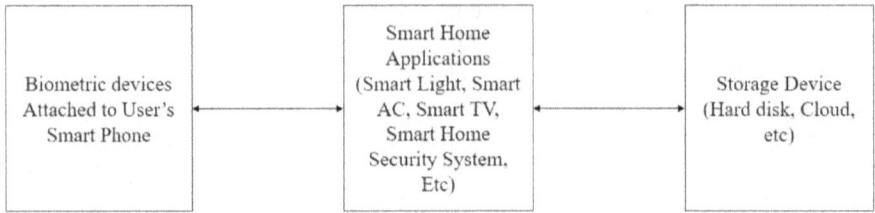

FIGURE 17.4 Biometrics in smart home applications.

If the biometrics are not managed and used properly, then there are privacy threats, so biometrics should be used with privacy-preserving techniques [30]. In future, the biometrics can be used efficiently with the privacy protecting factor in any green smart city applications. For example, biometric devices can be connected to the user's smart phone, which can be used to access the smart home applications like smart lights, smart AC, smart TV, smart home security system, etc. It is to ensure that no unauthorized user can access the biometric devices without the real user's attributes, and if there is an intrusion found, with the help of key exchange protocol, mutual or two-way authentication can be used to confirm with the real user and the intrusion can be avoided. All the data from the IoT applications are stored in a database or cloud that can also be accessed with user authentication to avoid a data breach, as shown in Figure 17.4.

17.5.2 CRYPTOGRAPHY

Cryptography is the backbone of the privacy and security protection mechanism for communication and information, as the cryptographic mechanism is capable of denying access to doubted or unauthenticated parties in the whole life cycle of data (process, store and share). Due to power consumption and computational complexity, traditional cryptography is not enough to protect the huge amount of data generated from all the IoT devices. So lightweight encryption has now been made a mandatory and basic requirement to be used for the protection of data. End-to-end lightweight protection for the communication channel was developed by Mahmood et al. [13] to be used in the IoT applications; this mechanism can protect the system from DDoS attacks. Public-key encryption protocols like ciphertext-policy attribute-based encryption (CP-ABE) and hybrid protocols can be used to protect the applications of the green smart city. Homomorphic encryption (HE) gains attention while securing green smart city data, since when an encrypted data is sent to a user, the receiver first performs some calculations on the cipher text and a result is obtained. Then, when decrypting the data, it is verified whether the same operations are performed that were done previously on the cipher text. So sensitive data can be more secure if the HE algorithm is used for security. The HE algorithm can be implemented in areas like security of cloud computing, protecting sensitive data in the health care sector, protecting the smart energy consumption system, etc. The only disadvantage of the HE algorithm is that it requires high computational cost. In cryptography, the zero-knowledge protocol can be also used when handling sensitive data. In this protocol, user A can provide proof to another user, B, where user B has the value of X, even without decrypting any data or using any other data, but the user A need only know the X value. The authentication issues in green smart city applications like a smart card can be handled by using the zero-knowledge protocol.

17.5.3 Data Mining and Machine Learning

Various sensors and IoT devices are used in the green smart city, most of which are capable of identifying the user pattern with the help of data mining for providing improved smart services. As there are lots of sensitive data regarding the user's unique attributes, behavioral data and location data are at risk of being exposed. So they are to be protected with privacy-preserving data mining (PPDM) techniques. Data mining can also be used to deduct malware in databases, networks, operating systems and servers and then prevent them from leading the system into chaos or malfunction. Machine learning (ML) and AI can be used to train on lot of situations for improving the intrusion deduction system (IDS) and intrusion prevention system (IPS) efficiently. The networks are protected from attacks with the help of ML. As the wireless sensor network (WSN) connects all the IoT devices across the green smart city, all the data are transmitted through the network, and it needs protection. The advantages of using ML are security in sensing data, the fusion of WSNs, and the ability to detect attacks in the network of Wi-Fi. User-centric techniques in ML are for analyzing, predicting and making decisions more personalized for users. The smart phones and sensor networks have lots of security and privacy concerns; thus, multi-sensor authentication technologies can be used in the users' smart phones. The ML is used to strengthen the defense strategies. ML along with game theory can prevent intrusions very well. In the smart transportation system, the ML can be used to stop intruders from entering into the vehicles, thus protecting the citizens from hazards due to hacking.

17.5.4 Game Theory

In the cybersecurity field of privacy and security protection, one of the most powerful tools is game theory, which can be applied for protection in various green smart city applications. The advantages of game theory over other traditional defense techniques are reliable defense, proven mathematics, distributed solutions and timely action. Using game theory for addressing cyber security issues is increasing; it can be used in cloud storage to analyze strategies. The lightweight game theory for anomaly detections ensures low energy consumption and high accuracy. The attack and defense issues in the communication network such as networks enabled with honeypot need security, which can be used in green smart city applications like smart networks, smart buildings, smart health care, etc. The honeypot game and zero-sum game are used for deducting attacks like spoofing on wireless networks. Game theory that balances data utility and protection intensity is currently integrated with various privacy protection mechanisms such as differential privacy and k-anonymity. When green smart cities are fully developed and all the smart devices are connected to the network, various new privacy issues would arise, and evolved game theory will play a role in the future.

17.5.5 BLOCKCHAIN

Blockchain is a decentralized peer to peer distributed ledger that consists of chained blocks of records. It consists of immutable, tamper-resistant and permanent records in chronological order. Blockchain gives a huge security boost to the system. Data transparency can also be achieved. It can be used in many fields where the copy of the data is to be maintained regularly, and so if there is a change of data or hacking occurred in one node, then the data can be collected from other nodes and the hacked node can be recovered easily. Blockchain can help solve double spending problems; issues in supply chain management can be solved by the transparency property of blockchain, and a third party can be eliminated completely. For example, eliminating the supply chain in agriculture would eliminate illegally stocking up goods at ration, stores, etc. The supply chain in the smart health care system will provide all the history of records and patient real time data in a single place to provide a better treatment. In smart water management, smart parking and many smart applications, the blockchain network can be used to monitor each and every activity in the system; thus intruders and hackers can be completely eliminated from the system.

   Authenticity and security issues with the Internet can also be solved with blockchain. A few other features of blockchain are availability, confidentiality, cryptography, integrity, tamper resistance, interoperability, verifiability and unchangeability. Currently, blockchain is evolving, and the usage of the blockchain concept in every field is increasing. A framework developed with blockchain can improve a system's efficiency and reliability and ensures the security in the communication network across many IoT devices. The blockchain can be used to solve security issues in the VANET network. Even the cloud technology cannot satisfy all green smart city requirements, but blockchain will solve all the issues regarding privacy and security in the green smart city applications and scalable networks. When blockchain is integrated with various new technologies such as software-defined networking (SDN) and fog computing, the distributed architecture as shown in Figure 17.5 can also satisfy the principles such as adaptability, resilience, scalability, efficiency and security. But blockchain is still at its starting point and developing; in future it will play the biggest role in the privacy and security of the green smart city.

17.5.6 NONTECHNICAL SUPPLEMENTS

Apart from technical solutions to protect privacy and security of the green smart city, there are also nontechnical supplements. The limitations of the existing technologies can be serious due to the reinforcement of government, policies, regulation, etc. The government has a huge responsibility to select which information should be kept open to the public and which information should be kept secure, only giving access permission to corresponding users and whoever else is not allowed

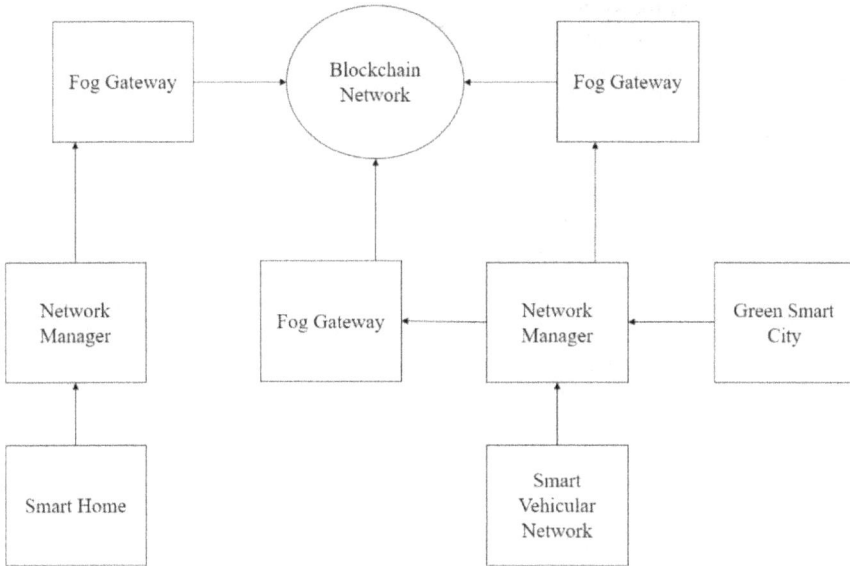

FIGURE 17.5   Blockchain in the green smart city.

to access the data. For example, the health details of a patient is sensitive data that should not be open to the public, but rather visible to the patient concerned and their doctor. The government should protect citizens' data by implementing all the protection mechanisms in the framework of green smart city. Efficient training should be given to developers, service providers, manufacturers and citizens of the green smart city to develop, maintain and use the green smart city applications efficiently. Developers should produce a stable system, vendors should fix all the vulnerabilities by using the updated firewalls, manufacturers should ensure that the standard quality and safety should be enhanced. Education and knowledge is given to the citizen on the working of the green smart city and how to be safe and protect themselves.

17.6   CONCLUSION AND FUTURE DIRECTION

The smart application and IoT devices have caused lots of privacy and security issues around the world. Even in 2020, accounts are hacked and cyber criminals demand ransoms and bitcoins in various platforms like YouTube and Twitter. So it is essential to develop advanced protection frameworks in all the fields. In this chapter, overview of the green smart city is explained, and then privacy and security issues that are present in the current applications of the green smart city are studied, and also the corresponding requirements that are needed to build a secure and stable green smart city are discussed. Finally, various existing technologies that are growing in recent years for protection are discussed. In the future, these technologies have a big role to play in the development of the green smart city and privacy protection. But citizens have to understand the system and should take all the safety measures to be safe in the green smart city, so that hackers and intruders can be completely eliminated from the system.

REFERENCES

[1] 50 Sensor Applications for the Smarter World, Libelium. Available at: www.libelium. com/resources/top_50_iot_sensor_applications_ranking/

[2] Artificial Intelligence Threats and Security Issues, Identity Management Institute. Available at: www.identitymanagementinstitute.org/artificial-intelligence-threats-and-security-issues/#:~:text=AI%20Can%20Backfire%20in%20the%20Hands%20of%20 Hackers&text=Attacks%20executed%20with%20AI%20tend,evade%20even%20 sophisticated%20threat%20detection

[3] Bijalwan, A., Solanki, V. K. and Pilli, E. S., Botnet Forensic: Issues, Challenges and Good Practices, *Security Protocols and Algorithms*, Vol 10(2), 2018. Available at: www.research-gate.net/publication/326058876_Botnet_Forensic_Issues_Challenges_and_Good_ Practices/citation/download

[4] Bowles, J., Autonomous Vehicles and the Threat of Hacking, *CPO Magazine*, 2018. Available at: www.cpomagazine.com/cyber-security/autonomous-vehicles-and-the-threat-of-hacking/

[5] Cimpanu, C., Emotet Botnet Returns after a Five-Month Absence, ZDNet, 2020. Available at: www.zdnet.com/article/emotet-botnet-returns-after-a-five-month-absence/

[6] Eckhoff, D. and Wagner, I., Privacy in the Smart City: Applications, Technologies, Challenges and Solutions, IEEE Communications Surveys & Tutorials, 2017.

[7] Elezaj, R., Autonomous Cars: Safety Opportunity or Cybersecurity Threat? Machine Design, 2019. Available at: www.machinedesign.com/mechanical-motion-systems/article/21837958/autonomous-cars-safety-opportunity-or-cybersecurity-threat#:~:text=Just%20like%20any%20other%20computer,and%20jeopardize%20 the%20passenger's%20safety

[8] ETGovernment, AI-Enabled Metering and Surveillance to Dominate Smart City Market: Report. ETGovernment.com, 2020. Available at: https://government. economictimes.indiatimes.com/news/digital-india/ai-enabled-metering-and-surveillance-to-dominate-smart-city-market-report/73432774

[9] Fineman, B. and Lewis, N., Securing Your Reality: Addressing Security and Privacy in Virtual and Augmented Reality Applications, Educause Review, 2018. Available at: https://er.educause.edu/articles/2018/5/securing-your-reality-addressing-security-and-privacy-in-virtual-and-augmented-reality-applications

[10] Gharaibeh, A., Salahuddin, M. A., Hussini, S. J., Khreishah, A., Khalil, I., Guizani, M. and Al-Fuqaha, A., Smart Cities: A Survey on Data Management, Security, and Enabling Technologies, *IEEE Communications Surveys & Tutorials*, Vol 19(4), 2017, 2456–2501.

[11] Glasco, J., Smart Education for Smart Cities: Visual, Collaborative & Interactive, Bee Smart City, 2019. Available at: https://hub.beesmart.city/en/solutions/smart-people/smart-education/viewsonic-smart-education-for-smart-cities#:~:text=Smart% 20education%20is%20a%20key,what%20defines%20a%20smart%20city

[12] Jamrozik, N., What Is Smart Waste Management? IoT for All, 2019. Available at: www.iotforall.com/smart-waste-management/#:~:text=Smart%20waste%20manage-ment%20solutions%20use,are%20ready%20to%20be%20emptied.&text=The%20 cost%20of%20these%20sensors,more%20attractive%20to%20city%20leaders

[13] Mahmood, Z., Ning, H. and Ghafoor, A., Lightweight Two-Level Session Key Management for End User Authentication in Internet of Things, in Internet of Things (iThings) and IEEE Green Computing and Communications (GreenCom) and IEEE Cyber, Physical and Social Computing (CPSCom) and IEEE Smart Data (SmartData), 2016 IEEE International Conference on IEEE, Chengdu, China, 2016, 323–327.

[14] McGee, M. K., Virtual Reality: Real Privacy and Security Risks, Data Breach Today, 2016. Available at: www.databreachtoday.com/interviews/virtual-reality-real-privacy-security-risks-i-3221

[15] Nagesh, R., Rao, A. and Haribabu, P., Smart Water Distribution Network IoT. C-DAC. Available at: www.cdac.in/index.aspx?id=pe_iot_Smart_Water_Distr_Network#:~: text=Distribution%20Network%20IoT,Smart%20Water%20Distribution%20 Network,Consumer%20points%20with%20flow%20meters

[16] Knorr, E., The Internet of Things in 2020: More Vital Than Ever, Network World, 2020. Available at: www.networkworld.com/article/3542891/the-internet-of-things-in-2020-more-vital-than-ever.html

[17] Kochar, M., Smart Transportation: A Key Building Block for a Smart City, ForbesIndia, 2018. Available at: www.forbesindia.com/blog/who/smart-transportation-a-key-building-block-for-a-smart-city/

[18] Korolov, M., What Is a Botnet? When Armies of Infected IoT Devices Attack, CSO, 2019. Available at: www.csoonline.com/article/3240364/what-is-a-botnet.html

[19] Leprince-Ringuet, D., Self-Driving Cars: The Hunt for Security Flaws Steps Up a Gear, ZDNet, 2020. Available at: www.zdnet.com/article/the-hunt-for-security-flaws-in-self-driving-cars-steps-up-a-gear/

[20] Marr, B., Botnets: The Dangerous Side Effects of the Internet of Things, Forbes, 2017. Available at: www.forbes.com/sites/bernardmarr/2017/03/07/botnets-the-dangerous-side-effects-of-the-internet-of-things/#c4e6d1333043

[21] Nia, A. M. and Jha, N. K., A Comprehensive Study of Security of Internet of Things, IEEE Transactions on Emerging Topics in Computing, 2017.

[22] Philipp, R., Study: Augmented Reality and Privacy Concerns, 2019. Available at: www. philipprauschnabel.com/2019/02/research-on-augmented-reality-privacy/

[23] Powell, L., The Problem with Artificial Intelligence in Security, Dark Reading, 2020. Available at: www.darkreading.com/threat-intelligence/the-problem-with-artificial-intelligence-in-security/a/d-id/1337854

[24] Ranger, S., What Is the IoT? Everything You Need to Know about the Internet of Things Right Now, ZDNet, 2020. Available at: www.zdnet.com/article/what-is-the-internet-of-things-everything-you-need-to-know-about-the-iot-right-now/

[25] Sarkar, A. N., Smart Cities: A Futuristic Vision, *The Smart City Journal*, Available at: www.thesmartcityjournal.com/en/articles/1333-smart-cities-futuristic-vision

[26] Secure, Sustainable Smart Cities and the IoT, Thalesgroup. Available at: www.thales group.com/en/markets/digital-identity-and-security/iot/inspired/smart-cities

[27] Sicari, S., Rizzardi, A., Grieco, L. A. and Coen-Porisini, A., Security, Privacy and Trust in Internet of Things: The Road Ahead, *Computer Networks*, Vol 76, 2015, 146–164.

[28] Smart Transportation, HereMobility, Available at: https://mobility.here.com/learn/smart-transportation/introduction-smart-transport

[29] Verma, S., Smart Energy Management: First Step towards IoT Adaptation, Digital Utility Group, 2019. Available at: https://energycentral.com/c/iu/smart-energy-management-first-step-towards-iot-adaptation#:~:text=Smart%20energy%20management%20systems% 20allow%20businesses%20to%20detect%20any%20faulty,analyze%20the%20 power%20quality%20events

[30] Wang, Y., Wan, J., Guo, J., Cheung, Y. M. and Yuen, P. C., Inference Based Similarity Search in Randomized Montgomery Domains for Privacy Preserving Biometric Identification, IEEE Transactions on Pattern Analysis and Machine Intelligence, 2017.

[31] What Is a Botnet Attack? Akamai. Available at: www.akamai.com/us/en/resources/what-is-a-botnet.jsp

[32] Zhang, K., Ni, J., Yang, K., Liang, X., Ren, J. and Shen, X. S., Security and Privacy in Smart City Applications: Challenges and Solutions, *IEEE Communications Magazine*, Vol 55(1), 2017, 122–129.

[33] Saravanan, K., Golden Julie, E. and Herold Robinson, Y., Smart Cities & IoT: Evolution of Applications, Architectures & Technologies, Present Scenarios & Future Dream, for the upcoming book series *Intel. Syst. Ref. Library*, Vol. 154, Valentina E. Balas et al. (Eds.): Internet of Things and Big Data Analytics for Smart Generation, 978-3-030-04202-8, 467407_1_En, (7). Available at: www.springer.com/us/book/9783030042028

[34] Saravanan, K. and Srinivasan, P., Examining IoT's Applications Using Cloud Services, in P. Tomar and G. Kaur (Eds.), *Examining Cloud Computing Technologies through the Internet of Things* (pp. 147–163). Hershey, PA: IGI Global, 2017. doi:10.4018/978-1-5225-3445-7.ch008

[35] Brumfield, C., Attempted Cyberattack Highlights Vulnerability of Global Water Infrastructure, CSO online, 2020. Available at: www.csoonline.com/article/3541837/attempted-cyberattack-highlights-vulnerability-of-global-water-infrastructure.html

[36] Sinha, J., 5 Artificial Intelligence-Based Attacks That Shocked the World in 2018, Analytics India Mag, 2018. Available at: https://analyticsindiamag.com/5-artificial-intelligence-based-attacks-that-shocked-the-world-in-2018/

[37] Singal, N., Smart Lightbulbs Are Not So 'Smart'! They Can Be Hacked Using Just a Laptop, Antenna, *Business Today*, 2020. Available at: www.businesstoday.in/technology/news/smart-lightbulbs-not-smart-they-can-hacked-using-just-laptop-antenna/story/395620.html

[38] The Dark Side of Smart Lighting: Check Point Research Shows How Business and Home Networks Can Be Hacked from a Lightbulb, Check Point Blog. Available at: https://blog.checkpoint.com/2020/02/05/the-dark-side-of-smart-lighting-check-point-research-shows-how-business-and-home-networks-can-be-hacked-from-a-lightbulb/

[39] Greenberg, A., Crash Override: The Malware That Took Down a Power Grid, *Wired*, 2017. Available at: www.wired.com/story/crash-override-malware/

Index

For Product Safety Concerns and Information please contact our EU
representative GPSR@taylorandfrancis.com
Taylor & Francis Verlag GmbH, Kaufingerstraße 24, 80331 München, Germany

www.ingramcontent.com/pod-product-compliance
Lightning Source LLC
Chambersburg PA
CBHW060811220326
41598CB00022B/2586

9 7 8 0 3 6 7 5 5 4 9 9 6